# Linear Control Systems in Engineering

This textbook examines classical and modern control strategies toward systems' best performance, especially concerning design and operations. It simplifies control theory concepts through related mathematics and examples of real-life systems worldwide.

*Linear Control Systems in Engineering: Basics and Beyond* covers the fundamental principles of control systems and advanced topics providing a comprehensive resource for readers at different levels of ability. It is written in an infographic language as much as possible, making complex concepts in control systems accessible to a broad audience, including students and professionals. The textbook includes many examples and practical exercises to reinforce learning and demonstrate how control systems work in various engineering domains. The textbook focuses on both the conventional and contemporary control systems technologies and trends, such as digital control, automation, and robust control. It also highlights analysis, stability, and optimization techniques for control systems in a sole source.

The textbook is written for both undergraduate and graduate courses dealing with the subjects of electrical, mechanical, mechatronics, chemical, and aerospace engineering. It will take the reader from basic concepts and applications to advanced topics, and the book will be the sole source to reach knowledge and explore future possibilities related to control design techniques, methodologies, and operations from basic to beyond.

A solutions manual and PowerPoint slides are available for qualified textbook adoption.

# Linear Control Systems in Engineering

## Basics and Beyond

Noman Jafri, Imran Shafi, Uzair Khaleeq uz Zaman, and Atal Anil Kumar

CRC Press
Taylor & Francis Group
Boca Raton London New York

CRC Press is an imprint of the
Taylor & Francis Group, an **informa** business

Designed cover image: Shutterstock

First edition published 2026
by CRC Press
2385 NW Executive Center Drive, Suite 320, Boca Raton FL 33431

and by CRC Press
4 Park Square, Milton Park, Abingdon, Oxon, OX14 4RN

*CRC Press is an imprint of Taylor & Francis Group, LLC*



ISBN: 978-1-032-77628-6 (hbk)
ISBN: 978-1-032-77654-5 (pbk)
ISBN: 978-1-003-48420-2 (ebk)

DOI: 10.1201/9781003484202

Typeset in Times
by Apex CoVantage, LLC

Access the Instructor and Student Resources: www.Routledge.com/9781032776286

*Dedicated to all who seek knowledge and contribute to the understanding and advancement of control systems.*

# Contents

# Foreword

In the ever-evolving landscape of science and engineering, certain principles stand as timeless cornerstones, guiding the creation of technologies that define our modern world. *Linear Control Systems in Engineering: Basics and Beyond* invites readers on a transformative journey into one such fundamental domain, where classical foundations meet cutting-edge innovation.

Every aspiring engineer envisions crafting systems that not only perform their tasks but do so with unparalleled precision and adaptability. This aspiration demands more than just creativity—it requires a profound grasp of both the basic and advanced principles governing system control. By beginning with linear time-invariant systems and expanding into the complexities of nonlinearities, this book equips readers with the intellectual tools to tackle real-world challenges. Moreover, it bridges the gap between classical methodologies and emerging techniques like robust control and AI-driven approaches, ensuring relevance in today's rapidly advancing technological era.

What sets this book apart is its clear and accessible presentation of control theory. Through carefully chosen real-world examples, insightful explanations, and essential mathematics, the authors transform a complex subject into an engaging and relatable experience. Infographic-style visuals enrich understanding, while in-depth discussions provide readers with the confidence to apply these principles to real-life systems.

This work is more than just a textbook—it is a companion for those eager to explore the boundaries of knowledge. Designed for both undergraduate and graduate students, it caters to learners at every level. Whether you are just beginning your journey or are an experienced practitioner, this book will guide you from fundamental concepts to advanced techniques, in control design and operations.

At its heart, this monograph emphasizes practical understanding over abstract mathematics. The authors thoughtfully introduce proofs only when they illuminate key ideas, ensuring that the focus remains on comprehension and application. Every chapter brims with carefully chosen examples, offering not only clarity but also inspiration for new ideas and possibilities.

This book is a wonderful effort to the enduring relevance of control systems in engineering, celebrating both the classical wisdom that forms its foundation and the forward-looking innovations shaping its future. As you turn these pages, may you find not only knowledge but also the spark to contribute meaningfully to the incredible journey of technological progress.

**Prof. Srdjan Stanković**
*Full Professor, Faculty of Electrical Engineering,*
*University of Montenegro, Montenegro*

# Preface

Control theory is the queen of engineering and design. Manufacturing and analysis of systems can only be carried out with its knowledge. This monograph presents a simplified control theory knowledge and the associated relevant examples. The topics cover the syllabus of undergraduate students of electrical, mechanical, chemical, and aerospace systems, and the course is designed to be used as a text for the first course in control systems. The prerequisite on the part of students is the introductory knowledge of different equations, vector-matrix analysis, circuit analysis, mechanics, hydraulics, and Laplace transform.

This textbook is organized into 11 chapters. The outline of the monograph is as follows. Chapter 1 presents an introduction to systems and linear control systems. Chapter 2 briefly deals with Laplace transforms of commonly encountered time functions and useful theorems on Laplace transforms. Chapter 3 describes the modeling of control systems using transfer function approach. Different types of systems such as mechanical, electrical circuits, simple hydraulic circuits, thermal systems, etc. are modeled using block diagrams and the associated transfer functions. Chapter 4 employs Mason's gain formula to model the control systems using signal flow graph approach. Chapter 5 deals with system types and steady-state error of different systems with different inputs. Time response analysis is presented in Chapter 6 wherein controller elements and their effect on control systems is described. Chapter 7 introduces the absolute stability of control systems along with the study of special cases using the Routh-Hurwitz criterion. Chapter 8 discusses control system analysis and design by Root Locus analysis. Chapter 9 briefly presents the frequency method, which includes the Bode diagram and analysis and design by the Nquist theorem and sampling techniques. Chapter 10 presents a fundamental analysis of control systems in state space. The concepts of controllability and observability of control systems are discussed in detail, along with the transformation of system models from transfer function model to state space model and vice versa. This chapter also includes design techniques in state space. Finally, Chapter 11 briefly presents different modern methods of analysis and design to familiarize the students with advances in control system design and analysis.

Last but not least, we would like to convey our gratitude to Ms. Aisha Qamar, Ms. Sher Bano, and Dr. Shahzeb Iqbal for helping us prepare the manuscript of the book.

Conclusively, in this monograph, the basic concepts are emphasized, and highly mathematical arguments are carefully avoided. Mathematical proofs are provided as and when they contribute to understanding the subject presented to students. Each chapter carefully chooses examples so that the reader will have a clear understanding of the subject matter.

PowerPoint slides and a solutions manual for qualified textbook adoption are available on the Routledge website (www.Routledge.com/9781032776286).

# About the Authors

**Dr. Noman Jafri** is a distinguished academic and engineering professional, currently serving as the dean of engineering at Abasyn University in Pakistan. His academic journey has been marked by a commitment to excellence and a passion for engineering innovation. He obtained his master's and PhD in electrical engineering in 1967 from the University of Ottawa in Canada. With his extensive educational background and expertise, Dr. Jafri has made significant contributions to the field of engineering. He has not only been a diligent educator but has also played a pivotal role in shaping the curriculum and research initiatives. Under his leadership, the engineering department has thrived and grown, attracting aspiring engineers from all over Pakistan.

**Dr. Imran Shafi** is a distinguished Pakistani computer engineer and researcher, known for his pioneering contributions to time-frequency analysis, particularly through the integration of neural networks to enhance signal processing techniques. His research focuses on improving time-frequency distributions (TFDs) for the analysis of nonstationary signals, addressing key challenges such as blurring and cross-term interference that commonly hinder the accuracy of traditional methods. By developing innovative approaches that utilize neural networks, Dr. Shafi has significantly advanced the resolution and energy concentration of TFDs, enabling more precise representation of instantaneous signal frequencies. With over 120 research publications to his name and a substantial citation count, his work has made a lasting impact in the fields of signal processing and computational intelligence, earning recognition within the global academic community.

**Dr. Uzair Khaleeq uz Zaman** is a highly accomplished academic and globally engaged educator, holding a PhD in industrial engineering from Arts et Métiers ParisTech, France, with over ten years of experience in university teaching, curriculum development, and applied research. Currently serving as associate professor and head of the Mechanical Engineering Department at NUST, Pakistan, he is recognized for his expertise in smart manufacturing, industry 4.0, and digital transformation. Dr. Uzair's international teaching footprint spans France, China, and Luxembourg, where he has delivered lectures and workshops at leading institutions. As a published author, keynote speaker, and subject matter expert, his work focuses on industrial automation, additive manufacturing, and smart manufacturing systems. Moreover, with a strong commitment to bridging academia and industry, he has successfully led industry-funded research projects and consultancy assignments in areas such as manufacturing optimization, warehouse automation using AGVs, and process planning. Beyond his technical expertise, Dr. Uzair is a strong advocate for global academic collaboration and student-centered, outcome-based learning, ensuring that his teaching and research contribute meaningfully to both academic excellence and real-world industrial innovation.

**Dr. Atal Anil Kumar** is currently working as a professor at the Laboratory of Design, Fabrication, and Control (LCFC), Université de Lorraine as the chair of Robotics for Sustainable and Agile Industry. He has a bachelor's degree in mechatronics and an Erasmus master's in advanced robotics (with scholarship) from France and Italy. He has been awarded a research internship grant by the Indian Institute of Technology, Bombay (IIT-B), for performing research during his bachelor's degree. Before joining as professor, Dr. Atal worked as a research scientist in the Department of Engineering at the University of Luxembourg. Dr. Atal has also worked as an R&D engineer in the automotive and steel manufacturing industries to gain significant hands-on experience to apply his theoretical knowledge. His research interests include nonlinear control, collaborative robots, robot kinematics and dynamics, and human-robot interaction.

# 1 Linear Control System

## 1.1 BACKGROUND

The real world depends upon declaring variables and measuring quantities. However, the requirement is to control them to behave desirably over time. Many engineering applications of control can be observed daily, such as holding a mobile phone for a selfie, driving a bicycle, or walking down the stairs. The air conditioners in our homes are fitted with a thermostat to regulate their operation to control the temperature at a specified value. Moreover, to maintain the speed of a car independent of road bumps and inclines, cruise control mode is provided in almost all cars these days. Yet another example of automatic control is an aircraft's autopilot, which tries to maintain variables like heading, altitude, and speed as per the pilot's desire. Similarly, the human body is an ideal combination of sensors, controllers, and actuators. There are various control strategies in action within our bodies to ensure their perfect functioning in adverse conditions, ranging from controlling blood pressure and blood sugar to balancing body temperature. Further, to manage a country's economy, governments take fiscal measures to control inflation and unemployment. The ingress of technology in today's world has, hence, given rise to highly challenging and widely varying control issues. From autonomous vehicles to spaceships, manufacturing processes to the chemical environment, intelligent machines to collaborative robots, communication networks to cyber security and blockchains, every aspect shows a range of problems. Figure 1.1 shows some common examples of everyday control.

It is evident that the universe is controlled by nature, wherein corrective actions are applied to prevent disturbances to keep the universe within certain bounds. Similarly, control engineering revolves around comprehending, designing, fabricating, and managing nature's materials and forces to enhance humanity's well-being. Engineers in control systems focus on understanding and directing specific aspects of their surroundings to deliver valuable economic products for society by controlling machines, plants, and processes using specific mechanisms to provide the best products or services. Figures 1.2 and 1.3 show uncontrolled and controlled phenomena in the system environment, respectively.

Linear control systems are founded on the principles of feedback theory and linear system analysis. It incorporates the principles of network theory and communication theory, as well. Consequently, control engineering is an interdisciplinary field that extends beyond traditional engineering disciplines and is equally relevant to electrical, mechanical, chemical, environmental, civil, political, and economic systems.

A control phenomenon frequently involves electrical, mechanical, and chemical systems or components. In essence, a control system is a network of interconnected components arranged in a system configuration designed to deliver enhanced performance in terms of improved speed of response, accuracy, and stability. Classical or

DOI: 10.1201/9781003484202-1

(a) (b) (c)

**FIGURE 1.1** (a) Holding a mobile phone for a selfie, (b) driving a bicycle, and (c) walking down the stairs (everyday control).

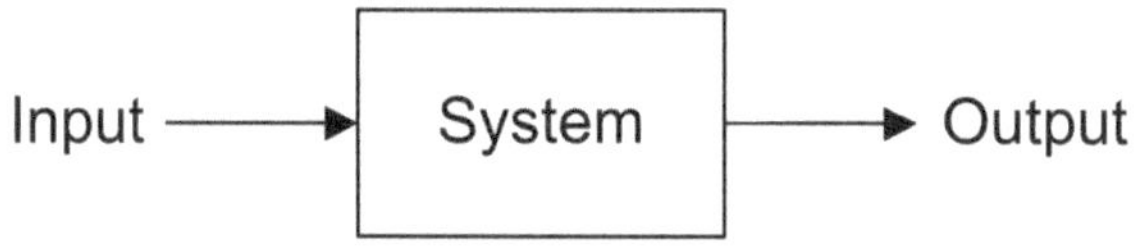


**FIGURE 1.2** Uncontrolled phenomenon.

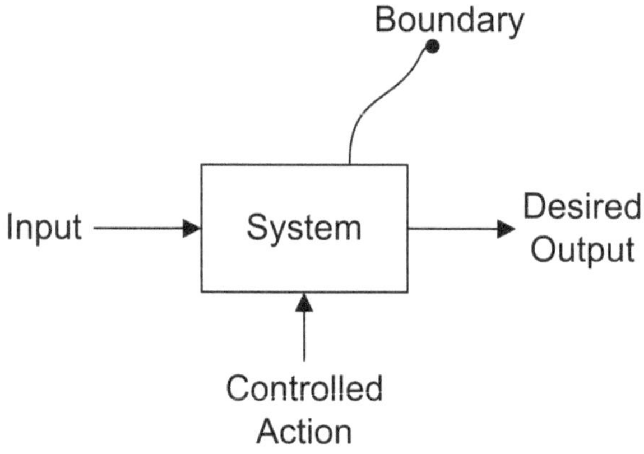


**FIGURE 1.3** Controlled phenomenon.

modern methods are used to control systems, dividing the approaches into two types for controlling a particular system, plant, or product. Figure 1.4 gives the general process of solving the control problem for a layman.

## 1.2 TYPES OF CONTROL ENGINEERING

Classical control engineering starts with continuous or discrete mathematical modelling of the system in time, frequency or complex domain using differential

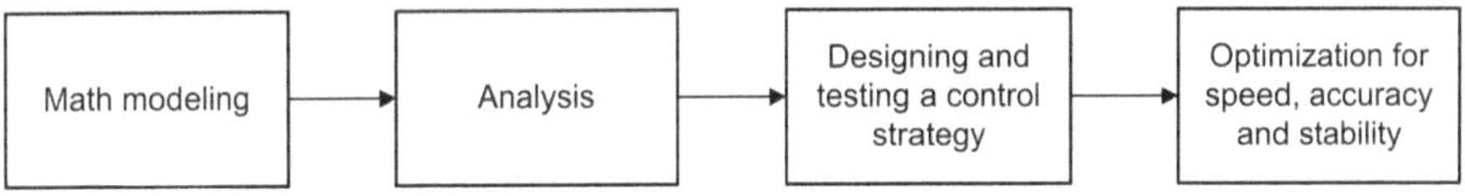


**FIGURE 1.4** General process of solving a control problem.

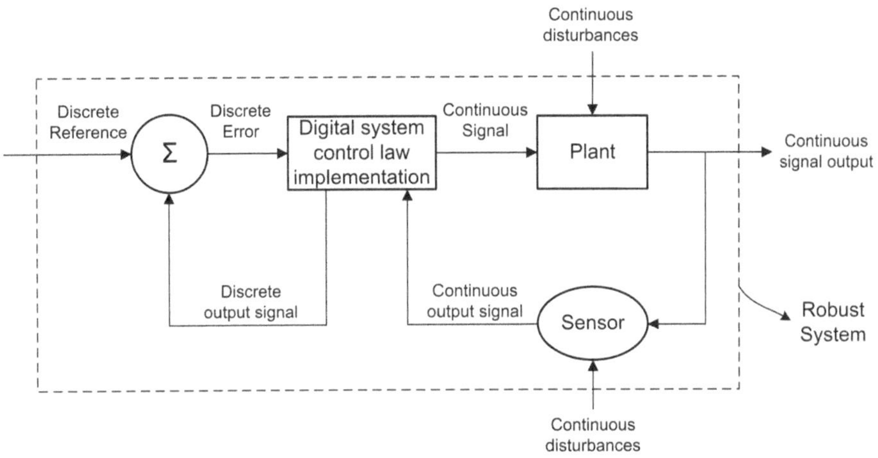


**FIGURE 1.5** Robust control engineering.

equations, Laplace transform, Fourier transform, or z-transform methods. Responses to various inputs, including impulse, step function and others, are mathematically modelled to subsequently find and analyze the related parameters like settling time, percentage overshoot, open loop gain, phase margin, bandwidth, etc. Furthermore, classical control engineering covers concepts like sampling, poles and zeros, transfer function, Nyquist plots, Bode plots, etc., which are generally used for single input and single output (SISO) systems.

On the other hand, modern control engineering approaches the problem by simplifying higher-order equations to lower-order differential equations and using matrix algebra to solve them. It extensively uses the state space approach with the help of Eigenvalues and Eigenvectors instead of complex ordinary differential equations to deal with multiple input and multiple output (MIMO) systems.

Additionally, there can be other sub-types of control engineering based on methodologies. These include robust control engineering, whose aim is to widen stability and performance in the presence of internal and external disturbances and faults in the system. Moreover, 'optimal control engineering' minimizes the cost function under physical and performance constraints. Game theory is an example of robust and optimal control engineering with a relationship between conflict and cooperation, reducing cost function against disturbances and noise. Figure 1.5 shows a typical block diagram of an elementary type of plant being controlled in a robust manner. The system is said to be robust if it has good regulation in the presence of

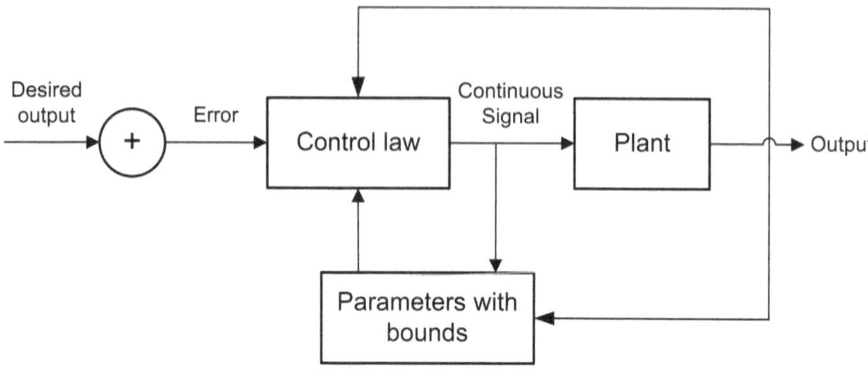


**FIGURE 1.6** Adaptive control engineering.

disturbances and changes in plant parameters, that is, it possesses good disturbance rejection and low sensitivity to changes in parameters.

Adaptive control engineering employs an adaptive controller by some mechanism for adapting certain parameters. Parameter adjustment uses an extra loop other than a regular feedback loop (see Figure 1.6).

Nonlinear control engineering handles nonlinearities by dividing the system into linear and nonlinear parts and analyzes extensive mathematical computations. The linear system theory and principles are generally applied to the analysis of any system. The input-output relationship can be suitably estimated using available mathematical methods in the form of impulse or frequency response and transfer function. These are then analyzed and optimized subsequently to get better control over the system functions.

## 1.3 HISTORICAL DEVELOPMENT OF CONTROL SYSTEM

The deployment of automatic control systems was first observed somewhere in the third century BC in Egypt, where a water clock was being used as the first feedback control device to regulate the water level in a vessel. Due to its effectiveness, this feedback control device was still used in Baghdad until the late twelfth century AD. On careful analysis, it is evident that a plethora of automatic devices have been observed being used in history for accomplishing various tasks, including entertainment. The automata was one example where the dancing figures repeating a particular task were used in Europe during the seventeeth and eighteenth centuries. A selected historical development of control system theory and application can be described as follows in Table 1.1.

## 1.4 EXISTING CHALLENGES AND EMERGING TRENDS IN CONTROL ENGINEERING

The use of high-tech equipment in ordinary people's daily lives is increasing, imposing highly efficient, accurate, but less expensive control solutions for more complex problems

**TABLE 1.1**
**Historical Evolution of Control Systems**

| Year | Historical Developments of Control Systems |
|---|---|
| 1258 | Water clocks use water as a regulator to hold time. |
| 1620 | Temperature regulator of a furnace attributed to Drebbel, an example of a closed-loop automatic control system. |
| 1769 | James Watt's steam engine was a pivotal development regulated by a centrifugal flyball governor. The Watt steam engine is recognized as a milestone marking the commencement of the Industrial Revolution in Great Britain. This era witnessed significant advancements in mechanization, a technological precursor to automation. |
| 1800 | Eli Whitney's pioneering idea of interchangeable parts manufacturing was demonstrated in the production of muskets. Whitney's innovation is widely regarded as the initiation of mass production, marking a significant milestone in manufacturing history. |
| 1868 | J.C. Maxwell formulated a mathematical model for controlling a steam engine using a flyball governor and highlighted its instabilities, which marked the inaugural use of differential equations in modelling control systems, signifying the commencement of mathematical control and systems theory. Maxwell's analytical assessment compellingly laid the foundation for control theory, catalyzing significant advancements over the next 150 years. |
| 1913 | Henry Ford introduced Assembly lines and automatic assembly mechanisms in automotive production setups. |
| 1927 | Feedback amplifiers were developed and tested by H. W. Bode. |
| 1932 | H. Nyquist gave a methodical analysis of systems' stability. |
| 1952 | The mechanism for controlling the machine tools' axes was developed numerically. |
| 1954 | George Devol proposed robotic mechanism theory and design. |
| 1960 | The robotic mechanism was developed and tested based on programmed article transfer given by George Devol. |
| 1970 | Theoretical and mathematical foundations for state space and optimal control were developed. |
| 1980 | Theoretical and mathematical foundations for robust control. |
| 1990 | Global attention towards automation in manufacturing industries. |
| 1994 | Feedback control is widely utilized in various industries, including automobiles, aviation, and mass transit applications. Manufacturing setups moved towards robust control systems to improve reliability in manufacturing. |

with little information about their dynamic behavior. Examples include autonomous road vehicles, space vehicles, underwater remotely operated vehicles, and unmanned aerial vehicles. Today, we need more novel control strategies and mixed interdisciplinary methods to meet these challenges and optimize existing control methods.

Moreover, due to widely expanding research in the computer science area, the twenty-first century has witnessed significant advances in computational power at low cost, influencing the control engineering domain. Such influence has given rise to intelligent control methods in the form of expert systems as decision-support processes. Moreover, a heavy interest is observed towards understanding and controlling discrete event systems that cannot be described by differential/difference equations in modern manufacturing processes, leading to interdisciplinary research. The rise of artificial intelligence (AI), machine learning (ML) approaches, artificial neural networks (ANNs), and fuzzy and hybrid control of sequential machines are a few emerging trends being investigated by control engineers to control the variety of complicated systems.

## 1.5 DEFINITION OF A SYSTEM

A system 's' is a partially interconnected set of abstract objects '$a_1$, $a_2$ $a_3$ . . .', that are termed as the components of 's'. The components of 's' may be oriented or non-oriented, finite or infinite, and each may be associated with a finite or infinite number of variables called input or output. Subsequently, a system may range from SISO to single input multiple outputs (SIMO), and multiple inputs single-output (MISO) to MIMO. A system's desirable purpose is to generate useful output(s) by aggregating different components under action from various input(s). This definition covers all the systems in the universe, for example, electrical, mechanical, chemical, nuclear, astronomical, ecological, economical, astrological, political, etc. Many possible inputs can be given to a system, as shown in Figure 1.7.

The system can be categorized into various types based on its physical properties, as given in Figure 1.8. Continuous and discrete-time systems are the two most popular system classes encountered in real life based on temporal characteristics. Continuous time systems are mostly systems in nature, whereas discrete-time systems require computer programs that work on specific clock pulses.

Nonlinear systems may be approximated with piecewise linear elements, which necessitate the learning of linear system theories, properties, and mathematics.

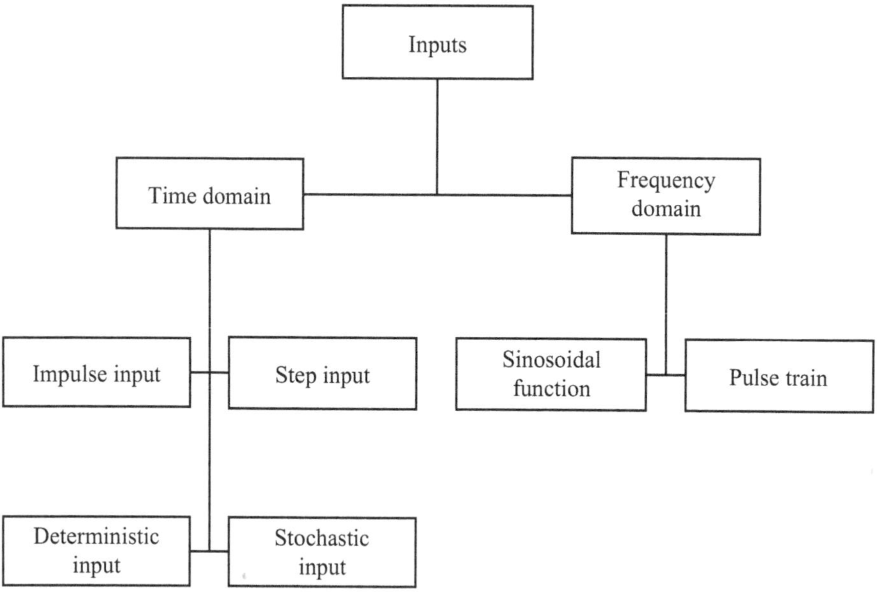


**FIGURE 1.7** Possible inputs to the system.

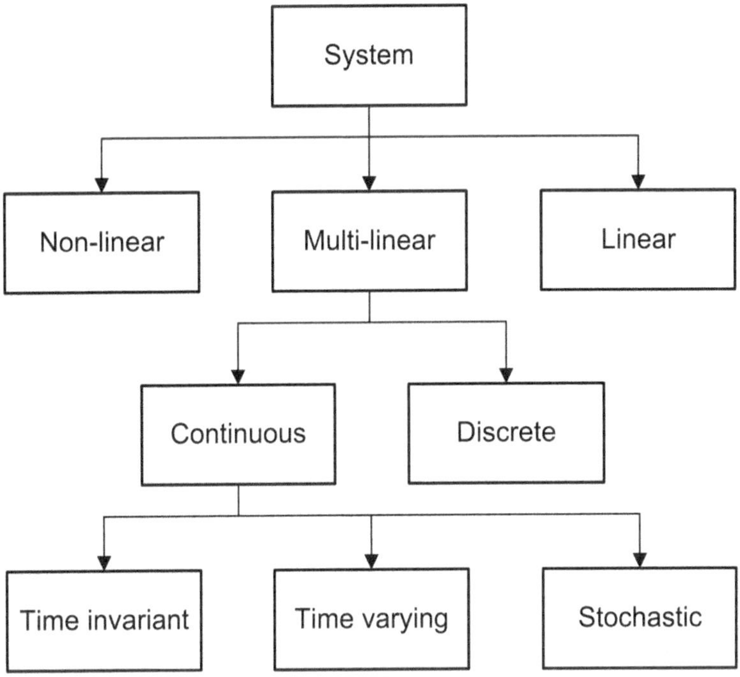


**FIGURE 1.8** Types of physical systems.

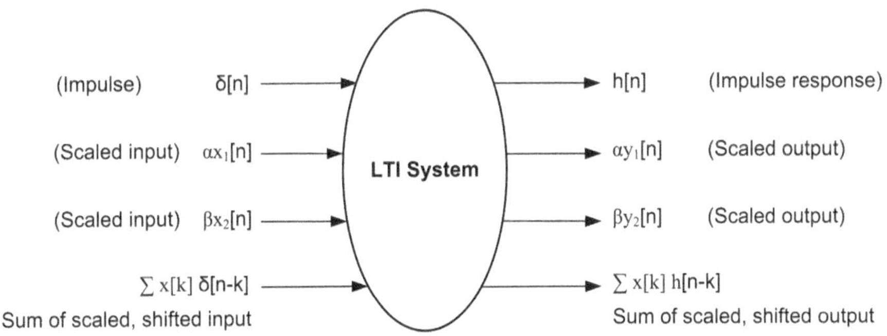


**FIGURE 1.9** Linear time invariant system.

A linear system, on the other hand, follows the principle of superposition by upholding scaling and additivity properties. A linear time-invariant (LTI) system also follows time invariance in addition to a linear system's scaling and additivity properties as shown in Figure 1.9.

## 1.6 TYPES OF CONTROLLERS

The controllers can be divided into two main classes, open loop and closed loop controllers, to achieve either of the two main objectives, that is, regulation or tracking. A regulator is a controller that tries to maintain a constant value of the desired parameter under control, whereas if the desired parameter varies and is to be maintained, then the controller becomes a tracker. Modern approaches to control engineering try to optimize for both types. Broadly, the control system classification is given in Figure 1.10.

## 1.7 CONTROL SYSTEM MODELLING AND DESIGN

Control system modelling first entails identifying the system and selecting suitable mathematical relationships to fit the observed input/output experimental data. It is then followed by analytical modelling, which consists of the systematic application of basic physical laws to system components and the interconnection of these components. Once the system is modelled, its experimentation in the time domain is carried

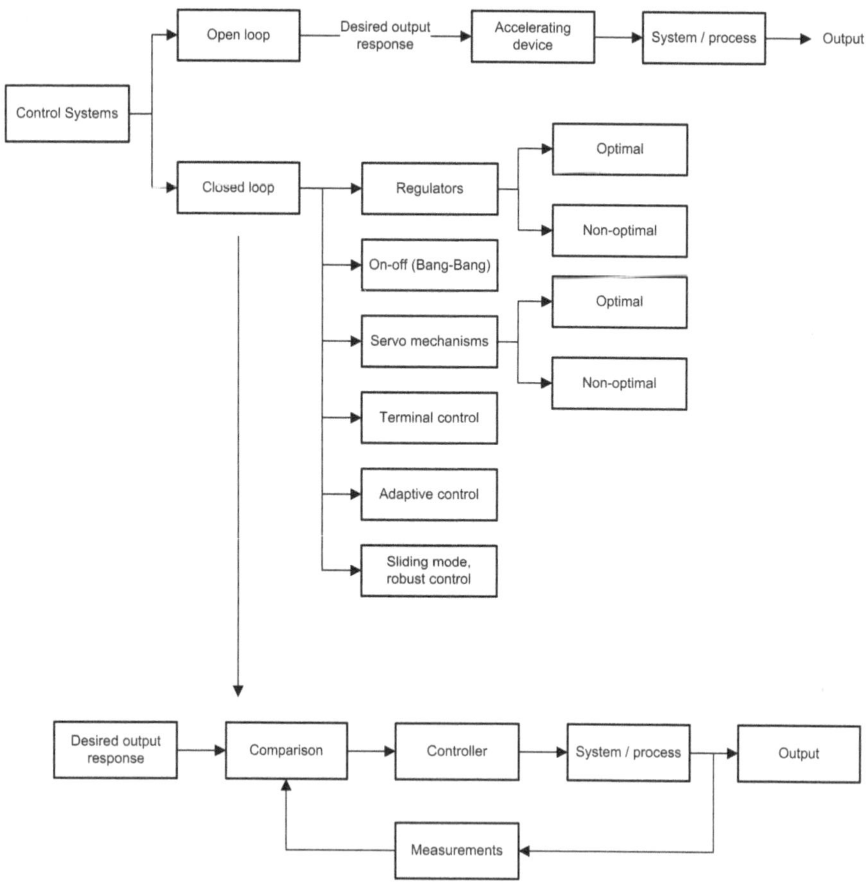


**FIGURE 1.10** Open loop and closed loop controllers and classification of control system.

out using impulse, series of pulses and step input (it has the maximum rate of change) to check out the controller's performance, as shown in Figure 1.11. The experimental performance can also be confirmed in the frequency domain or stochastically.

The techniques in control system analysis are broadly divided into two categories, as shown in Figure 1.12. Each category includes respective methods that will be discussed in the respective chapters of the book.

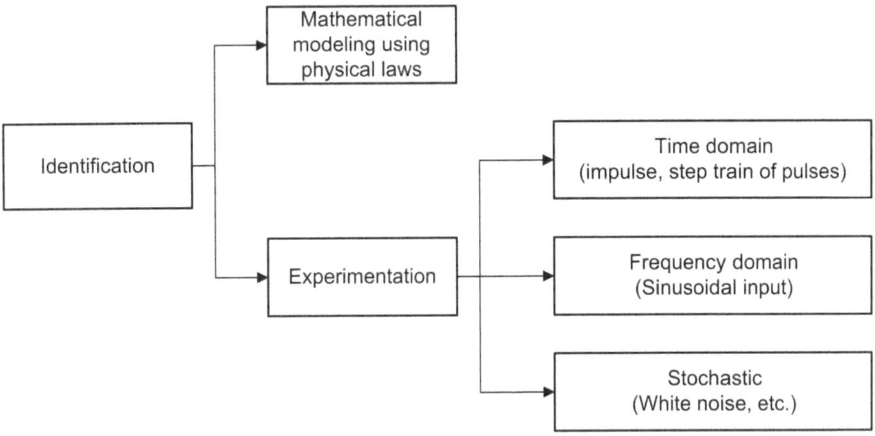


**FIGURE 1.11** Identification of control system.

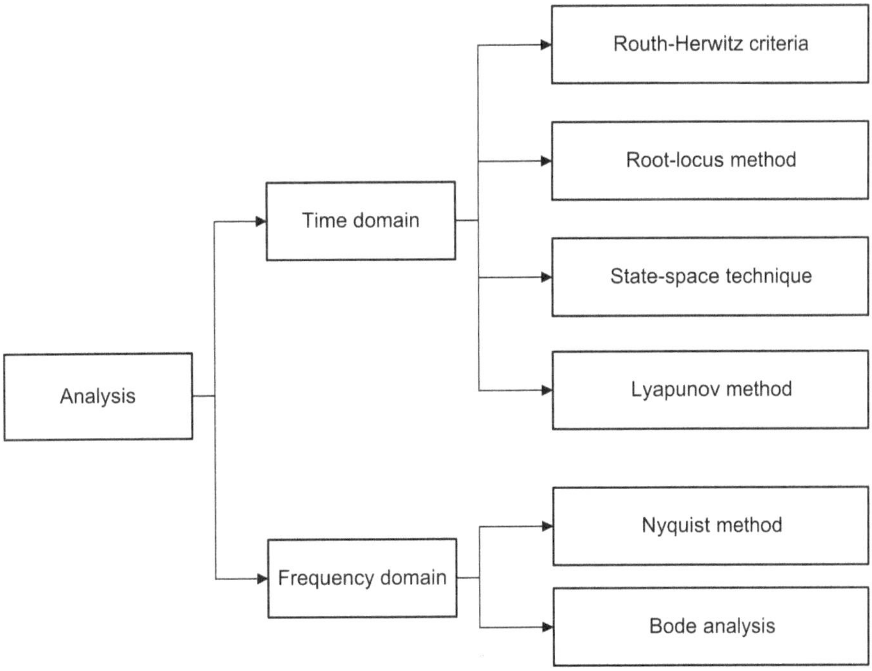


**FIGURE 1.12** Analysis techniques of control system.

Moreover, the techniques for the synthesis and design of feedback control systems are again divided into two categories. The first category is based on classical methods using transforms, including Laplace, Fourier, or z-transforms, etc. The second category is based on state space technique, also known as modern methods, where higher order coupled differential equations are de-coupled into lower order differential equations using state space. The matrix algebra takes a central seat once the state space method is used. The same is depicted in Figure 1.13.

**Example 1.1: Design a controller using an operational amplifier (Opamp) for Figure 1.16**

For any Opamp, following equations are used for its modeling. First, it can be modeled using Eq. 1.1 and Figure 1.14 in the time domain:

$$V_{out}(t) = -\frac{1}{RC}\int V_{in}(t)\,dt \tag{1.1}$$

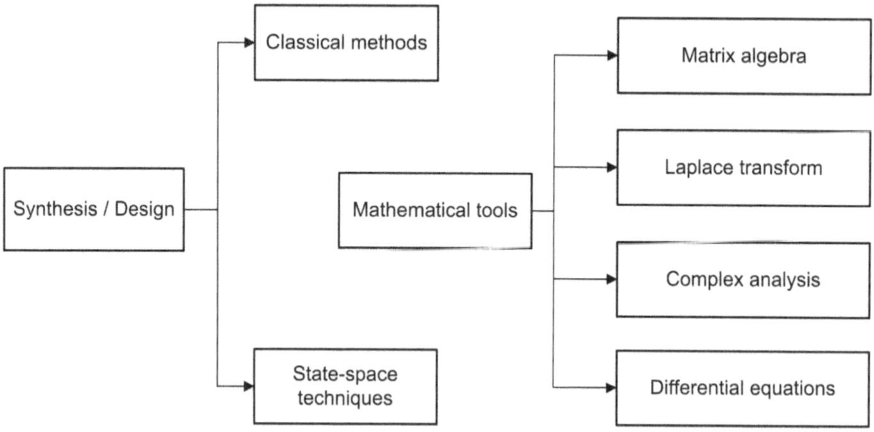


**FIGURE 1.13** Design and synthesis of control system.

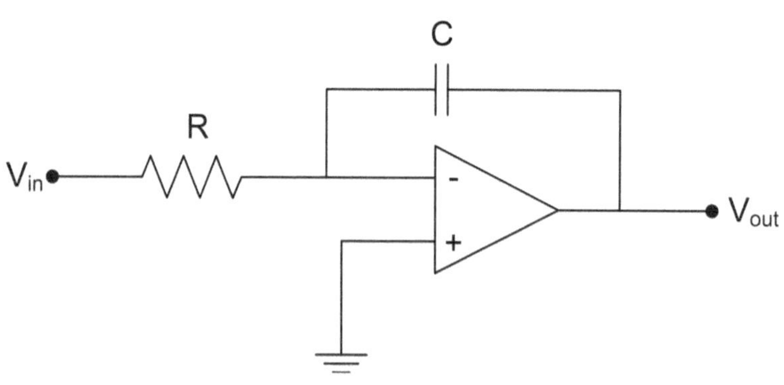


**FIGURE 1.14** An operational amplifier RC circuit.

If values of 'R' and 'C' are unity in Eq 1.1, then opamp is simply an integrator.

Applying Laplace transform to Eq 1.1 gives:

$$V_{out}(s) = -\frac{1}{RC}\frac{V_{in}(s)}{s} \tag{1.2}$$

Simplifying Eq 1.2 would give:

$$\frac{V_{out}(s)}{V_{in}(s)} = \frac{-\frac{1}{Cs}}{R} = -\frac{Z_c}{Z_R} = -\frac{Z_b}{Z_a} \tag{1.3}$$

Furthermore, it is known that impedance term, (Z), in Laplace domain is defined for resistor, capacitor, and inductor as:

$Z_R = R$ (Resistor)

$Z_C = \frac{1}{C_s}$ (Capacitor with 'C' being the capacitance value)

$Z_L = sL$ (Inductor with 'L' being the inductance value)

So the following figure can be generated and drawn (see Figure 1.15) for an operational opamp.

Moving on, the transfer function of the controller in Figure 1.16 can be given as under:

$$C(s) = \frac{s+b}{s^2+as+c} \tag{1.4}$$

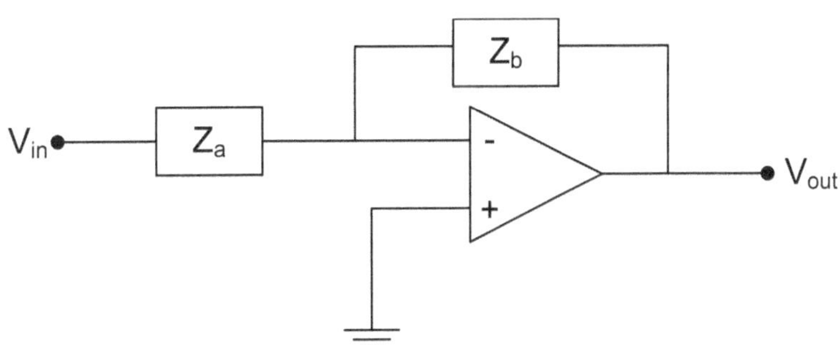


**FIGURE 1.15** An operational amplifier impedance circuit.

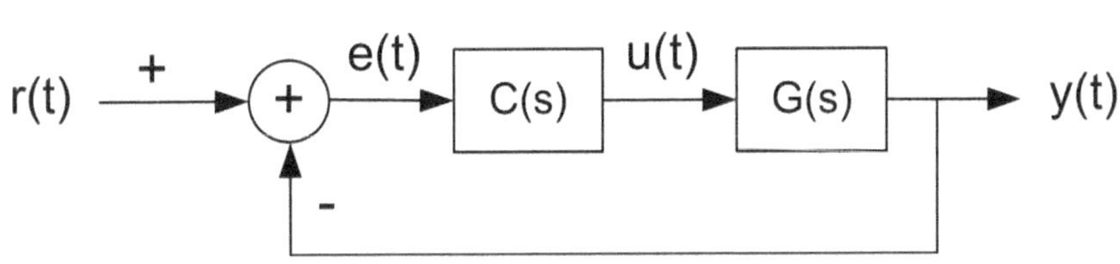


**FIGURE 1.16** The block diagram for a closed loop controller system.

where a, b, and c are constants.

$$\because\ C(s) = \frac{U(s)}{E(s)},$$

$$s^2U(s) + asU(s) + cU(s) = sE(s) + bE(s) \tag{1.5}$$

Taking inverse Laplace transform of Eq 1.5 would give:

$$u''(t) + au'(t) + cu(t) = e'(t) + be(t)$$

Simplifying,

$$u''(t) = -au' - cu(t) + e'(t) + be(t)$$

Taking double integral on both sides gives:

$$u(t) = \int[-au(t) + e(t) - c\int u(t)dt + b\int e(t)dt]dt \tag{1.6}$$

An opamp can be used as an 'integrator' due to its physical properties. Based on the results, a controller using an opamp can be drawn as shown in Figure 1.17.

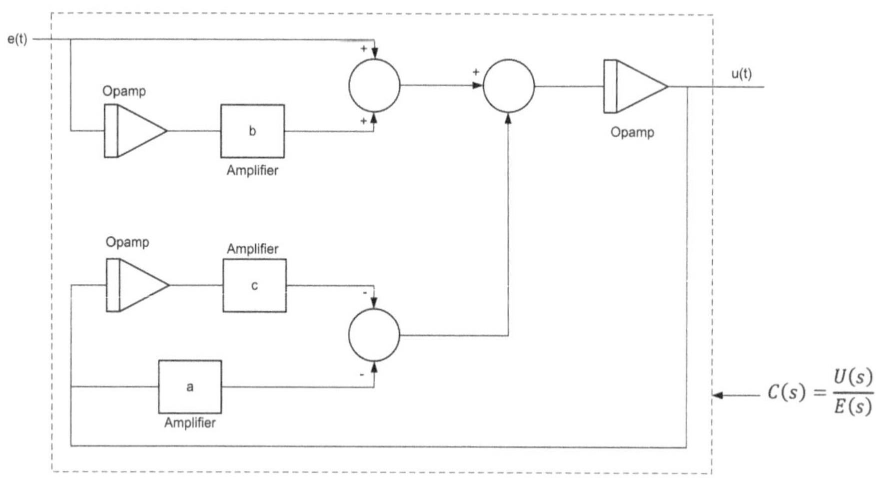


**FIGURE 1.17** Controller using an opamp.

# 2 Laplace Transform

## 2.1 DEFINITION

Laplace transform is a method of solving differential equations by converting them to an algebraic equation that is easier to solve. By applying the Laplace transform, a time-domain system is transformed to s-domain (also known as Laplace domain). The inverse Laplace transform is used to revert to the original time-domain system as shown in Figure 2.1.

Let f(t) be a function of t with following properties:

- f(t) is identically zero for $t < 0$
- f(t) is continuous from the right at $t = 0$ ($\lim_{t \to 0} f(t) = f(0)$) for $t > 0$)
- f(t) is piece wise continuous

Then the Laplace transform of f(t) denoted by $\mathcal{L}$ [f(t)] is defined as

$$\mathcal{L}[f(t)] = F(s) = \int_0^\infty e^{-st} f(t)\,dt \tag{2.1}$$

where, $s = \sigma + j\omega$ is a complex number. The Laplace transform shall exist if the integral converges for some value of $\sigma$.

## 2.2 LAPLACE TRANSFORM OF SOME SIMPLE FUNCTIONS

### 2.2.1 Unit Impulse

The unit impulse is given by Eq. 2.2:

$$\delta(t) = \begin{cases} \dfrac{1}{\epsilon}, & 0 < t < \epsilon \\ 0, & t > \epsilon \end{cases} \tag{2.2}$$

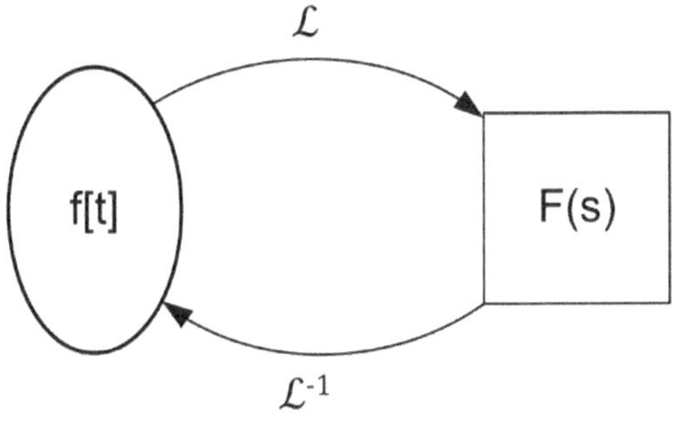


**FIGURE 2.1** Laplace transform.

DOI: 10.1201/9781003484202-2

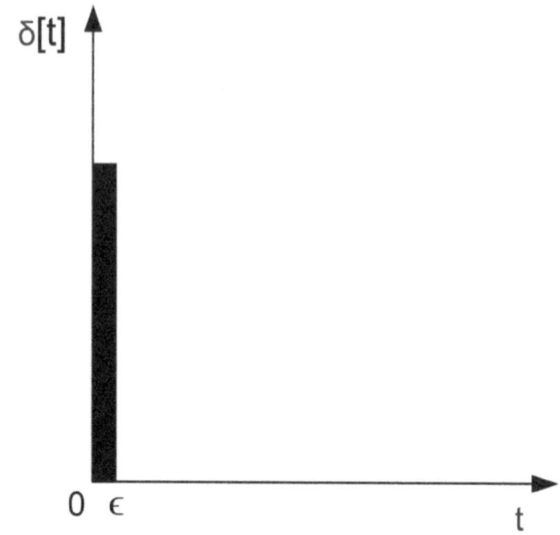


**FIGURE 2.2** Distribution of δ(t) with time.

'∈' is a very small number such that $\epsilon \approx 0$ and $1/\epsilon$ is very large. The distribution of δ(t) with time is given in Figure 2.2 and is depicted in Eq. 2.3.

$$\int_0^{\infty} \delta(t)dt = \int_0^{\epsilon} \delta(t)dt + \int_{\epsilon}^{\infty} \overset{0}{\delta(t)}dt = \int_0^{\epsilon} \frac{1}{\epsilon} dt = \frac{1}{\epsilon} t \Big|_0^{\epsilon} = \frac{1}{\epsilon} \times \epsilon = 1 \tag{2.3}$$

There is a sifting property associated with δ(t) as shown in Eq. 2.4.

$$\int_0^{\infty} \delta(t) g(t) dt = g(0), \text{ and}$$

$$\int_0^{\infty} \delta(t - t_0) g(t) dt = g(t_0) \tag{2.4}$$

The previous equation in view, the functional form can be written as follows:

$$f(t) = \delta(t) = \begin{cases} 0 & t \neq 0 \\ 1 & t = 0 \end{cases} \tag{2.5}$$

$$\int_a^b \delta(t) dt = \begin{cases} 1 & t = 0 \\ 0 & t \neq 0 \end{cases} \tag{2.6}$$

$$\delta(t - t_0) = 0 \qquad t \neq t_0$$

$$\int_a^b \delta(t - t_0) d(t) = 1 \qquad a \leq t_o \leq b \tag{2.7}$$

In addition

$$\int_a^b f(t)\delta(t-t_0)d(t)=1 \qquad \begin{cases} f(t_0) & a \le t_0 \le b \\ 0 & otherwise \end{cases}$$

$$\int_a^b e^{-st}\delta(t)d(t)=1 \tag{2.8}$$

Now,

$$\mathcal{L}[\delta(t)] = \int_0^\infty e^{-\delta t}\delta(t)dt = \int_0^\in e^{\delta t}\delta(t)dt + \int_\in^\infty e^{\delta t}\overset{0}{\not\delta}(t)dt = \frac{1}{\in}\int_0^\in e^{\delta t}dt = e^0 = 1$$

$$\text{So, } \mathcal{L}[\delta(t)]=1 \tag{2.9}$$

This means the inverse transformation will be, $\mathcal{L}^{-1}[1]=\delta(t)$

## 2.2.2 Step Function

$$u(t)=\begin{cases} 1 & t \ge 0 \\ 0 & t < 0 \end{cases} \tag{2.10}$$

$$F(s)=\int_a^\infty 1.e^{-st}\,d(t)=\frac{e^{-st}}{s}\Bigg|_0^\infty=\frac{1}{s}$$

So

$$\mathcal{L}[u(t)]=\frac{1}{s} \tag{2.11}$$

Eq. 2.11 can be written as:

$$\mathcal{L}[C]=\frac{C}{s}$$

where, C is a constant.

## 2.2.3 Ramp Function

$$f(t)=t \quad t \ge 0$$

$$F(s)=\int_a^\infty t.e^{-st}\,d(t)=\frac{t.e^{-st}}{s}\Bigg|_0^\infty+\frac{e^{-st}}{s^2}\Bigg|_0^\infty=\frac{1}{s^2} \tag{2.12}$$

So

$$\mathcal{L}[t] = \frac{1}{s^2} \tag{2.13}$$

Subsequently,

$$\mathcal{L}\left[t^2\right] = \frac{2!}{s^3}$$

And

$$\mathcal{L}\left[t^3\right] = \frac{3!}{s^4}$$

In simple terms, the previous equation can be generalized as:

$$\mathcal{L}\left[t^n\right] = \frac{n!}{s^{n+1}}$$

### 2.2.4 Exponential Function

$$f(t) = e^{at}$$

$$F(s) = \int_a^{\infty} e^{at}.e^{-st}d(t) = \int_0^{\infty} e^{(a-s)t} \quad dt = \frac{e^{-(S-a)t}}{a-s}\Bigg|_0^{\infty} = \frac{1}{a-s}\left[\frac{1}{\infty} - 1\right] \tag{2.14}$$

$$F(s) = \frac{1}{s-a}$$

### 2.2.5 Sinusoidal Functions

$$f(t) = \sin(at) = \frac{e^{jat} - e^{-jat}}{2j} \tag{2.15}$$

$$F(s) = \frac{1}{2j}\left[\int_0^{\infty}\left(e^{jat} - e^{-jat}\right)e^{-st}dt\right] = \frac{1}{2j}\left[\int_0^{\infty} e^{-j(a+s)t}\,dt - \int_0^{\infty} e^{-(ja+s)t}dt\right.$$

$$= \frac{1}{2j}\left[\frac{1}{-ja+s} - \frac{1}{ja+s}\right] = \frac{a}{s^2+a^2} \tag{2.16}$$

Similarly, for

$$f(t) = \cos(at)$$

$$F(s) = \frac{s}{s^2+a^2} \tag{2.17}$$

### 2.2.6 Laplace Transforms of Derivatives

$$f(\mathrm{t}) = \dot{\mathrm{y}}(\mathrm{t}) \tag{2.18}$$

$$\mathcal{L}[\dot{\mathrm{y}}(\mathrm{t})] = \int_0^\infty \dot{\mathrm{y}}(\mathrm{t})\mathrm{e}^{-\mathrm{st}}\mathrm{dt} = \left|-\mathrm{y}(\mathrm{t})\mathrm{e}^{-\mathrm{st}}\right|_0^\infty - \int_0^\infty -\mathrm{sy}(\mathrm{t})\mathrm{e}^{-\mathrm{st}}\mathrm{dt} \tag{2.19}$$

$$L[\dot{\mathrm{y}}(\mathrm{t})] = -y(0) + sY(\mathrm{s}) \tag{2.20}$$

Similarly,

$$L\left[\frac{d^2y}{dt^2}\right] = s^2Y(s) - sy(0) - \frac{dy}{dt}(0) \tag{2.21}$$

$$L\left[\frac{d^ny}{dt^n}\right] = s^nY(s) - s^{n-1}y(0) - s^{(n-2)}\frac{dy}{dt}(0)\ldots\frac{dy^{n-1}}{dt^{n-1}}(0) \tag{2.22}$$

## 2.3 INVERSE LAPLACE TRANSFORM

Following methods are used to obtain inverse Laplace transform:

a) Laplace transform tables
b) Partial fraction method
c) Inversion Integral

The Laplace transform table is given in Table 2.1.

**TABLE 2.1**
**Laplace Transform Table**

| f(t) | F(s) |
|---|---|
| Step function u(t) | $\frac{1}{s}$ |
| $e^{-\alpha t}$ | $\frac{1}{s+a}$ |
| $\sin wt$ | $\frac{w}{s^2+w^2}$ |
| $\cos wt$ | $\frac{s}{s^2+w^2}$ |
| $e^{-at}f(t)$ | $F(s+a)$ |
| $t^n$ | $\frac{n!}{s^n+1}$ |
| $\int_{-\infty}^{t} f(t)dt$ | $\frac{F(s)}{s} + \frac{1}{s}\int_{-\infty}^{0} f(t)dt$ |
| Impulse function | 1 |

### 2.3.1 Partial Fraction Method

**Example 2.1**

$$Y(s) = \frac{s^3 + 8s^2 + 26s + 22}{s^3 + 7s^2 + 14s + 8} = 1 + \frac{s^2 + 12s + 14}{s^3 + 7s^2 + 14s + 8} = 1 + Y_1(s) \tag{2.23}$$

Now,

$$Y_1(s) = \frac{A}{s+1} + \frac{B}{s+2} + \frac{C}{s+4}$$

Assuming values for A, B, and C as

$$A = 1, B = 3, C = -3$$

Now,

$$Y(s) = 1 + \frac{1}{s+1} + \frac{3}{s+2} - \frac{3}{s+4} \tag{2.24}$$

Finally,

$$y(t) = \delta(t) + e^{-t} + 3e^{-2t} - 3e^{-4t} \tag{2.25}$$

**Example 2.2**

$$Y(s) = \frac{s^2 + 3s + 1}{s(s+1)^3} = \frac{A}{s} + \frac{B}{(s+1)^3} + \frac{C}{(s+1)^2} + \frac{D}{s+1} \tag{2.26}$$

$$A = \frac{(s^2 + 3s + 1).s}{s(s+1)^3}\Big|_{s=0} = 1$$

$$B = \frac{(s^2 + 3s + 1).(s+1)^3}{s(s+1)^3}\Big|_{s=-1} = 1$$

To calculate C,

$$\frac{d}{ds}\left[(s+1)^3 Y(s)\right] = \frac{d}{ds}\left(\left[\frac{(s+1)^3 (s^2 + 3s + 1)}{s(s+1)^3}\right]\right) \tag{2.27}$$

$$= \frac{d}{ds}\left[\frac{\left(s^2+3s+1\right)}{s}\right]$$

$$= \frac{s(2s+3)-(s^2+3s+1)}{s^2}\Big|_{s=-1}$$

$$C = \frac{s^2-1}{s^2}\Big|_{s=-1} = 0 \tag{2.28}$$

To Calculate D,

$$D = \frac{1}{2!}\left(\frac{d}{ds}\left(\frac{s^2-1}{s^2}\right)\right)\Big|_{s=-1}$$

$$D = \frac{d}{ds}1\Big|_{s=-1} - \frac{d}{ds}\left(\frac{1}{s^2}\right)\Big|_{s=-1}$$

$$D = -1 \tag{2.29}$$

Therefore, $Y(s) = \frac{1}{S} - \frac{1}{S+1} + \frac{1}{(S+1)^3}$

$$y(t) = 1 - e^{-t} + \frac{t^2}{2}e^{-t} \tag{2.30}$$

## 2.3.2 Laplace Transform Theorems

### a. Linearity

If ‘a’ and ‘b’ are constants and f(t) and g(t) are transformable, then

$$L\left[a{*}f(t)+b{*}g(t)\right] = a*L\left[f(t)\right]+b*L\left[g(t)\right] \tag{2.31}$$

### b. Complex Translation

If f(t) is transformable with transform F(s), then $e^{at}f(t)$ is also transformable where ‘a’ is real constant and has transform $F(s-a)$. Similarly, $e^{-at}f(t)$ has transform $F(s+a)$.

For example,

$$e^{-at}f(t) \rightarrow \frac{1}{s-a} \tag{2.32}$$

**c. Real Transformation to Right**

If f(t) is a transformable function with transform F(s) and 'a' is a nonnegative real number, then $f(t-a)$ is also a transformable function with transform $e^{-as}F(s)$.
Translation to left

$$f(t+a) \rightarrow e^{-as}F(s) \tag{2.33}$$

$$u(t-a) \rightarrow \frac{e^{-as}}{s} \tag{2.34}$$

$$u(t+a) \rightarrow \frac{e^{as}}{s} \tag{2.35}$$

**d. Real Differentiation**

$$\mathcal{L}\left[\frac{df}{dt}\right] = -f(0) + sF(s)$$

$$f(t) = e^{at}, f(0) = 1 \tag{2.36}$$

$$\frac{d}{dt}[f(t)] = \frac{s}{s-a} - 1 \tag{2.37}$$

**e. Real Integrations**

If f(t) is transformable, then its definite integral is also transformable.

$$L\left[\int_0^t f(u)\,du\right] = \frac{1}{s}F(s) \tag{2.38}$$

**f. Complex Differentiation**

If f(t) is transformable with transform F(s), tf(t) is also transformable and has transform, $-\frac{d}{ds}[F(s)]$. For example,

$$L[u(t)] = \frac{1}{s} \tag{2.39}$$

$$L[t] = -\frac{d}{ds}\left(\frac{1}{s}\right) = \frac{1}{s^2} \tag{2.40}$$

$$L[t^2] = -\frac{d}{ds}\left(\frac{1}{s^2}\right) = \frac{2!}{s^3} \tag{2.41}$$

$$L[t^n] = (-1)^n \frac{d^n}{ds^n}\left[\frac{1}{s}\right] = \frac{n!}{s^{n+1}} \tag{2.42}$$

$$L[sinwt] = \frac{w}{s^2 + w^2}$$

$$L[tsinwt] = -\frac{d}{ds}\left(\frac{w}{s^2 + w^2}\right) = \frac{2ws}{\left(s^2 + w^2\right)^2} \tag{2.43}$$

**g. Periodic Functions**

$$\text{If } f(t+T) = f(t)$$

$$\text{and } f(t+kT) = f(t) \tag{2.44}$$

Then it is a periodic function. If f(t) is a transformable function of period T, then its transform may be found by integration over the first period.

$$L[f(t)] = \frac{\int_0^T f(t)e^{-st}dt}{1-e^{-sT}} \tag{2.45}$$

$$L[f(t)] = \frac{\int_0^T f(t)e^{-st}dt}{1-e^{-2s}} = \frac{\int_0^T e^{-st}dt}{1-e^{-2s}} \tag{2.46}$$

$$= \frac{1-e^{-s}}{s\left(1-e^{-2s}\right)} = \frac{1}{s\left(1+e^{-s}\right)} \tag{2.47}$$

**h. Real Convolution**

Suppose f(t) and g(t) are Laplace transformable having transforms F(s) and G(s) respectively, then product *F(s) G(s)* is the transform of

$$h(t) = \int_0^t f(\vartheta)g(t-\vartheta)d\vartheta \tag{2.48}$$

$$= \int_0^t f(t-\vartheta)g(\vartheta)d\vartheta$$

$$h(t) = f(t) * g(t) = g(t) * f(t) \tag{2.49}$$

**Example 2.3**

$$h(t) = e^{-at} * \cos t = \int_0^t e^{-a(t-u)} \cos u.du \tag{2.50}$$

$$h(t) = e^{-at}\int_0^t e^{au}\cos u\,du = \frac{1}{a^2+1}\left[a\cos t + \sin t - ae^{-at}\right] \tag{2.51}$$

**i. Initial Value theorem**

If f(t) and its derivatives are transformable then

$$\lim_{s\to\infty} sF(s) = f(0) \tag{2.52}$$

**j. Final Value Theorem**

If f(t) and its derivatives are transformable and if f(t) has a transform F(s) and if all the singularity of F(s) are in the left half plane, then

$$\lim_{s\to 0} sF(s) = \lim_{t\to\infty} f(t) \tag{2.53}$$

**Example 2.4**

**Find the initial and final value of the function having Laplace transform**

$$F(s) = \frac{s+9}{s^2+7s+3} \tag{2.54}$$

Initial value

$$\lim_{t\to 0} F(s) = \lim_{s\to\infty} sF(s) = \lim_{s\to\infty}\frac{s(s+9)}{(s^2+7s+3)}$$

$$= \lim_{s\to\infty}\left(\frac{s^2+9s}{s^2+7s+3}\right) \tag{2.55}$$

$$= \lim_{s\to\infty}\left(\frac{1+\frac{9s}{s^2}}{1+\frac{7s}{s^2}+\frac{3}{s^2}}\right) = 1 \tag{2.56}$$

Final value

$$\lim_{t\to\infty} f(t) = \lim_{s\to 0} sF(s) = \lim_{s\to 0}\frac{s(s+9)}{s^2+7s+3} = 0 \tag{2.57}$$

**Example 2.5**

**Find f(t) if** $F(s) = \frac{s+3}{(s+1)(s+2)(s+4)}$

$$F(s) = \frac{s+3}{(s+1)(s+2)(s+4)} = \frac{A}{s+1} + \frac{B}{s+2} + \frac{C}{s+4} \tag{2.58}$$

$$A = \frac{2}{3}, B = -\frac{1}{2}, C = -\frac{1}{6}$$

$$F(s) = \frac{2}{3(s+1)} - \frac{1}{2(s+2)} - \frac{1}{6(s+4)} \tag{2.59}$$

$$F(s) = \left(\frac{2}{3}\right)e^{-t} - \left(\frac{1}{2}\right)e^{-2t} - \left(\frac{1}{6}\right)e^{-4t} \tag{2.60}$$

**Example 2.6**

**Find f(t) if** $F(s) = \frac{s}{(s+1)^2(s+3)}$

$$F(s) = \frac{A}{(s+1)^2} + \frac{B}{s+1} + \frac{C}{s+3}$$

$$A = \frac{s(s+1)^2}{(s+1)^2(s+3)}\Big|_{s=-1} = -\frac{1}{2}$$

$$B = \frac{d}{ds}\left[\frac{s}{s+3}\right]\Big|_{s=-1} = \frac{3}{4}$$

$$C = \frac{s(s+3)}{(s+1)^2(s+3)}\Big|_{s=-3}$$

$$C = \frac{-3}{4}$$

Therefore,

$$F(s) = -\frac{1}{2(s+1)^2} + \frac{3}{4(s+1)} - \frac{3}{4(s+3)} \tag{2.61}$$

$$f(t) = \left(-\frac{1}{2}\right)te^{-t} + \frac{3e^{-t}}{4} - \frac{3e^{-3t}}{4} \tag{2.62}$$

**Example 2.7**

**Response of second-order system under differential initial conditions**

$$\ddot{y} + 3\dot{y} + 2y = u(t) \tag{2.63}$$

$$s^2Y(s) - sy(0) - \dot{y}(0) + 3sY(s) - 3y(0) + 2Y(s) = \frac{2}{s}$$

$$Y(s) = \frac{2 + s^2y(0) + \dot{y}(0) + 3y(0)}{s^2 + 3s + 2}$$

$$Y(s) = \frac{2 + s^2y(0) + sy(0)}{s(s+1)(s+2)} \tag{2.64}$$

(a) We can vary the initial conditions for Equation 2.64 and find the time-response as under

$$\dot{y}(0) = 0, y(0) = 0 \tag{2.65}$$

$$Y(s) = \frac{2}{s(s+1)(s+2)} = \frac{1}{s} - \frac{2}{s+1} + \frac{1}{s+2}$$

$$y(t) = u(t) - 2e^{-t} + e^{-2t} \tag{2.66}$$

(b) Another set of initial conditions can be considered as under

$$\dot{y}(0) = 1, y(0) = 0 \tag{2.67}$$

$$Y(s) = \frac{2 + s^2}{s(s+1)(s+2)} = \frac{1}{s} - \frac{3}{s+1} + \frac{3}{s+2}$$

$$y(t) = u(t) - 3e^{-t} + 3e^{-2t} \tag{2.68}$$

(c) The third set of initial conditions may be

$$\dot{y}(0) = 1, y(0) = 1 \tag{2.69}$$

$$Y(s) = \frac{2 + s^2 + 4s}{s(s+1)(s+2)} = \frac{1}{s} + \frac{1}{s+1} - \frac{1}{s+2}$$

$$y(t) = u(t) + e^{-t} - e^{-2t} \tag{2.70}$$

Figure 2.3 shows the response of second order system under different initial conditions.

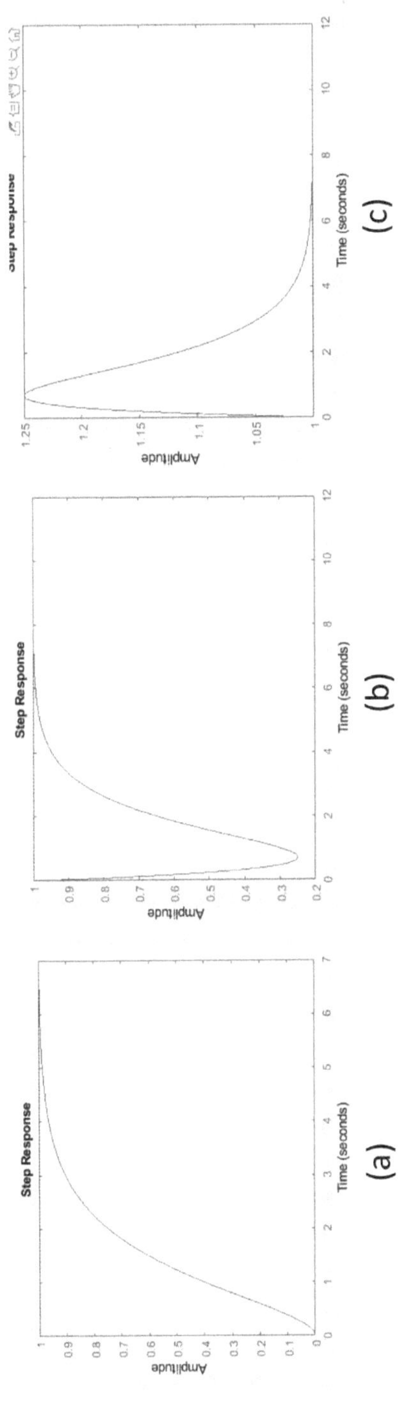


**FIGURE 2.3** Response of second-order system under different initial conditions.

## Problems

P2.1 Find the Laplace transform of the following

$$5 + e^{-3t} + te^{-2t}$$

$$2t + 2e^{-3t} + e^{-t} sint$$

$$e^{-t} cost + 2e^{-3t} + t$$

P2.2 Obtain the Laplace transform of the gate function given here:

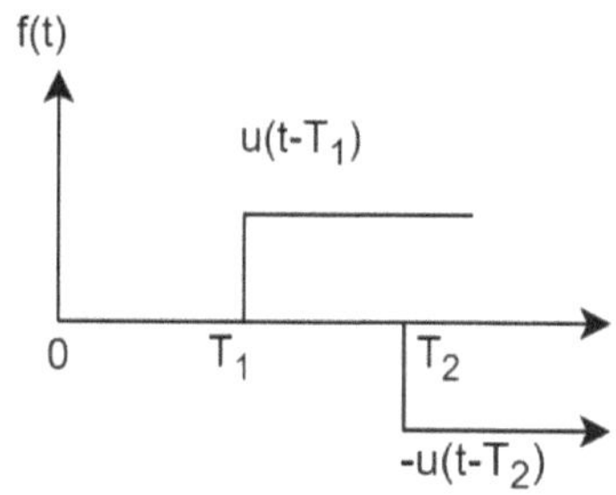


P2.3 Find the initial and final values of f(t) whose Laplace transform is

$$F(s) = \frac{1}{s\left(s^2 + 2s + 3\right)}$$

P2.4 Determine f(t) for the system whose $F(s) = \frac{s+3}{s(s+1)^2(s+2)}$

P2.5 Find Y(s) if $\frac{d^2y}{dt^2} + \frac{8dy}{dt} + 5y = 6$

P2.6 A laser jet printer uses a laser beam to print copies rapidly for a computer. The laser is positioned by a control input r(t), so we have

$$Y(s) = \frac{5(s + 100)}{s^2 + 60s + 500} R(s)$$

The input r(t) represents the desired position of the laser beam.

a. If r(t) is a unit step input, find the output y(t)
b. What is the final value of y(t)?

P2.7 A second-order differential equation is given here:

$$\frac{d^2y}{dt^2} + \frac{4dy}{dt} + 3y = 2r(t)$$

$$r(t) = u(t), y(0) = 1, \dot{y}(0) = 0$$

Find the steady-state value of y(t).

P2.8 Find the initial and final values of the function having Laplace transform

$$F(s) = \frac{s+9}{s^2 + 7s + 3}$$

P2.9 Find f(t) if $F(s) = \frac{s+3}{(s+1)(s+2)(s+4)}$

P2.10 Find f(t) if $F(s) = \frac{s}{(s+1)^2(s+3)}$

P2.11 Find the response of the second order under different initial conditions

$$\ddot{y} + 3\dot{y} + 2y = 2u(t)$$

$$\dot{y}(0) = 0, y(0) = 0$$

$$\dot{y}(0) = 1,\ y(0) = 0$$

$$\dot{y}(0) = 1,\ y(0) = 1$$

P2.12 Find the inverse Laplace transform of

$$\frac{s^3 + 5s^2 + 9s + 7}{(s+1)(s+2)}$$

$$\frac{(s+1)}{s(s^2 + s + 1)}$$

$$\frac{1}{s^2(s^2 + \omega^2)}$$

P2.13 Find the solution of the following differential equation:

$$2\ddot{x} + 7\dot{x} + 3x = 0$$

# 3 Modeling of Control System-Transfer Function Approach

## 3.1 MODELLING OF MECHANICAL SYSTEMS

Newton's law states that for a translational system, the algebraic sum of forces acting on a body is equal to the mass time of the linear acceleration of the body. If the forces are balanced, then the mass will not move. For a rotational system, the law states that the sum of the torque acting on a body is equal to the moment of inertia times the angular accelerations of the body. Examples are provided later to explain the idea.

### 3.1.1 LINEAR TRANSLATIONAL SYSTEM

In Figure 3.1, $K$ is the spring constant, $m$ is the mass, and $f$ is the damping ratio. $F(t)$ is the applied force, and $x$ is displacement.

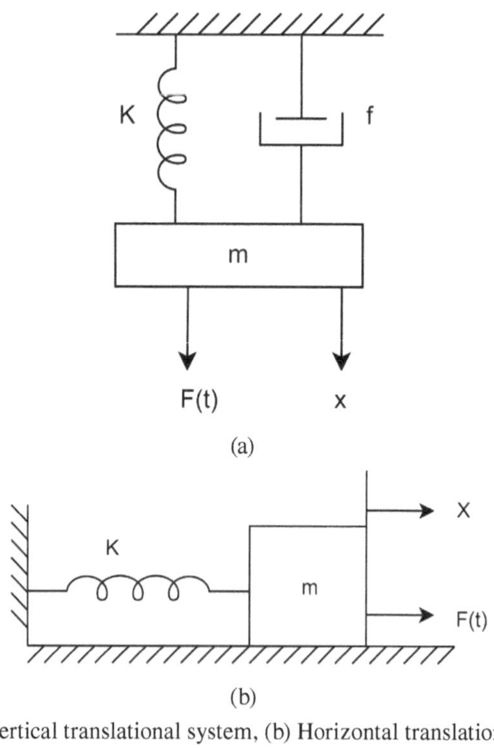


**FIGURE 3.1** (a) Vertical translational system, (b) Horizontal translational system.


DOI: 10.1201/9781003484202-3

$$\frac{md^2x}{dt^2} = -f\frac{dx}{dt} - kx + F(t)$$
$$\frac{md^2x}{dt^2} + f\frac{dx}{dt} + kx = F(t) \tag{3.1}$$

Assuming zero initial conditions

$$s^2mX(s) + sfX(s) + kX(s) = F(s)$$
$$X(s) = \frac{F(s)}{ms^2 + fs + k} \tag{3.2}$$

### 3.1.2 Linear Rotational System

Figure 3.2 shows a linear rotational system where K: Shaft stiffness, $\theta$: Angle of rotation, *J*: Moment of inertia, *f*: damping ratio, and T: Torque.

By applying Newton's law,

$$J\ddot{\theta} = -f\dot{\theta} - k\theta + T$$
$$J\ddot{\theta} + f\dot{\theta} + k\theta = T \tag{3.3}$$

$$s^2 + J\theta(s) + sf\theta(s) + k\theta(s) = T(s)$$

$$\theta(s) = \frac{T(s)}{Js^2 + fs + k} \tag{3.4}$$

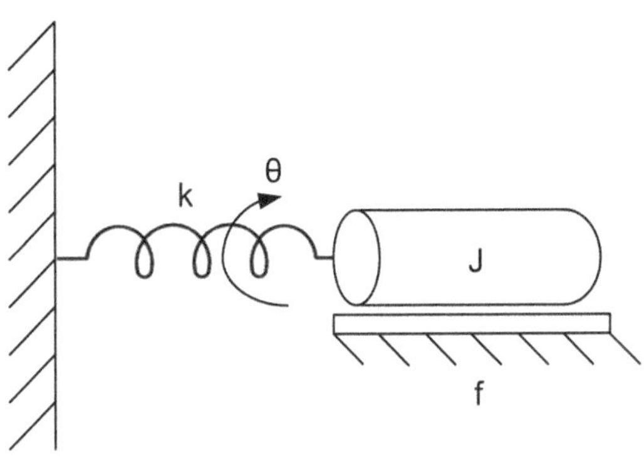


**FIGURE 3.2** Linear rotational system.

### 3.1.3 Complex Translational System

Figure 3.3 shows a complex translational system.

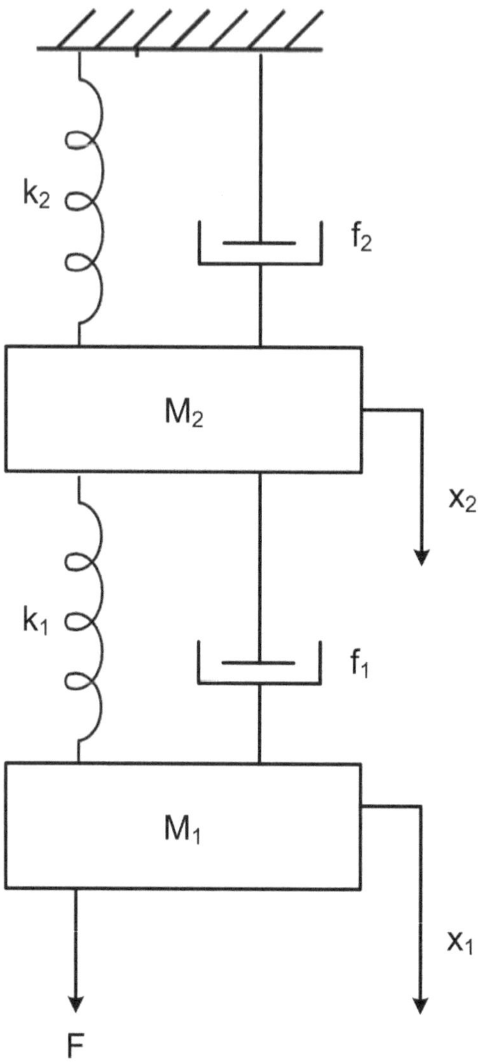


**FIGURE 3.3** Complex translational system.

By applying Newton's Law,

$$m_1\ddot{x}_1 = -f_1\left(\dot{x}_1 - \dot{x}_2\right) - k_1\left(x_1 - x_2\right) = F\left(t\right) \tag{3.5}$$

$$m_2\ddot{x}_2 = -k_2x_2 - f_2\dot{x}_2 + f_1\left(\dot{x}_1 - \dot{x}_2\right) + k_1\left(x_1 - x_2\right) \tag{3.6}$$

After Laplace transform,

$$m_1 s^2 X_1(s) + f_1 s X_1(s) + k_1 X(s) - f_1 s X_2(s) - k_1 X_2(s) = F(s) \tag{3.7}$$

$$m_2 s^2 X_2(s) + (f_1 + f_2) s X_2(s) + (k_1 + k_2) X_2(s) - f_1 s X_1(s) - k_1 X_1(s) = 0 \tag{3.8}$$

$$\begin{bmatrix} s^2 m_1 + s f_1 + k_1 & -s f_1 - k_1 \\ -s f_1 - k_1 & s^2 m_2 + (f_1 + f_2) s + (k_1 + k_2) \end{bmatrix} \begin{bmatrix} X_1(s) \\ X_2(s) \end{bmatrix} = \begin{bmatrix} F(s) \\ 0 \end{bmatrix} \tag{3.9}$$

**Example 3.1**

Write down the dynamical equation of the complex translational system given in Figure 3.4.

***Solution***

$$m_1 \ddot{x}_1 = -k_1 x_1 - k_2 (x_1 - x_2) - f(\dot{x}_1 - \dot{x}_2) + u \tag{3.10}$$

$$m_2 \ddot{x}_2 = -k_3 x_2 - k_2 (x_2 - x_1) - f(\dot{x}_2 - x_1)$$

$$\begin{bmatrix} m_1 s^2 + fs + k_1 + k_2 & -fs - k_2 \\ -fs - k_2 & m_2 s^2 + fs + k_2 + k_3 \end{bmatrix} \begin{bmatrix} X_1(s) \\ X_2(s) \end{bmatrix} = \begin{bmatrix} 1 \\ 0 \end{bmatrix} U(S)$$

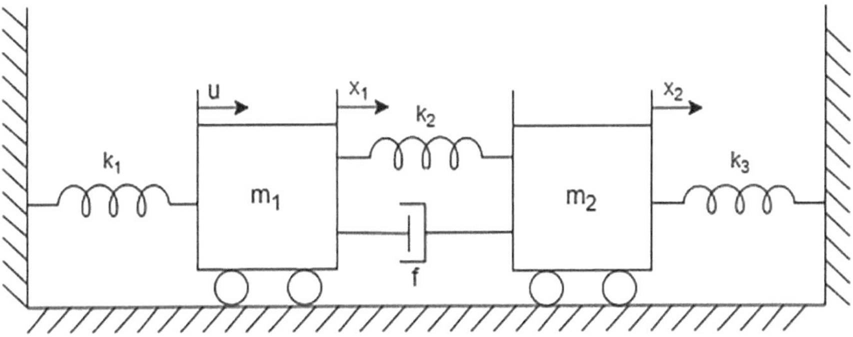


**FIGURE 3.4** Complex translational system.

**Example 3.2**

Write down the dynamical equations for the translational system given in Figure 3.5.

***Solution***

Applying Newton's Law,

$$m_1 \ddot{y}_1 + f \dot{y}_1 + k(y_1 - y_2) = 0 \tag{3.11}$$

$$m_2 \ddot{y}_2 + k(y_2 - y_1) = u \tag{3.12}$$

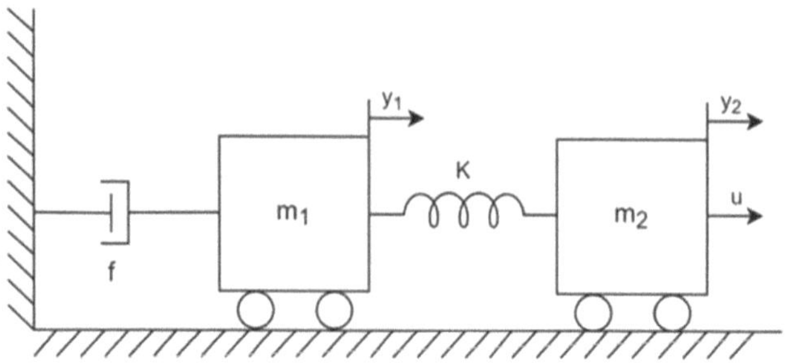


**FIGURE 3.5** Translational system.

## 3.2 MODELING OF ELECTRICAL CIRCUITS

Figure 3.6 shows an electrical system where L: Inductance, R: Resistance, C: Capacitance, I(t): Current, $e_1(t)$: Input Voltage, and $e_2(t)$: Output Voltage. Following equations govern electrical circuits.

$$L\frac{di}{dt} + Ri + \frac{1}{c}\int i\,dt = e_1(t) \tag{3.13}$$

$$\frac{1}{c}\int i\,dt = e_2(t) \tag{3.14}$$

$$sLI(s) + RI(s) + \frac{1}{Cs}I(s) = E_1(s) \tag{3.15}$$

$$\frac{1}{Cs}I(s) = E_2(s) \tag{3.16}$$

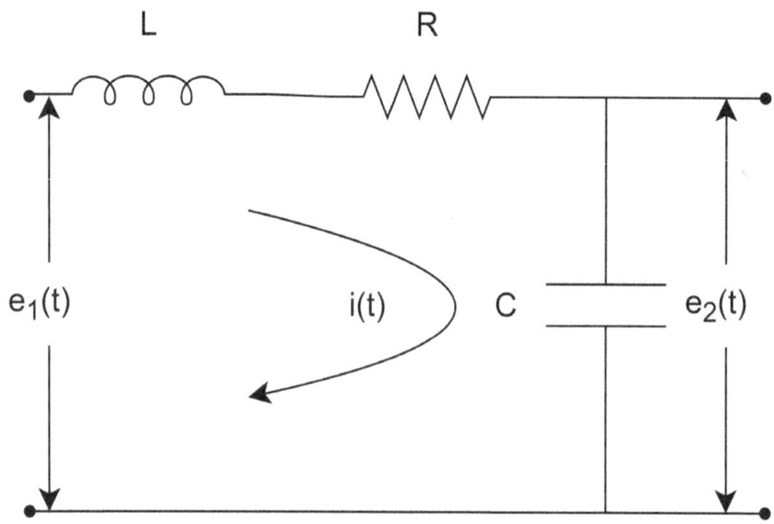


**FIGURE 3.6** Electrical system.

$$\frac{E_2(s)}{E_1(s)} = \frac{\left(\frac{1}{Cs}\right)I(s)}{\frac{s^2LCI(s)+CRI(s)+I(s)}{Cs}}$$

$$\frac{E_2(s)}{E_1(s)} = \frac{1}{s^2LC+RS+1} \tag{3.17}$$

**Example 3.3**

Write down the dynamic equations for the complex electrical system given in Figure 3.7.

***Solution***

$$\left(R_1+\frac{1}{C_1s}\right)I_1(s)-\frac{I_2(s)}{C_1s}=V(s) \tag{3.18}$$

$$\left(sL_1+\frac{1}{C_1s}+\frac{1}{C_2s}\right)I_2(s)-\frac{I_1(s)}{C_1s}-\frac{I_3(s)}{C_2s}=0 \tag{3.19}$$

$$\left(R_2+\frac{1}{C_2s}\right)I_3(s)-\frac{I_2(s)}{C_2s}=0 \tag{3.20}$$

$$\begin{bmatrix} R_1+\frac{1}{C_1s} & \frac{1}{C_1(s)} & 0 \\ -\frac{1}{C_1(s)} & sL_1+\frac{1}{C_1s}+\frac{1}{C_2s} & -\frac{1}{C_2(s)} \\ 0 & -\frac{1}{C_2(s)} & R_2+\frac{1}{C_2s} \end{bmatrix}\begin{bmatrix} I_1(s) \\ I_2(s) \\ I_3(s) \end{bmatrix}=\begin{bmatrix} V(s) \\ 0 \\ 0 \end{bmatrix} \tag{3.21}$$

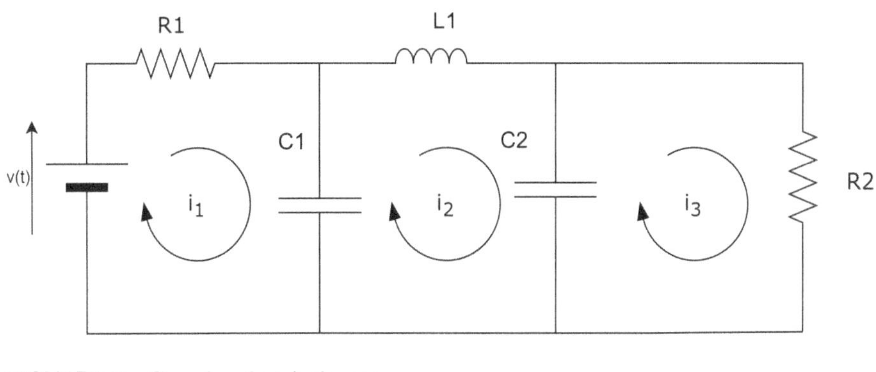


FIGURE 3.7 Complex electrical system.

## 3.3 MODELING OF A SIMPLE HYDRAULIC CIRCUIT

Figure 3.8 shows a hydraulic system where, $q_i$: input flow, $q_0$: output flow, A: cross-section area of tank, R: resistance in pipe, $\rho$: density of fluid, and h(t): fluid height.

$$q_0 = \frac{h}{R}$$

$$\rho q_i(t) - \rho q_0(t) = \frac{d}{dt}\left(\rho A h(t)\right) \tag{3.22}$$

$$q_i(t) - q_0(t) = \frac{Adh}{dt}$$

$$but\, q_0 = \frac{h}{R}$$

$$q_i - \frac{h}{R} = A\frac{dh}{dt} \tag{3.23}$$

Assuming initial conditions to be zero,

$$Q_i(s) = \frac{H(s)}{R} + sAH(s)$$

$$Q_i(s) = \left(\frac{1}{R} + As\right)H(s) \tag{3.24}$$

$$H(s) = \frac{RQ_i(s)}{1+\tau s}, \qquad \tau = AR \tag{3.25}$$

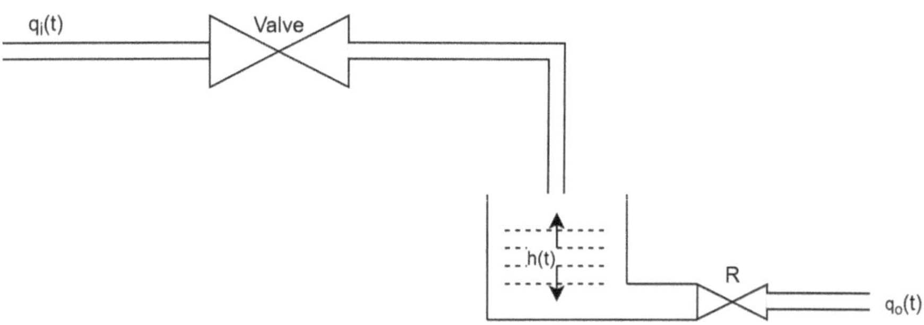


FIGURE 3.8 Hydraulic system.

## 3.4 MODELING OF A SIMPLE THERMAL SYSTEM

Figure 3.9 shows a thermal system. Newton's law of cooling states that the rate of change of temperature in a thermometer is proportional to the difference between bath and measured temperature. Following are governing equations for a simple thermal system.

$$\dot{\theta}(t) = \frac{1}{T(\mu - \theta)} \tag{3.26}$$

$$T\left[s\theta(t)-\theta(0)\right]+\theta(s)=\frac{\mu}{s} \tag{3.27}$$

$$Ts\theta(s)+\theta(s)=\frac{\mu}{s}+T\theta(s)$$
$$\theta(s)(1+Ts)=\frac{\mu T}{s(1+Ts)}+\frac{T\theta(0)}{1+Ts} \tag{3.28}$$

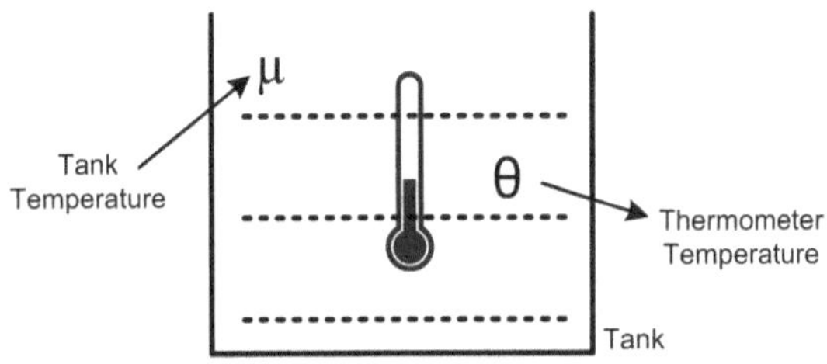


**FIGURE 3.9** A thermal system.

**Example 3.4**

Write down the dynamical equation for the mechanical system given in Figure 3.10.

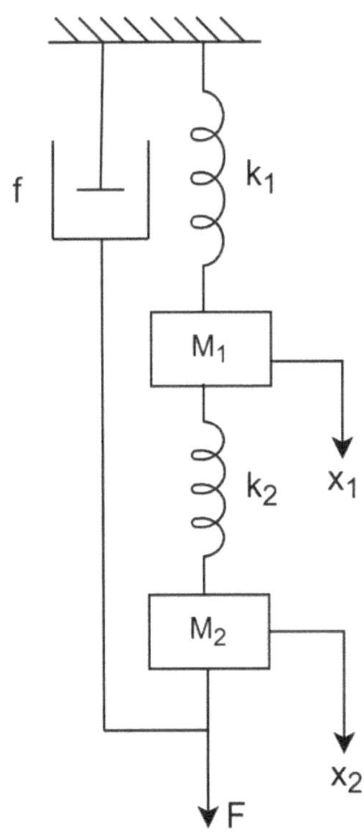


**FIGURE 3.10** A mechanical system.

*Solution*

$$m_1\ddot{x}_1 + k_1x_1 - k_2(x_2 - x_1) = 0$$

$$m_2\ddot{x}_2 + k_2(x_2 - x_1) + f\dot{x}_2 = F \tag{3.29}$$

**Example 3.5**

Write down the dynamical equation for the mechanical system given in Figure 3.11.

*Solution*

$$m_1\ddot{x}_1 + k_1x_1 + f(\dot{x}_1 - \dot{x}_2) = F_1$$

$$m_2\ddot{y}_2 + k_2x_2 + f(\dot{x}_2 - \dot{x}_1) = F_2 \tag{3.30}$$

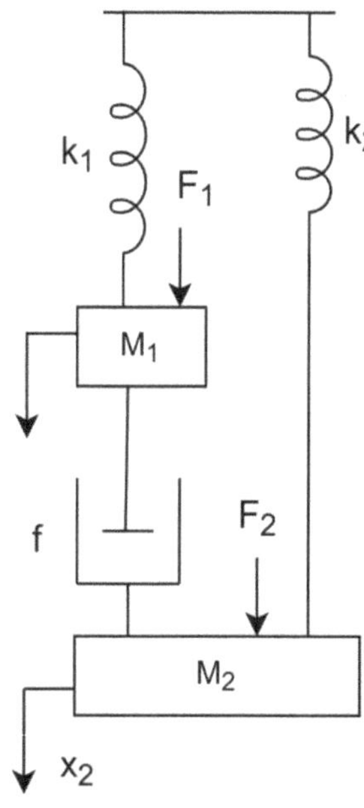


FIGURE 3.11 A mechanical system.

**Example 3.6**

Write down the dynamical equation for the mechanical system given in Figure 3.12.

*Solution*

$$x_0(f_2s + k_2) = (f_1s + k_1)(x_1 - x_0) \tag{3.31}$$

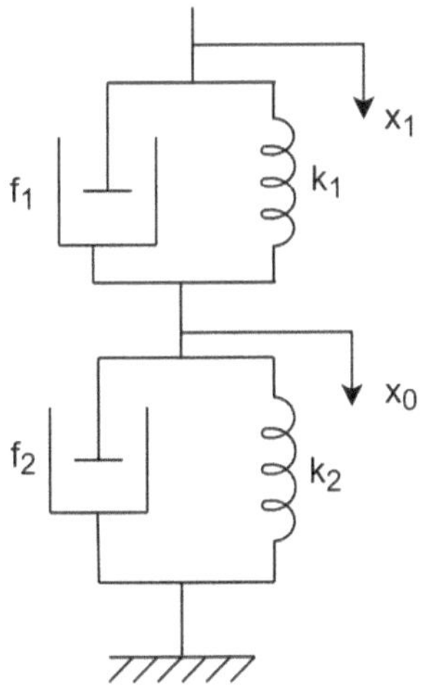


FIGURE 3.12 A mechanical system.

## 3.5 ARMATURE CURRENT CONTROLLED MOTOR

Figure 3.13 shows an armature current controlled motor where, $i_a$: armature current, $i_f$: field current, $e_i$: armature voltage, $T_m$: motor torque, $T_L$: load torque, $J_m$: motor inertia, $J_L$: load inertia, $\theta_m$: motor angle, and $\theta_L$: load angle.

$$L\frac{di_a}{dt}+i_aR_a+k_b\frac{d\theta_m}{dt}=e_i \tag{3.32}$$

$$(Ls+R)I_a(s)+k_bs\theta_m(s)=E_i(s) \tag{3.33}$$

$$J_m\ddot{\theta}_m+T=T_m=Ki_a \tag{3.34}$$

$$T\theta_m=T_L\theta_L \tag{3.35}$$

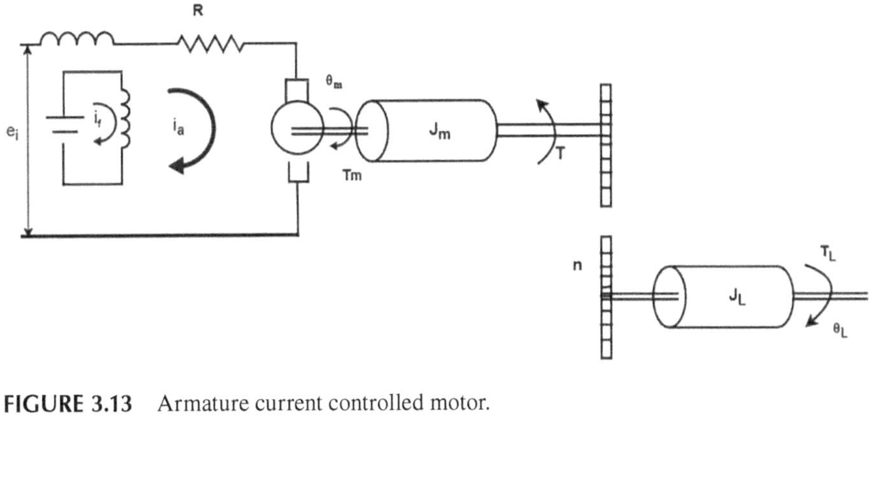


FIGURE 3.13 Armature current controlled motor.

$$T = nT_L, \qquad \frac{\theta_L}{\theta_m} = n \tag{3.36}$$

$$J_L \ddot{\theta}_L = T_L \tag{3.37}$$

Combining,

$$\left(J_m + n^2 J_L\right)\ddot{\theta}_L = nki_a \tag{3.38}$$

$$(J_m + n^2 J_L)S^2 0_L(s) = nkI_a(s) \tag{3.39}$$

$$I_a(s) = \frac{\left(J_m + n^2 J_L\right)s^2\theta_L(s)}{nk} \tag{3.40}$$

Putting in Eq. 3.33,

$$\frac{(Ls+R)\left(J_m + n^2 J_L\right)s^2\theta_L(s)}{nk} + \frac{k_b s\theta_L(s)}{n} = E_i(s)$$

$$\left[(Ls+R)\left(J_m + n^2 J_L\right)s^2 + k\,k_b s\right]\theta_L(s) = nkE_i(s)$$

$$\frac{\theta_L(s)}{E_i(s)} = \frac{nk}{s\left[(Ls+R)\left(J_m + n^2 J_L\right)s + k\,k_b\right]}$$

**Example 3.7: A Mechanical System**

Write down the dynamical equation for the mechanical system given in Figure 3.14.

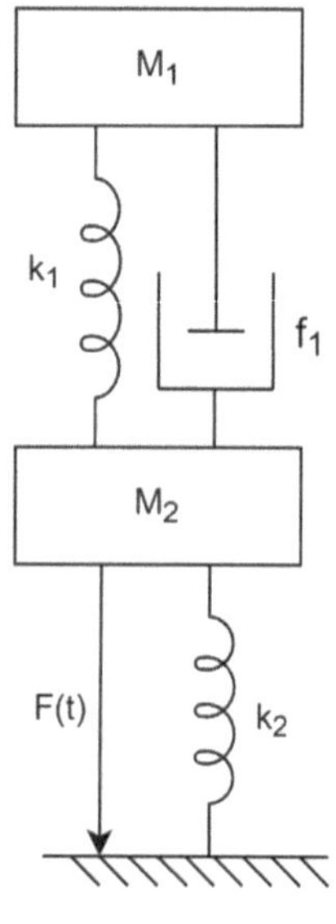


**FIGURE 3.14** A mechanical system.

***Solution***

$$m_1\ddot{x}_1 + f_1(\dot{x}_1 - \dot{x}_2) + k_1(x_1 - x_2) = 0 \tag{3.41}$$

$$m_2\ddot{x}_2 + f_1(\dot{x}_2 - \dot{x}_1) + k_1(x_2 - x_1) + k_2x_2 = F(t) \tag{3.42}$$

## 3.6 INVERTED PENDULUM

Figure 3.15 shows an inverted pendulum. The kinetic energy of the pendulum can be modeled as:

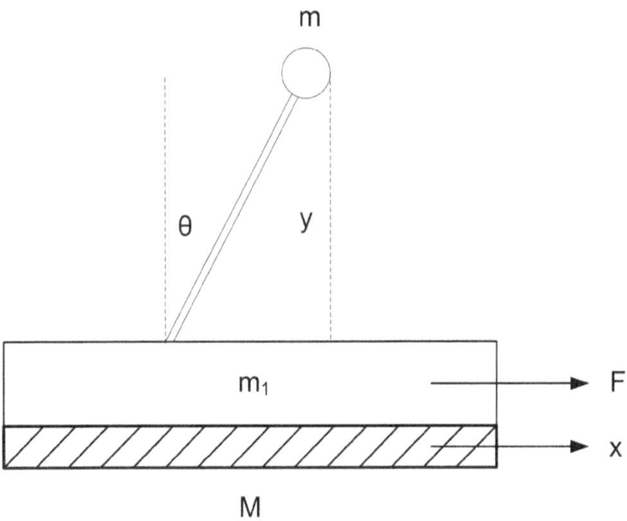


**FIGURE 3.15** Inverted pendulum.

$$T = \frac{1}{2}(m_1 + m)\dot{x}^2 + \frac{1}{2}m\dot{y}^2 \tag{3.43}$$

$$V = -mgy \tag{3.44}$$

$$L = T + V \tag{3.45}$$

$$L = \frac{1}{2}(m_1 + m)\dot{x}^2 + \frac{1}{2}m\dot{y}^2 - mgy \tag{3.46}$$

The dynamical equations are as follows:

$$\frac{d}{dt}\left(\frac{\partial L}{\partial \dot{x}}\right) - \frac{\partial L}{\partial x} = F \tag{3.47}$$

$$\frac{d}{dt}\left(\frac{\partial L}{\partial \dot{y}}\right) - \frac{\partial L}{\partial x} = 0 \tag{3.48}$$

Previous equations result in:

$$(m_1 + m)\ddot{x} = F \tag{3.49}$$

$$m\ddot{y} + mg = 0 \tag{3.50}$$

The state variables for the previous system are:

$$(x, \dot{x}, y, \dot{y}) \rightarrow (x_1, x_2, x_3, x_4) \tag{3.51}$$

$$\dot{x}_1 = x_2$$

$$\dot{x}_2 = x_3 = \frac{F}{M+m} \tag{3.52}$$

$$\dot{x}_3 = x_4 \tag{3.53}$$

$$\dot{x}_4 = -g \tag{3.54}$$

$$\begin{bmatrix} \dot{x}_1 \\ \dot{x}_2 \\ \dot{x}_3 \\ \dot{x}_4 \end{bmatrix} = \begin{bmatrix} 0 & 1 & 0 & 0 \\ 0 & 0 & 1 & 0 \\ 0 & 0 & 0 & 1 \\ 0 & 0 & 0 & 0 \end{bmatrix} \begin{bmatrix} x_1 \\ x_2 \\ x_3 \\ x_4 \end{bmatrix} + \begin{bmatrix} 0 \\ \frac{1}{M+m} \\ 0 \\ -1 \end{bmatrix} \begin{bmatrix} 0 & F & 0 & g \end{bmatrix} \tag{3.55}$$

**Example 3.8: Armature Current Controlled Motor**

Figure 3.16 shows an armature current controlled motor. It can be modeled as:

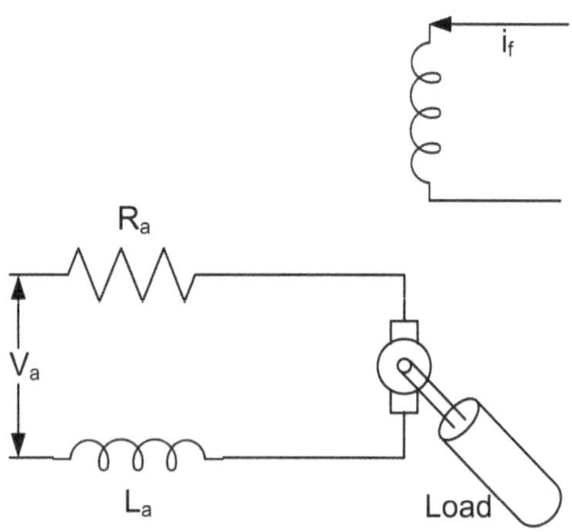


**FIGURE 3.16** Armature current controlled motor.

$$v_a = i_a R_a + L\frac{d_{ia}}{dt} + k_b\frac{d\theta}{dt} \tag{3.56}$$

The developed torque is:

$$K_1 I_a = T_d \tag{3.57}$$

The load torque is:

$$J\frac{d^2\theta}{dt^2} + \frac{fd\theta}{dt} + L\theta \tag{3.58}$$

Assuming no loss,

$$T_L = T_d \tag{3.59}$$

$$\frac{Jd^2\theta}{dt^2} + f\frac{d\theta}{dt} + L\theta = K_i i_a \tag{3.60}$$

$$\frac{d^2\theta}{dt^2} = \frac{f}{J}\frac{d\theta}{dt} - \frac{L}{J}\theta + K_i i_a$$

$$\dot{\theta}_1 = \theta_2$$

$$\dot{\theta}_2 = -\frac{L}{J}\theta_1 - \frac{f}{J}\theta_2 + \frac{K_1}{J}i_a \tag{3.61}$$

$$\begin{bmatrix}\dot{\theta}_1\\ \dot{\theta}_2\end{bmatrix} = \begin{bmatrix}0 & 1\\ -\frac{L}{J} & -\frac{f}{J}\end{bmatrix}\begin{bmatrix}\theta_1\\ \theta_2\end{bmatrix} + \begin{bmatrix}0\\ \frac{K}{J}\end{bmatrix}\begin{bmatrix}0 & i_a\end{bmatrix} \tag{3.62}$$

Also,

$$(Js^2 + f\theta + L)\theta(s) = \frac{K_1(V_a - K_b s\theta)}{R_a + sL_a} \tag{3.63}$$

$$(R_a + sL_a)(Js^2 + fs + L) \tag{3.64}$$

$$(R_a + sL_a)(Js^2 + fs + L)\theta = K_1(v_a - K_b s\theta) \tag{3.65}$$

$$(R_a Js^2 + R_a fs + R_a L + L_a Js^3 + L_a fs^2 + L_a fs^2 + L_a Ls)\theta = K_1 v_a k_b s\theta \tag{3.66}$$

$$\frac{d^3\theta}{dt^3} = -\frac{k_3}{k_2}\frac{d^2\theta}{dt^2} - \frac{k_4 d\theta}{dt} - \frac{k_5}{k_2} + \frac{k_1}{k_2}v_a \tag{3.67}$$

$$\dot{\theta}_1 = \theta_2, \dot{\theta}_2 = \theta_3, \dot{\theta}_3 = -\frac{k_3}{k_2}\theta_3 - \frac{k_4}{k_2}\theta_2 - \frac{k_5}{k_2}\theta_1 + \frac{k_1}{k}v_a \tag{3.68}$$

$$\begin{bmatrix} \dot{\theta}_1 \\ \dot{\theta}_2 \\ \dot{\theta}_3 \end{bmatrix} = \begin{bmatrix} 0 & 1 & 0 \\ 0 & 0 & 1 \\ -\frac{k_5}{k_2} & -\frac{k_4}{k_2} & -\frac{k_3}{k_2} \end{bmatrix} \begin{bmatrix} \theta_1 \\ \theta_2 \\ \theta_3 \end{bmatrix} + \begin{bmatrix} 0 \\ 0 \\ \frac{k_1}{k} \end{bmatrix} v_a \tag{3.69}$$

**Example 3.9: Simple Pendulum**

Figure 3.17 shows a simple pendulum. It can be modeled as:

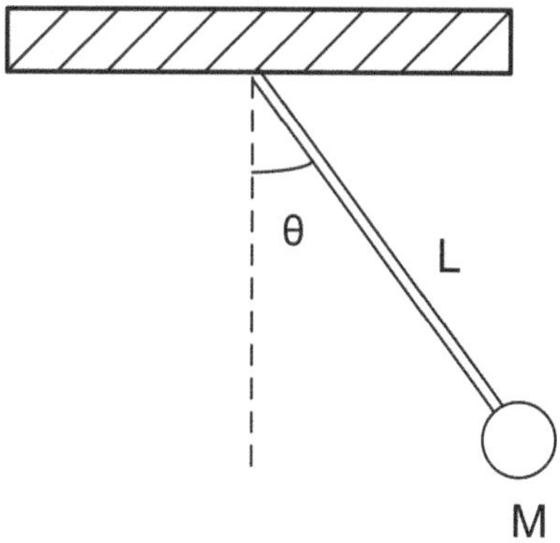


**FIGURE 3.17** Simple pendulum.

$$T = MgLsinq \tag{3.70}$$

The first derivative evaluated at equilibrium provides the linear application as follows:

$$T - T_0 = M_g L \frac{\partial sin\theta}{\partial \theta} |_{\theta=\theta_0} \tag{3.71}$$

$$T = MgL(cos\,0)(\theta - 0) = Mgl\theta \tag{3.72}$$

**Example 3.10: Field Current Controlled Motor**

Figure 3.18 shows a field current controlled motor. It can be modeled as:

$$\varphi = k_f I_f, T_m = k_1 \varphi i_a = k_1 k_f i_f i_a$$

Here, one current must be kept constant. Let $i_a$ be constant, then,

$$T_m(s) = (k_1 k_f I_a) i_f(s) = k_m I_f(s) \tag{3.73}$$

$$v_f(s) = (R_f + L_f s) I_f(s) \tag{3.74}$$

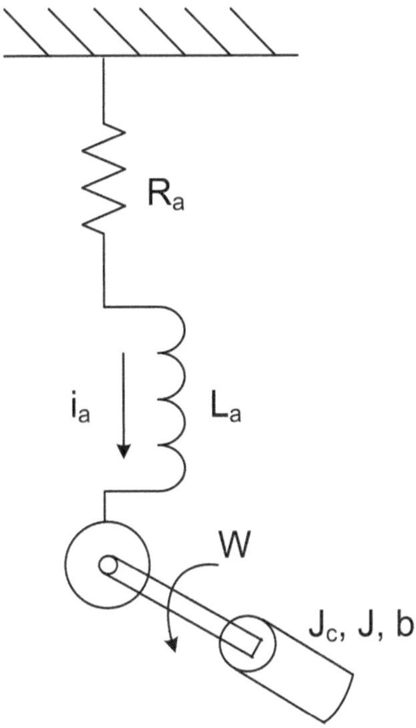


**FIGURE 3.18** Field current controlled motor.

$$T_m(s) = T_L(s) + T_d(s) \tag{3.75}$$

$$T_L(s) = Js^2\theta(s) + bs\theta(s) \tag{3.76}$$

$$T_L(s) = T_m(s) - T_d(s) \tag{3.77}$$

$$T_m(s) = k_m I_f(s) \tag{3.78}$$

$$I_f(s) = \frac{v_f(s)}{k_f + L_f s} \tag{3.79}$$

Therefore,

$$\frac{\theta(s)}{v_f(s)} = \frac{k_m}{s(Js+b)(L_f s + R_f)} \tag{3.80}$$

$$= \frac{\dfrac{K_m}{JL_f}}{s\left(s + \dfrac{b}{J}\right)\left(s + \dfrac{R_f}{L_f}\right)} \tag{3.81}$$

$$= \frac{\frac{k_m}{bR_f}}{s(\tau_f s+1)(\tau_L s+1)}, \quad \tau_f = \frac{L_f}{R_f}, \quad \tau_L = \frac{J}{b} \tag{3.82}$$

## 3.7 BLOCK DIAGRAM OF CLOSED LOOP SYSTEM AND ITS TRANSFER FUNCTION

Figure 3.19 shows the standard way of the canonical form of block diagram representation of a system with negative feedback. Here, R(s) is input or reference output, E(s) is an error, and G(s) is the system, which includes the controller, manipulator, and plant or process. C(s) is the output, and H(s) is the transducer. So the input and output have the same nature for comparison.

$$E(s) = R(s) - C(s)H(s)$$

$$C(s) = E(S)G(s)$$

$$E(s) = R(s) - E(S)G(s)H(s)$$

$$E(s)\left[1 + G(s)H(s)\right] = R(s)$$

$$E(s) = \frac{R(s)}{1 + G(s)H(s)}$$

$$C(s) = \left[R(s) = C(s)H(s)\right]G(s)$$

$$C(s) + C(s)G(s)H(s) = G(s)R(s)$$

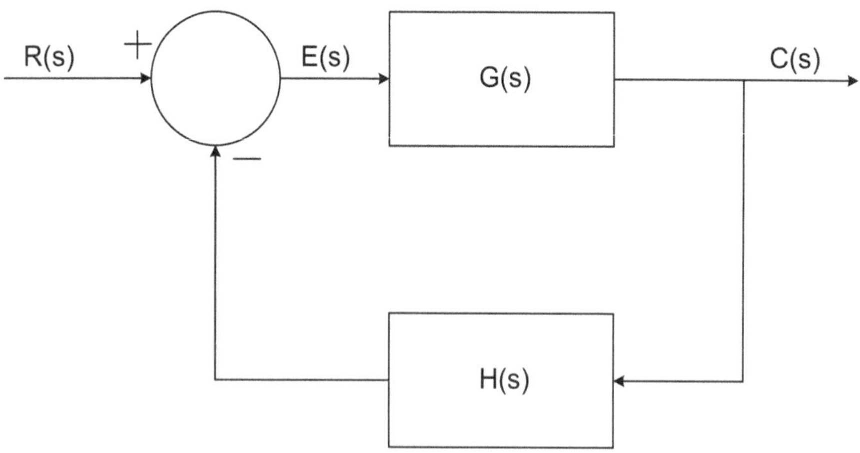


**FIGURE 3.19** The standard way of the canonical form of block diagram representation of a system with negative feedback.

$$C(s)\left(1+G(s)H(s)\right)=G(s)R(s)$$

$$C(s)=\frac{G(s).R(s)}{1+G(s)H(s)}$$

$$\frac{C(s)}{R(s)}=\frac{G(s).R(s)}{1+G(s)H(s)} \tag{3.83}$$

**Example 3.11**

Reduce the block diagram of Figure 3.20 and determine the transfer function.

***Solution***

$$C_1(s)=\frac{R(s)G_1(s)G_2(s)}{1+G_1(s)G_2(s)H(s}$$

$$N(s)-D(s)G_2(s)H(s)G_1(s)=D(s)$$

$$C_N(s)=D(s)G_2(s)$$

$$N(s)=D(s)\left(1+G_2(s)H(s)G_1(s)\right)$$

$$D(s)=\frac{N(s)}{1+G_1(s)G_2(s)H(s)}$$

$$C_N(s)=\frac{N(S)G_2(s)}{1+G_1(s)G_2(s)H(s)}$$

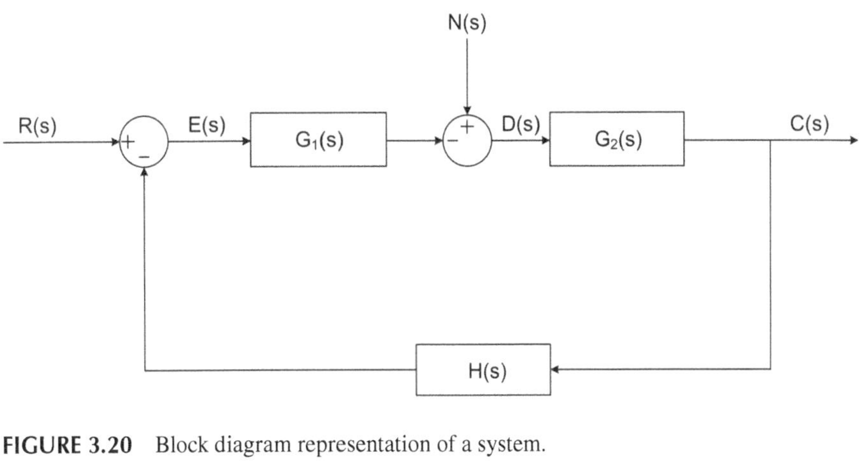


**FIGURE 3.20** Block diagram representation of a system.

$$C_1(S) + C_N(S) = \frac{G_2}{1 + G_1(s)G_2(s)H(s)}\left(G_i(s)R(s) + N(s)\right)$$

### 3.7.1 Rules of Reduction of Block Diagram

Figure 3.21 shows reduction of block diagram for various scenarios.

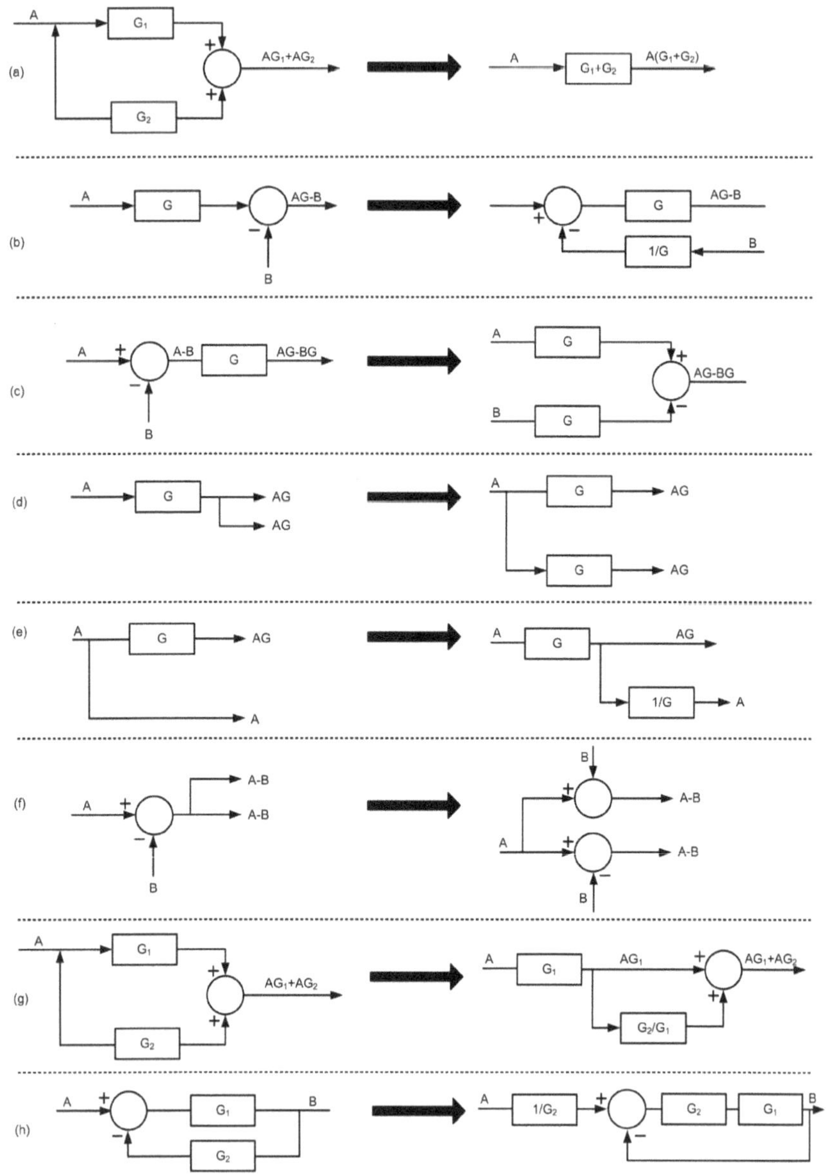


FIGURE 3.21 Rules of reduction of block diagram.

## Example 3.12

Figure 3.22 shows the block diagram representation of a system. Reduce the block diagram and write down the transfer function.

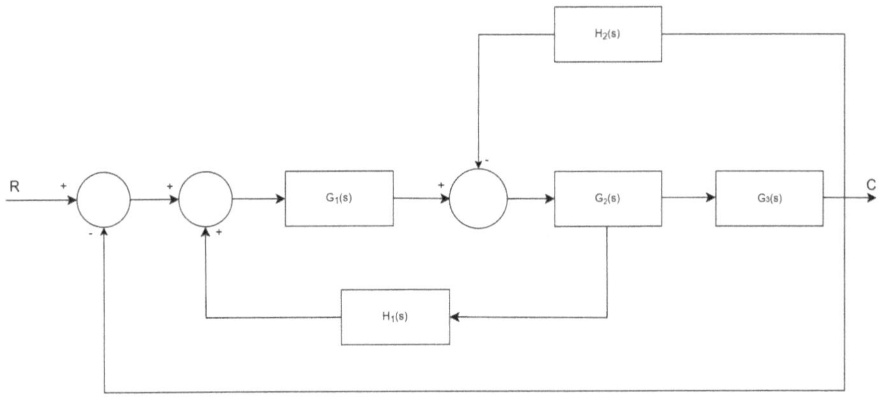


**FIGURE 3.22** Block diagram representation of a system.

***Solution***

Figure 3.23 shows the reduction of block diagram.

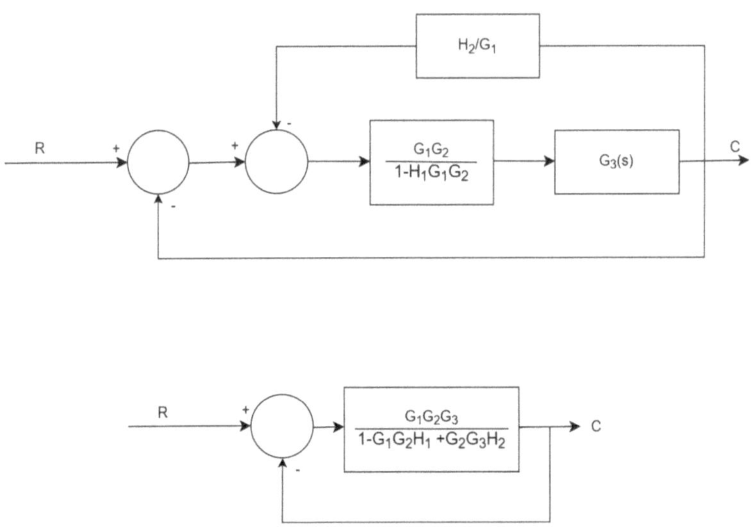


**FIGURE 3.23** Reduction of block diagram representation of a system.

**Example 3.13**

Reduce the block diagram of Figure 3.24 and write the transfer function.

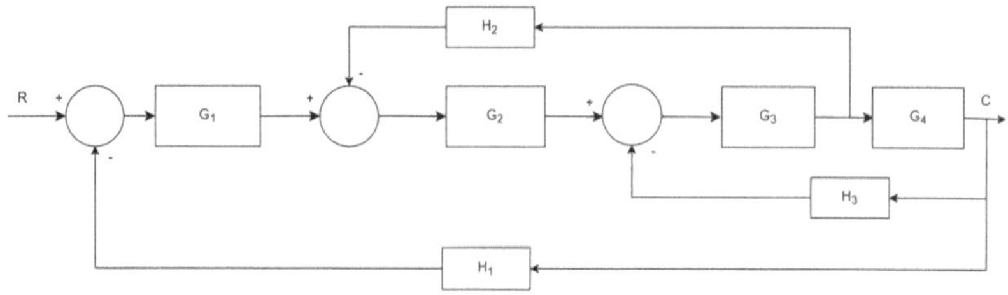


**FIGURE 3.24** Block diagram representation of a system.

***Solution***

Figure 3.25 shows the reduction of block diagram.

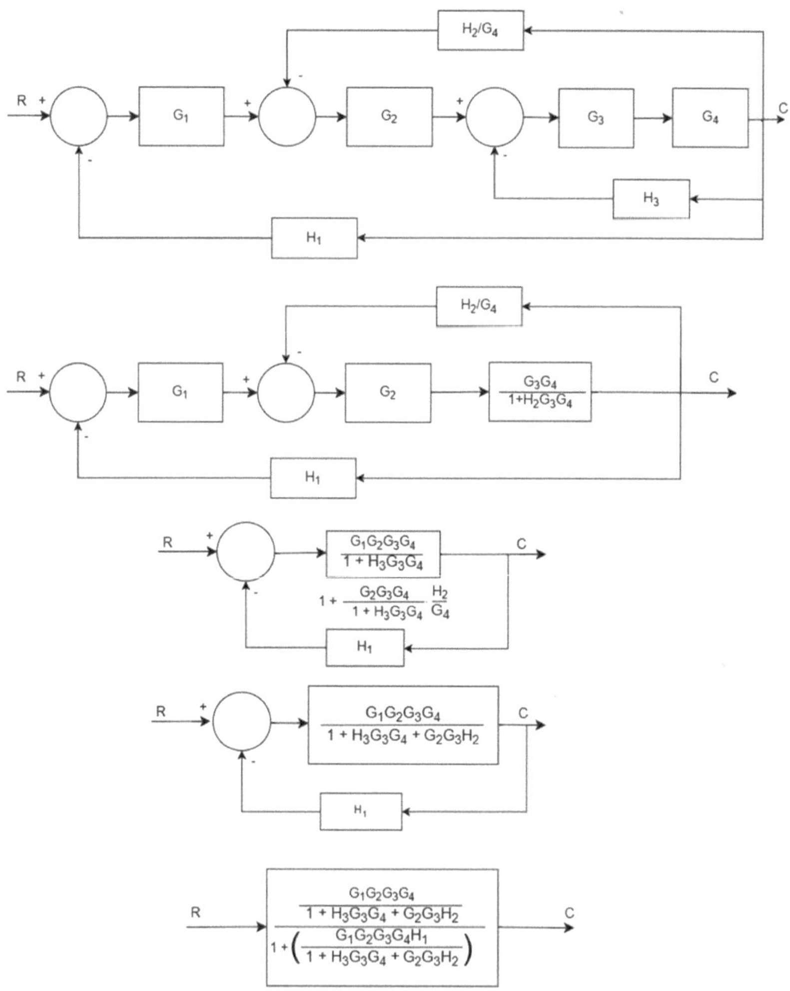


**FIGURE 3.25** Reduction of block diagram representation of a system.

**Problems**

P3.1. Write down the transfer function of the following mechanical systems:
Prob. 3.1 (a)

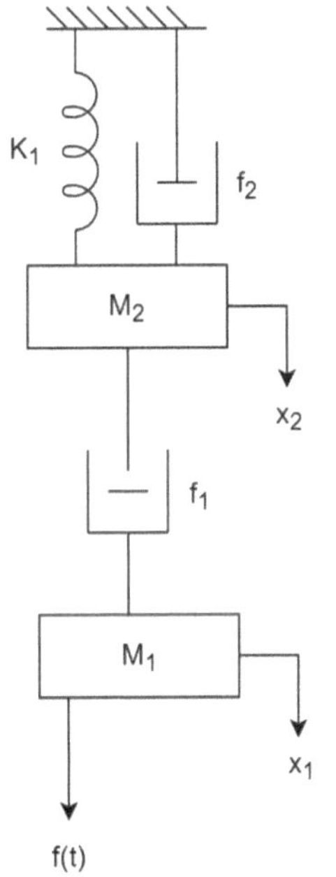


Prob. 3.1 (b)

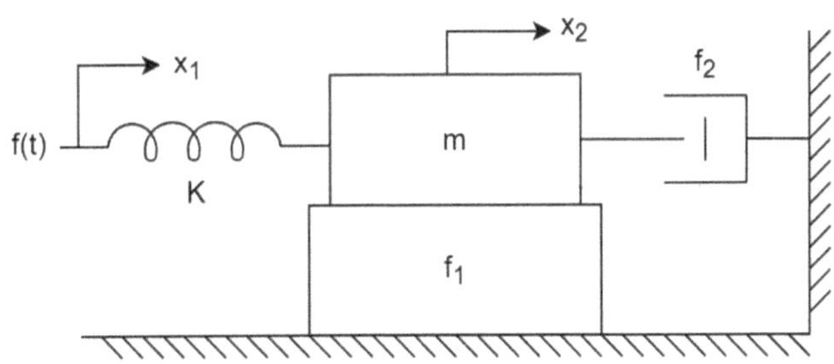


P3.2. Write down the equation of the following rotational mechanical systems:

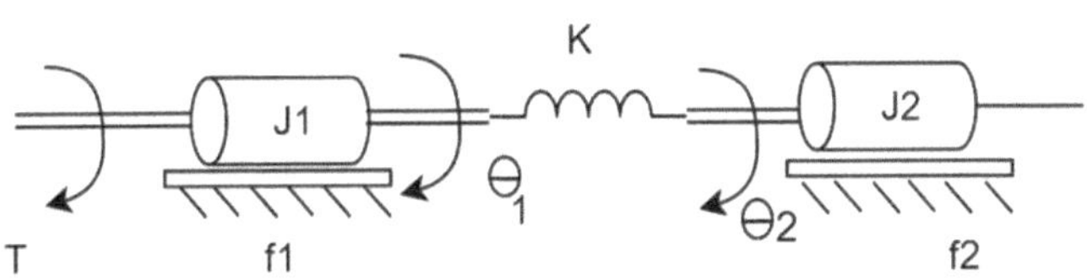

P3.3. Find the ratio $\frac{V_0(s)}{V_1(s)}$ for the following electrical network.

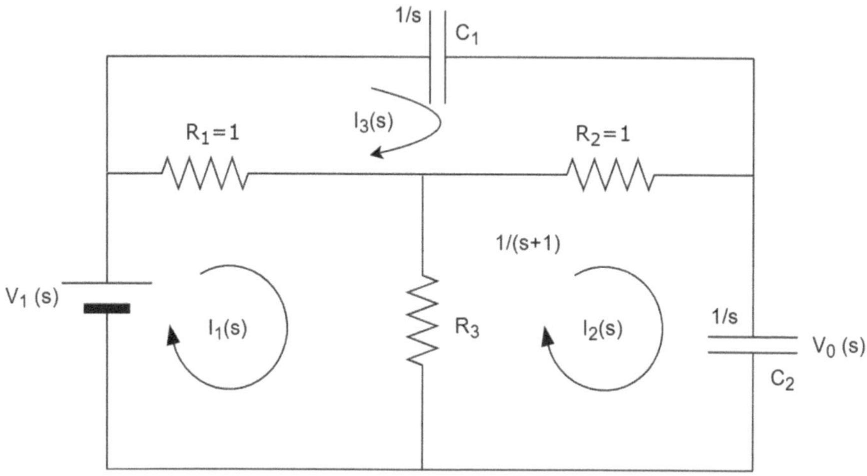


P3.4. Write down the equation for the following mechanical accelerometer.

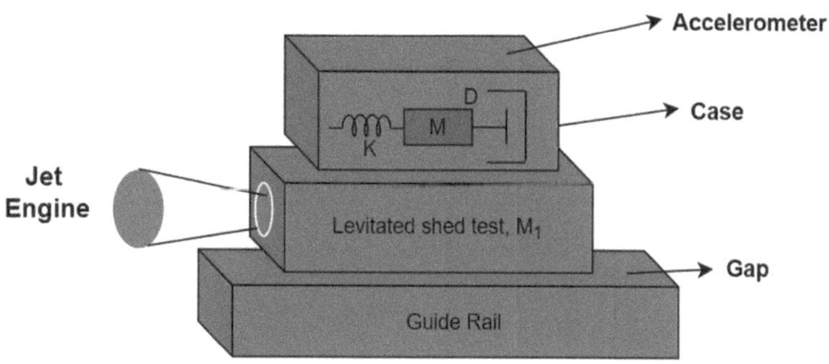


P3.5. Write down the equation for the following mechanical system given here.

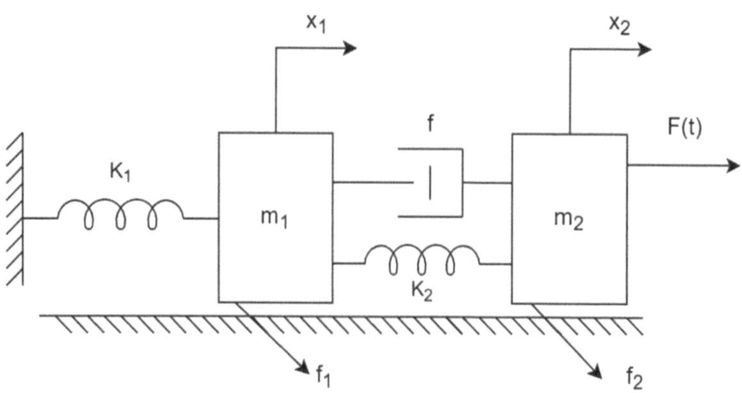

P3.6. Write down the equations for the following mechanical systems.

(a) Prob. 3.6 (a)

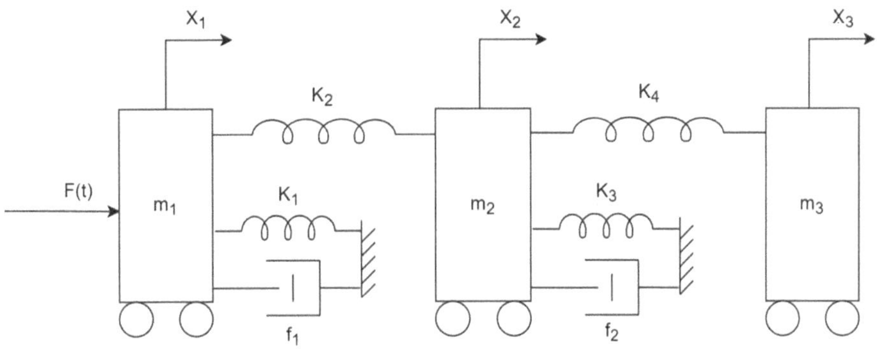


(b) Prob. 3.6 (b)

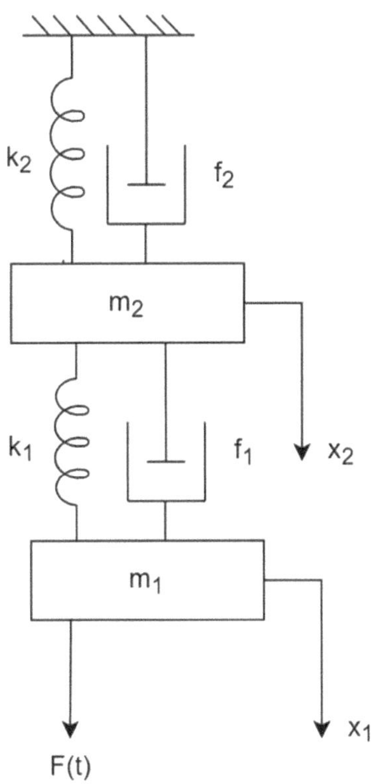

P3.7. Obtain the system equations for the mechanical system shown here.

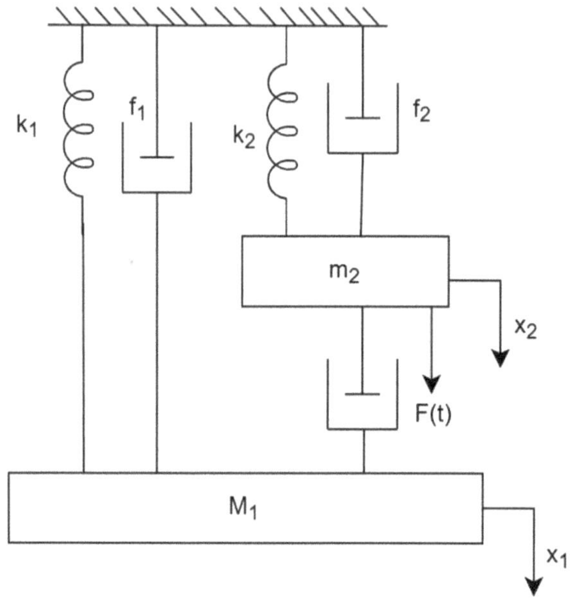


P3.8. A system block diagram is shown in Fig P.3.8. Obtain its transfer function by using the reduction method.

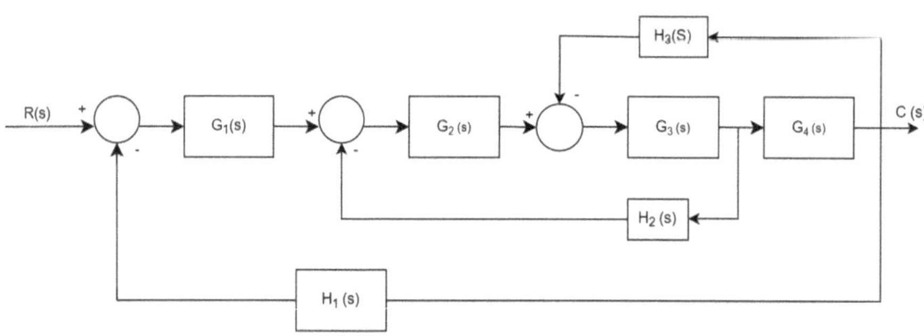


P3.9. Obtain the transfer function relating C(s) and R(s) for the block diagram shown here.

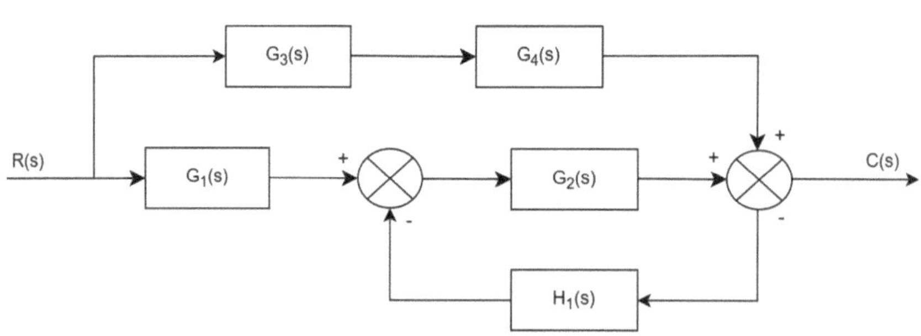

P3.10. Simplify the block diagram shown in Fig P3.10 and find the transfer function $\frac{C(s)}{R(S)}$.

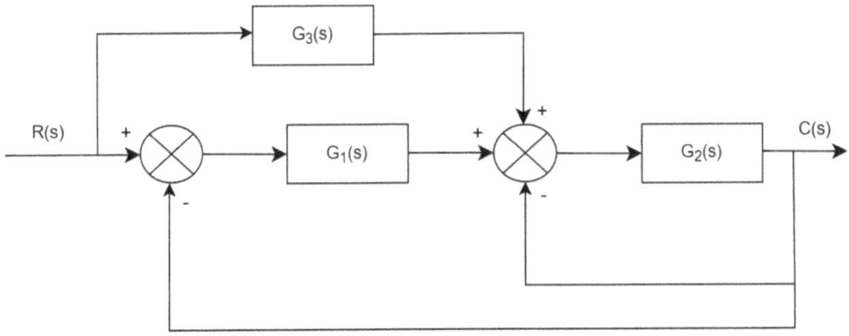


P3.11. A control system is shown in Fig P3.11. Determine the transfer function of the system.

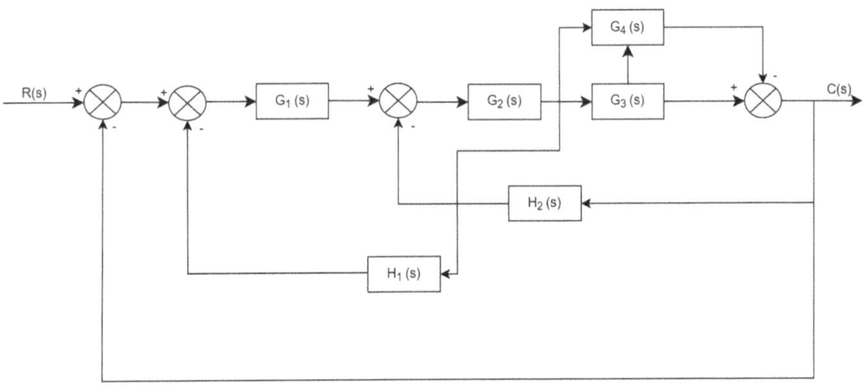


P3.12. A control system is shown in Fig P3.12. Determine the transfer function of the system.

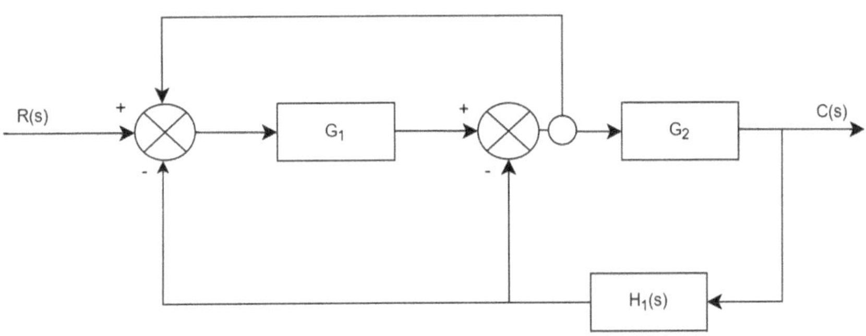

P3.13. A control system is shown in Fig P3.13. Determine the transfer function of the system.

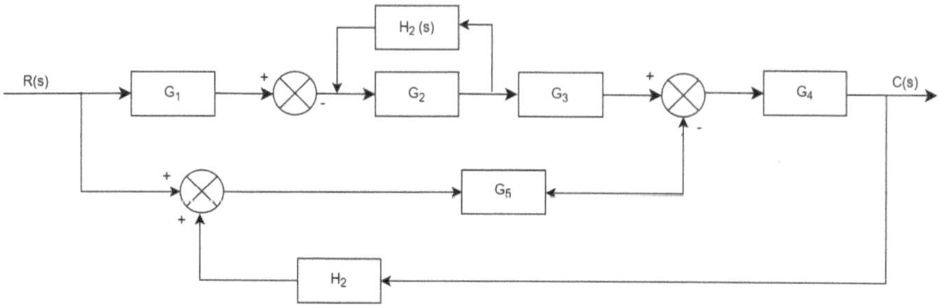


P3.14. A control system is shown in Fig P3.14. Determine the transfer function of the system.

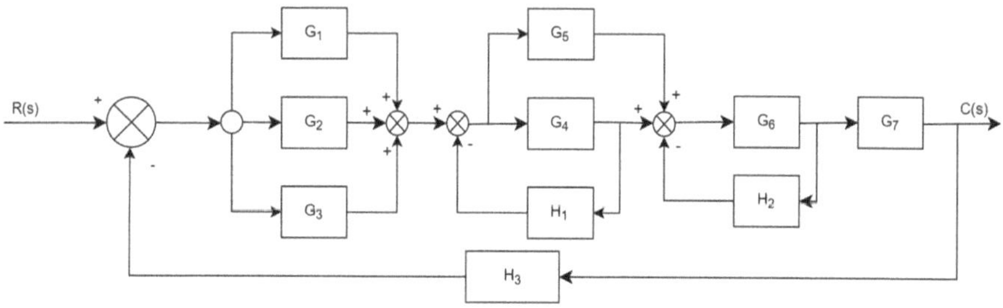

# 4 Modeling of Control Systems—Signal Flow Graph Approach

## 4.1 MASON'S GAIN FORMULA

Simple control systems are conveniently represented using block diagrams, but the process becomes challenging and error-prone once complex systems are encountered. The alternative approach is a signal flow graph that was developed by S. J. Mason, who used the variables of a set of linear algebraic equations derived from a signal flow graph for the analysis of control systems. It consists of a network in which directed branches connect nodes representing each system variable. A node is defined as an entity where incoming or converging signals are summed up to represent a system variable, whereas all exiting signals from it do not affect the node's value. The signal flow graph approach is relatively simpler and uses Mason's gain formula to model control systems, given as shown in Eq 4.1:

$$\mathrm{M} = \frac{1}{\Delta}\sum_{K=1}^{N} M_K \Delta_K \tag{4.1}$$

where,

$\Delta$ = 1 – (Sum of all individual loops) + Sum of gain products of all possible combinations of two non-touching loops – (Sum of products of three non-touching loops) + . . .

$\Delta_K$ = The value of $\Delta$ for that part of the graph not touching the Kth forward path.

$P_K$ = Gain of Kth forward path

The signal flow graph is shown in Figure 4.1, with R, E, and C as nodes.

Few signals converging at nodes and few exiting are not considered while defining the value of the node variable. The signal travels along a branch from one node to another node. The following nomenclature is used:

- Forward path: It is a path from the input node to the output node. R-E-C is a forward path, and its gain is G.
- Loop: It is a path that originates and terminates at the same node.
- Loop gain: It is the product of the branch gains encountered in traversing the loop.
- Non-touching loops: Loops are considered non-touching if they do not pass any common node.

DOI: 10.1201/9781003484202-4

**FIGURE 4.1** (a) Representation of a control system, and (b) signal flow graph.

**Example 4.1:** A signal flow graph for a control system is shown in Figure 4.2. How can its transfer function be obtained?

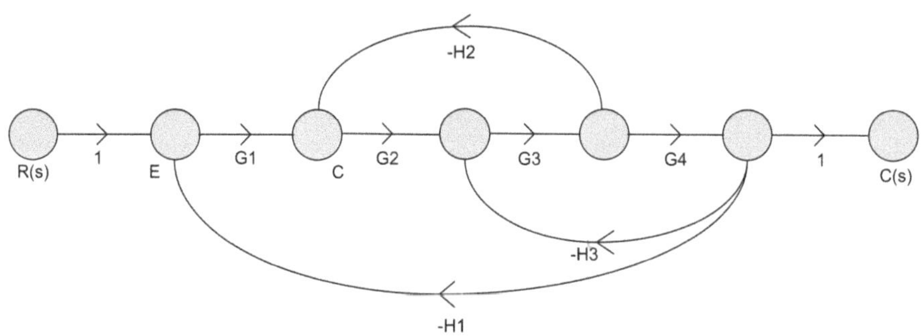


**FIGURE 4.2** Signal flow graph of a control system.

***Solution***

Forward paths

$$P_1 = G_1 G_2 G_3 G_4$$

Individual loops

$$L_1 = -G_1 G_2 G_3 G_4 H_1$$

$$L_2 = -G_2 G_3 H_2$$

$$L_3 = -G_3 G_4 H_3$$

$$\Delta_1 = 1$$

$$\Delta = 1 - \left(L_1 + L_2 + L_3\right)$$

$$\frac{C}{R} = \frac{P_1 \Delta_1}{\Delta} = \frac{G_1 G_2 G_3 G_4}{1 + G_1 G_2 G_3 G_4 H_1 + G_2 G_3 H_2 + G_3 G_4 H_3} \tag{4.2}$$

**Example 4.2:** Obtain the transfer function of the control system shown in Figure 4.3.

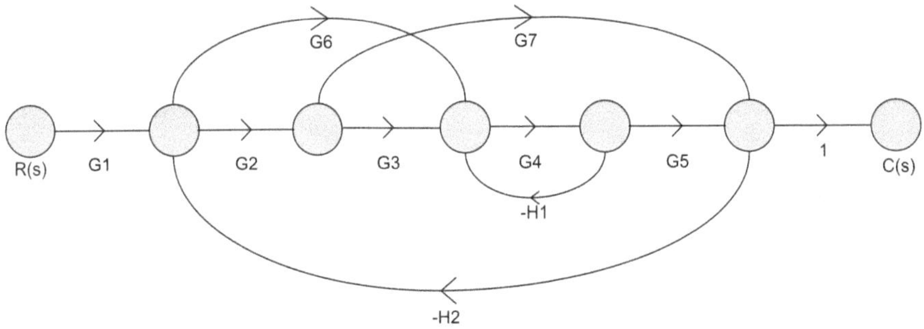


**FIGURE 4.3** Signal flow graph of a control system.

***Solution***

Forward paths

$$P_1 = G_1G_2G_3G_4G_5$$

$$P_2 = G_1G_6G_4G_5$$

$$P_3 = G_1G_2G_7$$

Individual loops

$$L_1 = -G_4H_1$$

$$L_2 = -G_2G_7H_2$$

$$L_3 = -G_6G_4G_5H_2$$

$$L_4 = -G_2G_3G_4G_5H_2$$

$L_1$ and $L_2$ do not touch each other. Therefore,

$$L_1L_2 = G_4H_1G_2G_7H_2$$

Also, $P_3$ does not touch $L_1$.

$$\Delta_1 = 1, \qquad \Delta_2 = 1, \qquad \Delta_3 = 1 - L_1 = 1 + G_4H_1$$

$$\frac{C}{R} = \frac{P_1\Delta_1 + P_2\Delta_2 + P_3\Delta_3}{1 + L_1 + L_2 + L_3 + L_4 + L_1L_2} \tag{4.3}$$

**Example 4.3:** Find the transfer function of the following system given in Figure 4.4.

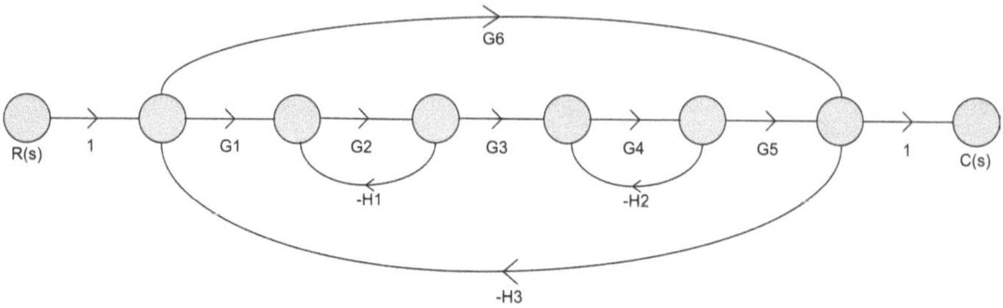


**FIGURE 4.4** Signal flow graph of a control system.

***Solution***

Forward Paths

$$P_1 = G_1G_2G_3G_4G_5$$
$$P_2 = G_6$$

Individual loops

$$L_1 = -G_2H_1$$
$$L_2 = -G_4H_2$$
$$L_3 = -G_6H_3$$
$$L_4 = -G_1G_2G_3G_4G_5H_3$$

Two non-touching loops

$$L_1L_2 = G_2G_4H_1H_2$$
$$L_1L_3 = G_2G_6H_1H_3$$
$$L_2L_3 = G_4G_6H_2H_3$$

Three non-touching loops

$$L_1L_2L_3 = -G_2G_4G_6H_1H_2H_3$$
$$\Delta = 1-\left(L_1+L_2+L_3+L_4\right)+\left(L_{12}+L_{13}+L_{23}\right)-L_{123}$$

Loops 1, 2, 3, and 4 touch path $P_1$

$$\Delta_1 = 1$$

Loops 1 and 2 do not touch $G_6$

$$\Delta_2 = 1 - \left(-G_2H_1 - G_4H_2\right) + G_2G_4H_1H_2$$

$$\frac{C}{R} = \frac{P_1\Delta_1 + P_2\Delta_2}{\Delta} \tag{4.4}$$

**Example 4.4:** See Figure 4.5 and try finding its transfer function using the signal flow graph.

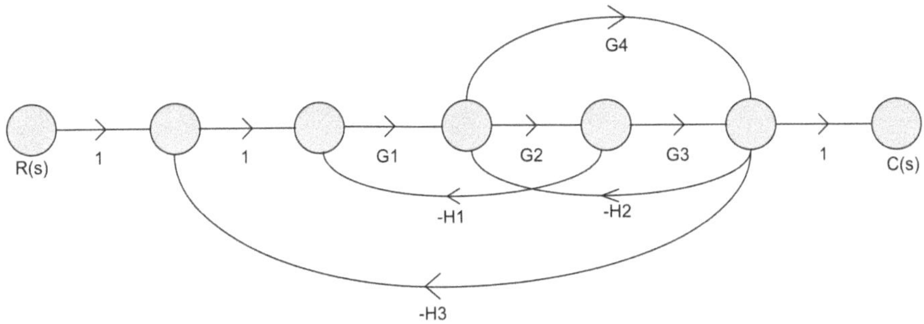


**FIGURE 4.5** Signal flow graph of a control system.

***Solution***

Forward paths

$$P_1 = G_1G_2G_3$$

$$P_2 = G_1G_4$$

Individual loops

$$L_1 = -G_1G_2G_3$$

$$L_2 = -G_1G_2H_1$$

$$L_3 = -G_2G_3H_2$$

$$L_4 = -G_1G_5$$

$$L_5 = -G_4H_2$$

All loops touch $P_1$ and $P_2$, therefore,

$$\Delta_1 = 1$$

$$\Delta_2 = 1$$

$$\Delta = 1 - \left(-G_1G_2G_3 - G_1G_2H_1 - G_2G_3H_2 - G_1G_4 - G_4H_2\right)$$

$$\frac{C}{R} = \frac{G_1G_2G_3 + G_1G_4}{1 - \left(-G_1G_2G_3 - G_1G_2H_1 - G_2G_3H_2 - G_1G_4 - G_4H_2\right)} \tag{4.5}$$

**Example 4.5:** In Figure 4.6, explore and figure out the transfer function of the control system.

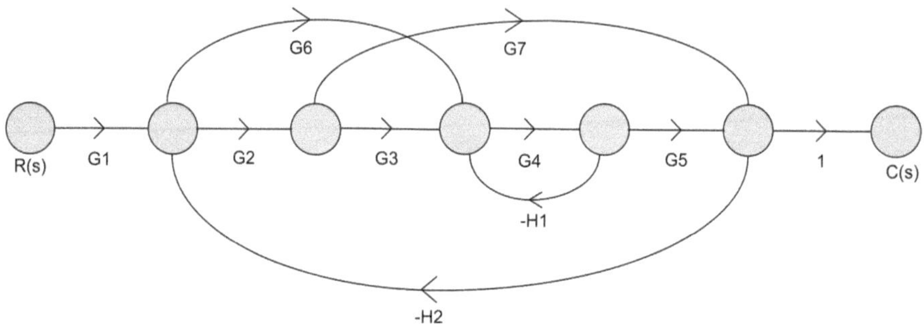


**FIGURE 4.6** Signal flow graph of a control system.

***Solution***

Forward Paths

$$P_1 = G_1G_2G_3G_4G_5$$

$$P_2 = G_1G_6G_4G_5$$

$$P_3 = G_1G_2G_7$$

Individual loops

$$L_1 = -G_4H_1$$

$$L_2 = -G_2G_7H_2$$

$$L_3 = -G_6G_4G_5H_2$$

$$L_4 = -G_2G_3G_4G_5H_2$$

$L_1$ and $L_2$ do not touch each other. Therefore,

$$L_1L_2 = G_4H_1G_2G_7H_2$$

Also $P_3$ does not touch $L_1$

$$\Delta_1 = 1$$

$$\Delta_2 = 1$$

$$\Delta_3 = 1 - L_1 = 1 + G_4 H_1$$

$$\frac{C}{R} = \frac{P_1\Delta_1 + P_2\Delta_2 + P_3\Delta_3}{\Delta}$$

$$\frac{C}{R} = \frac{G_1G_2G_3G_4G_5 + G_1G_6G_4G_5 + G_1G_2G_7\left(1 + G_4H_1\right)}{1 + G_4H_1 + G_2G_7H_2 + G_6G_4G_5H_2 + G_2G_3G_4G_5H_2 + G_4H_1G_2G_7H_2} \tag{4.6}$$

**Example 4.6:** In Figure 4.7, explore and figure out the transfer function of the control system.

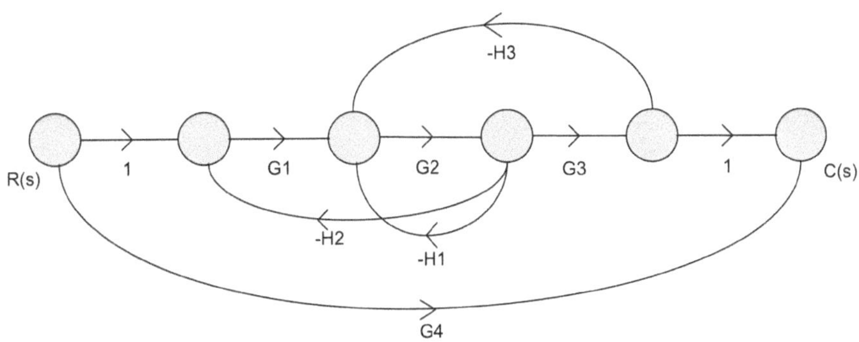


**FIGURE 4.7** Signal flow graph of a control system.

***Solution***

Forward paths

$$P_1 = G_1 G_2 G_3$$

$$P_2 = G_4$$

Individual loops

$$L_1 = -G_2 H_1$$

$$L_2 = -G_1 G_2 H_1$$

$$L_3 = -G_2 G_3 H_2$$

$$\Delta_1 = 1$$

$$\Delta_2 = 1 - \left(-G_2H_1 - G_1G_2H_1 - G_2G_3H_2\right) = 1 + G_2H_1 + G_1G_2H_1 + G_2G_3H_2)$$

$$\frac{C}{R}=\frac{G_1G_2G_3+G_4(1+G_2H_1+G_1G_2H_1+G_2G_3H_2)}{1+G_2H_1+G_1G_2H_1+G_2H_2G_3} \tag{4.7}$$

**Example 4.7:** In Figure 4.8, explore and figure out the transfer function of the control system.

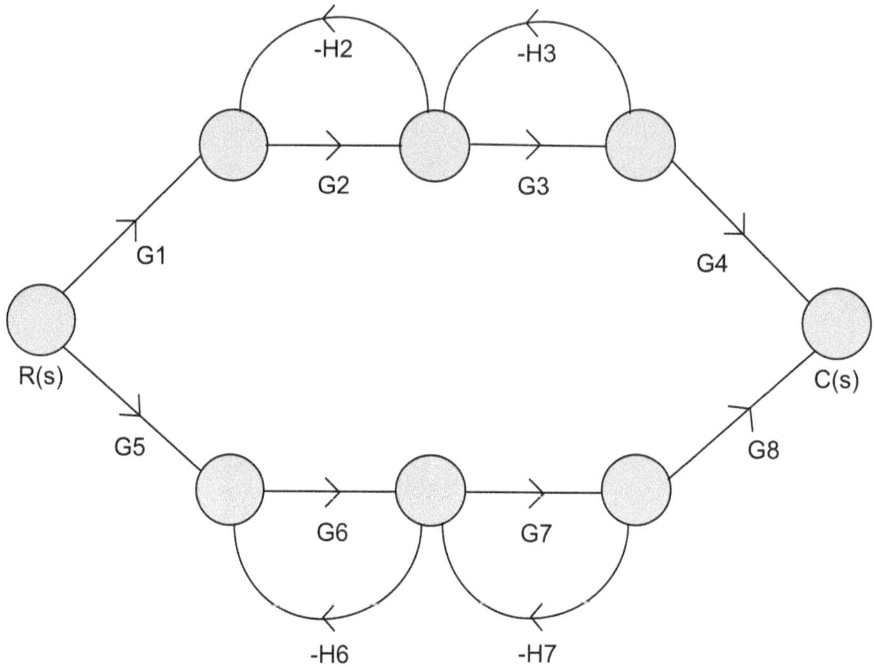


**FIGURE 4.8** Signal flow graph of a control system.

***Solution***

Forward Paths

$$P_1 = G_1G_2G_3G_4$$

$$P_2 = G_5G_6G_7G_8$$

Individual loops

$$L_1 = G_2H_2$$

$$L_2 = G_3H_3$$

$$L_3 = G_6H_6$$

$$L_4 = G_7H_7$$

Loops $L_1$ and $L_2$ do not touch $L_3$ and $L_4$. Therefore, the determinant is

$$\Delta = 1 - (L_1 + L_2 + L_3 + L_4) + (L_1L_3 + L_1L_4 + L_2L_3 + L_2L_4)$$

$$\Delta_1 = 1 - (L_3 + L_4)$$

$$\Delta_2 = 1 - (L_1 + L_2)$$

$$\frac{C}{R} = \frac{P_1\Delta_1 + P_2\Delta_2}{\Delta} \tag{4.8}$$

**Example 4.8:** In Figure 4.9, explore and figure out the transfer function of the control system.

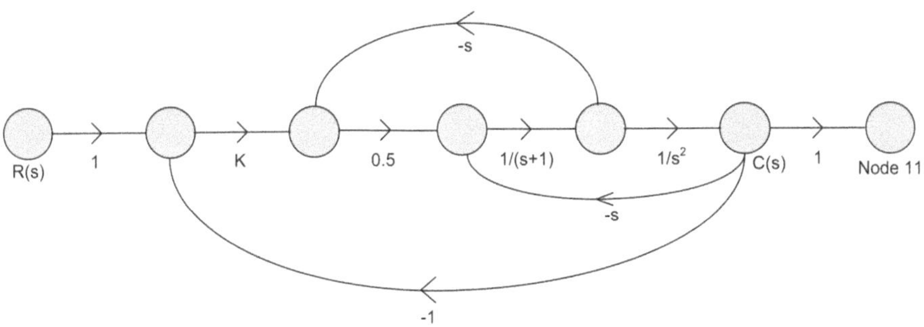


**FIGURE 4.9** Signal flow graph of a control system.

***Solution***

Forward Paths

$$P_1 = (1)(K)(0.5)\left(\frac{1}{s+1}\right)\left(\frac{1}{s^2}\right) = \frac{0.5K}{s^2(s+1)}$$

Individual loops

$$L_1 = (0.5)\left(\frac{1}{s+1}\right)(-5) = \frac{-2.5}{s+1}$$

$$L_2 = \left(\frac{1}{s+1}\right)\left(\frac{1}{s^2}\right)(-5) = \frac{-5}{s^2(s+1)}$$

$$L_3 = K(0.5)\left(\frac{1}{s+1}\right)\left(\frac{1}{s^2}\right)(-1) = \frac{-0.5K}{s^2(s+1)}$$

$$\Delta = 1 - (L_1 + L_2 + L_3)$$

$$\frac{C}{R} = \frac{P_1\Delta_1}{\Delta} = \frac{\dfrac{0.5K}{s^2(s+1)}}{1 - \left(\dfrac{-2.5}{s+1} - \dfrac{5}{s^2(s+1)} - \dfrac{0.5K}{s^2(s+1)}\right)}$$

$$\frac{C}{R} = \frac{0.5K}{s^3 + 3.5s^2 + s + 0.5K} \tag{4.9}$$

**Example 4.9:** In Figure 4.10, find out the closed-loop transfer function of the control system.

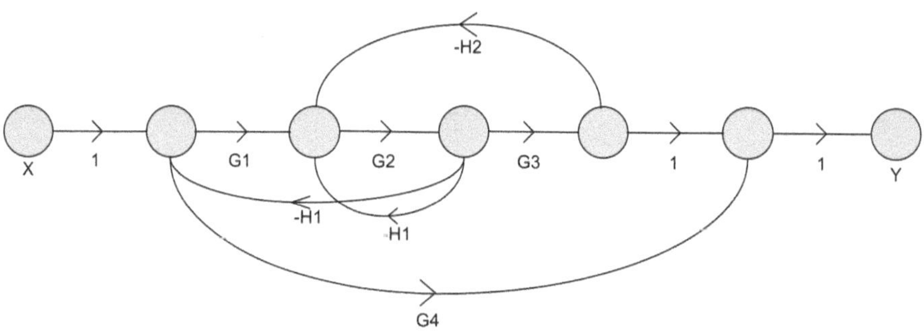


**FIGURE 4.10** Signal flow graph of a control system.

***Solution***

Forward Paths

$$P_1 = G_1G_2G_3$$

$$P_2 = G_4$$

Individual loops

$$L_1 = -G_2H_1$$

$$L_2 = G_1G_2H_1$$

$$L_3 = -G_2G_3H_2$$

$$\Delta = 1 - (L_1 + L_2 + L_3) = 1 + G_2H_1 - G_1G_2H_1 + G_2G_3H_2$$

$$\Delta_1 = 1$$

$$\Delta_2 = \Delta$$

$$\frac{Y}{X} = \frac{G_1G_2G_3 + G_4\left(1 + G_2H_1 - G_1G_2H_1 + G_2G_3H_2\right)}{1 + G_2H_1 - G_1G_2H_1 + G_2G_3H_2} \tag{4.10}$$

**Example 4.10:** Find $\frac{C}{R}$ using Mason's rule for the flow graph shown in Figure 4.11.

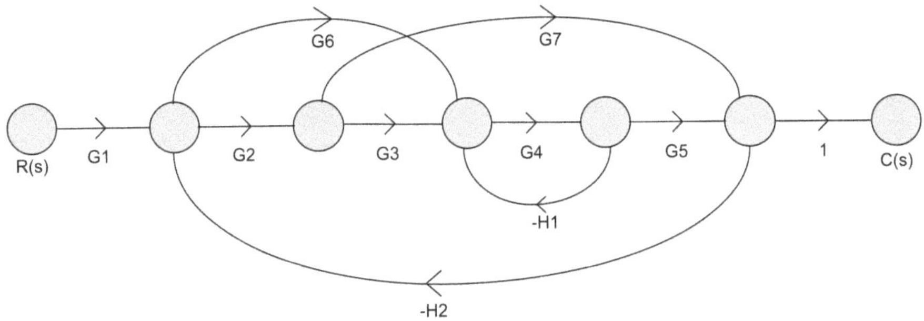


**FIGURE 4.11** Signal flow graph of a control system.

***Solution***

Forward Paths

$$P_1 = G_1G_2G_3G_4G_5$$

$$P_2 = G_1G_6G_4G_5$$

$$P_3 = G_1G_2G_7$$

Individual loops

$$L_1 = -G_4H_1$$

$$L_2 = G_2G_3G_4G_5H_2$$

$$L_3 = -G_2G_7H_2$$

$$L_4 = -G_6G_4G_5H_2$$

$$L_1L_3 = H_1H_2G_2G_3G_7$$

$$\frac{C}{R} = \frac{P_1\Delta_1 + P_2\Delta_2 + P_3\Delta_3}{\Delta}$$

$$\Delta = 1 - \left(L_1 + L_2 + L_3 + L_4\right) + L_1L_3$$

$$\Delta_1 = 1$$

$$\Delta_2 = 1$$

$$\Delta_3 = 1 - (-G_4H_1) = 1 + G_4H_1$$

$$\frac{C}{R} = \frac{P_1 + P_2 + P_3(1 + G_4H_1)}{1 - (L_1 + L_2 + L_3 + L_4) + L_1L_3} \tag{4.11}$$

**Example 4.11:** In Figure 4.12, explore and figure out the transfer function of the control system.

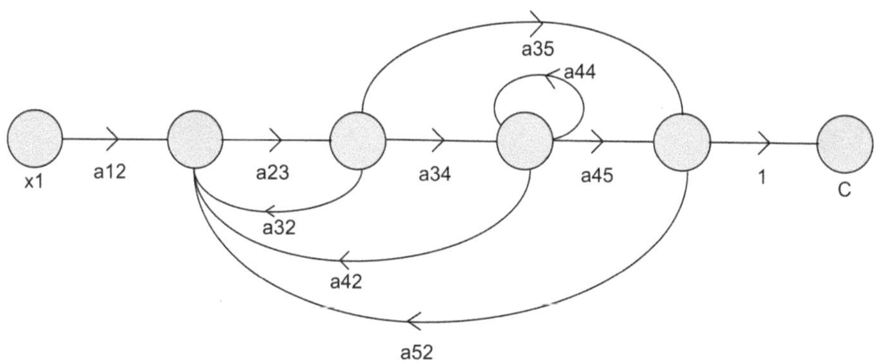


**FIGURE 4.12** Signal flow graph of a control system.

***Solution***

Forward Paths

$$P_1 = a_{12}a_{23}a_{34}a_{45}$$

$$P_2 = a_{12}a_{23}a_{35}$$

Individual loops

$$L_1 = a_{23}a_{32}$$

$$L_2 = a_{44}$$

$$L_3 = a_{23}a_{34}a_{45}$$

$$L_4 = a_{23}a_{34}a_{45}a_{52}$$

$$L_5 = a_{23}a_{35}a_{45}a_{52}$$

$$L_1L_2 = a_{23}a_{32}a_{44}$$

$$L_2L_5 = a_{44}a_{23}a_{35}a_{52}$$

$$\Delta_1 = 0$$

$$\Delta_2 = 1 - a_{44}$$

$$\Delta = 1 - (L_1 + L_2 + L_3 + L_4 + L_5) + L_1L_2 + L_2L_5$$

$$\frac{C}{R} = \frac{a_{12}a_{23}a_{35}a_{45} + a_{12}a_{23}a_{35}(1 - a_{44})}{\Delta} \tag{4.12}$$

**Example 4.12:** In Figure 4.13, explore and figure out the transfer function of the control system. The block diagram and signal flow of the block diagram in Figure 4.13 is given in Figure 4.14.

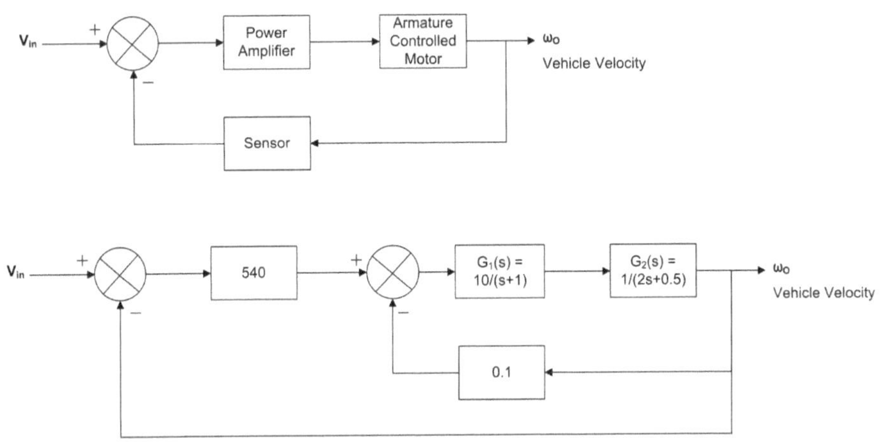


FIGURE 4.13 Block diagram and signal flow graph of a control system.

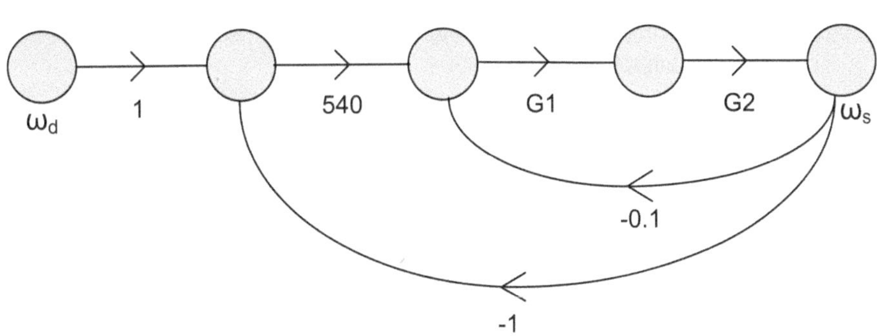


FIGURE 4.14 Signal flow graph for block diagram in Figure 4.13.

***Solution***

Electric traction motor control

$$\frac{\omega_s}{\omega_d}=\frac{540G_1(s)G_2(s)}{1+0.1G_1(s)G_2(s)+540G_1(s)G_2(s)}=\frac{540G_1(s)G_2(s)}{1+540.1G_1(s)G_2(s)}$$

$$\frac{\omega_s}{\omega_d}=\frac{5400}{(s+1)(2s+0.5)+540.1}$$

$$\frac{\omega_s}{\omega_d}=\frac{5400}{2s^2+2.5s+5401.5}$$

$$\frac{\omega_s}{\omega_d}=\frac{2700}{s^2+1.25s+2700.75}$$

**Problems**

P4.1. By constructing and analyzing signal flow graphs, determine the system transfer function for Prob. 4.1.

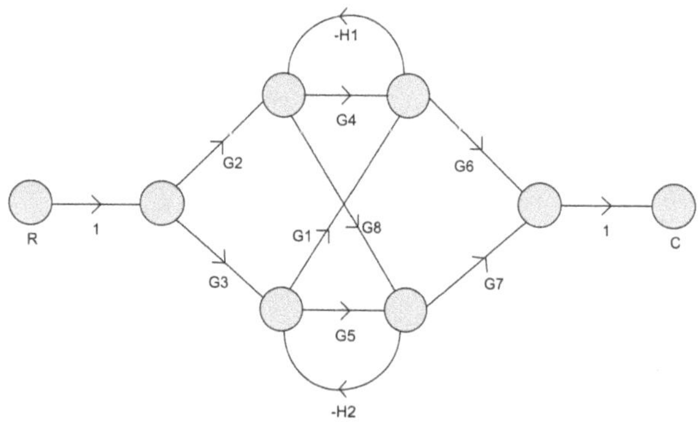


P4.2. By constructing and analyzing signal flow graphs, determine the system transfer function for Prob. 4.2.

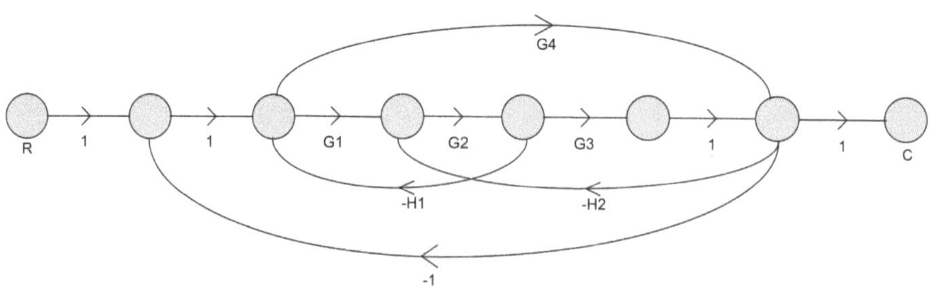

P4.3. By constructing and analyzing signal flow graphs, determine the system transfer function for Prob. 4.3.

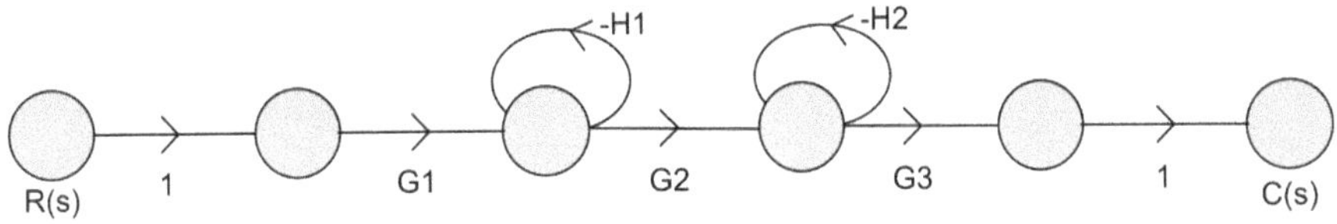


P4.4. By constructing and analyzing signal flow graphs, determine the system transfer function for Prob. 4.4.

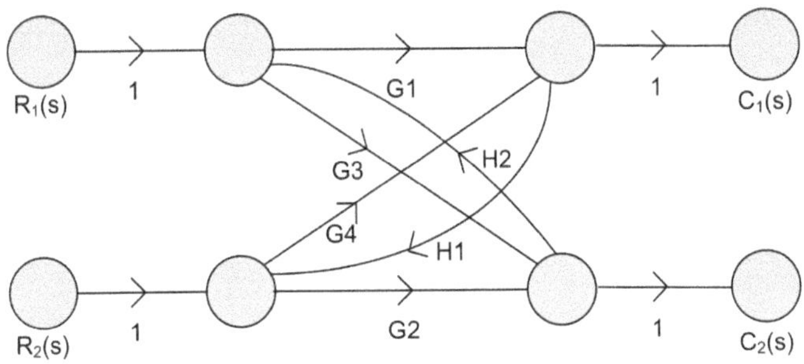


P4.5. By constructing and analyzing signal flow graphs, determine the system transfer function for Prob. 4.5.

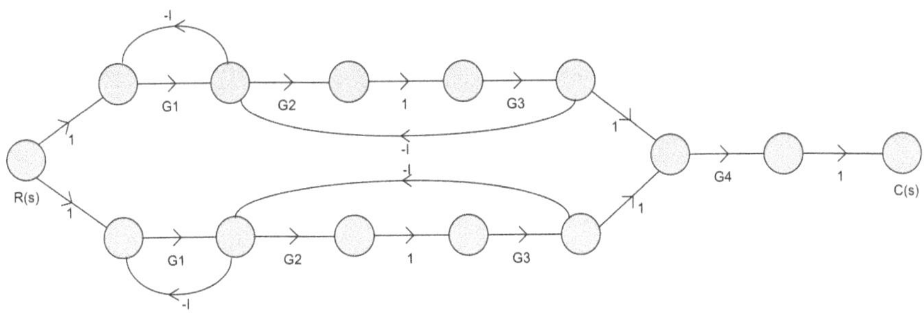

# 5 System Types and Steady-State Error Description

## 5.1 STEADY-STATE ERROR

The number of poles at the origin and the input type dictate the steady-state error of a feedback control system.

$$G(s) = \frac{\prod_{i=1}^{m}(s+z_i)}{s^l \prod_{j=1}^{n}(s+p_j)} \tag{5.1}$$

$$l + n > m$$

$$l = 0,1,2$$

where, $G(s)$ is the transfer function of the system and $l$ is the number of poles at the origin on the s-plane. The Type of system categorized by $l$ is as given here:

$l = 0$, Type 0
$l = 1$, Type 1
$l = 2$, Type 2

### 5.1.1 Steady-State Error and Error Constraints

Consider Figure 5.1.

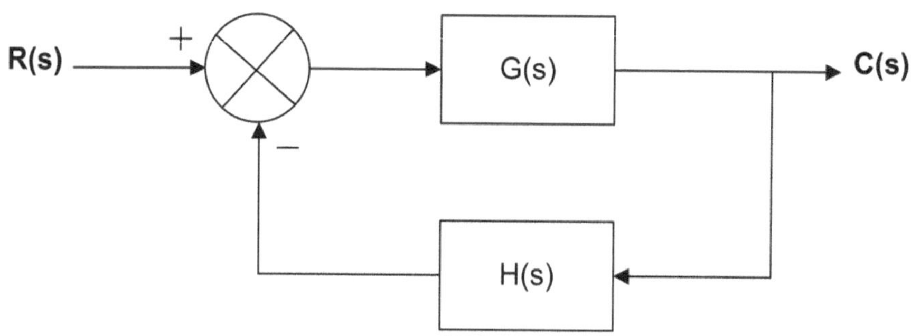


**FIGURE 5.1** A close loop control system.


DOI: 10.1201/9781003484202-5

Now, consider a feedback control system,

$$E(s) = R(s) - C(s)H(s) \tag{5.2}$$

$$C(s) = G(s)E(s) \tag{5.3}$$

Putting Eq. 5.3 in Eq. 5.2,

$$E(s) = R(s) - E(s)G(s)H(s)$$

$$E(s)(1 + G(s)H(s)) = R(s)$$

$$E(s) = \frac{R(s)}{1 + G(s)H(s)}$$

$$H(s) = 1$$

$$E(s) = \frac{R(s)}{1 + G(s)} \tag{5.4}$$

a) **Steady-state error of control systems with step input (Positional error constant)**

$$R(t) = u(t)$$

$$R(s) = \frac{1}{s}$$

$$e_{ss} = \lim_{s \to 0} \frac{sR(s)}{s(1 + G(s))} = \lim_{s \to 0} \frac{R(s)}{1 + G(s)} \tag{5.5}$$

Let

$$k_p = \lim_{s \to 0} G(s)$$

Then,

$$e_{ss} = \frac{R}{1 + k_p}$$

$$k_p = \frac{R - e_{ss}}{e_{ss}} = \frac{Desired \quad output - Allowable \quad e_{ss}}{Allowable \quad e_{ss}} \tag{5.6}$$

For Type 0 system, $e_{ss}$ is finite. Therefore, $k_p$ is also finite. For Type 1 and Type 2 systems, $e_{ss}$ is zero. Therefore, $k_p$ is infinite. Figure 5.2 shows step input given to a control system and its resultant transient response.

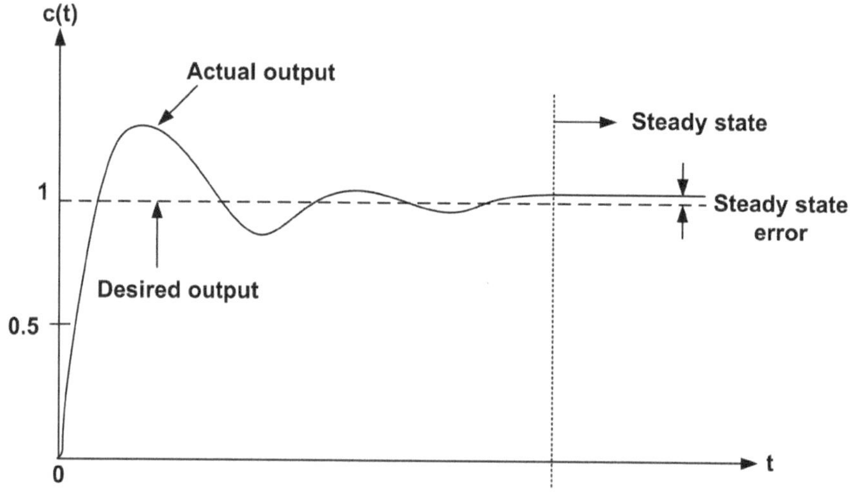


**FIGURE 5.2** Step input given to a control system and its transient response.

### b) Steady-state error of systems with velocity input

$$R(s) = \frac{R}{s^2}$$

$$e_{ss} = \lim_{s \to 0} \frac{s}{1 + G(s)} \cdot \frac{R}{s^2} = \lim_{s \to 0} \frac{R}{sG(s)} \tag{5.7}$$

Define

$$e_{ss} = \frac{R}{k_v}$$

$$\lim_{s \to 0} \quad sG(s) = k_v$$

Here, $k_v$ is the velocity error constant.

For Type 0, $e_{ss} = \infty$, $k_v = 0$

For Type 1, $e_{ss} = \dfrac{R}{k_v}$

For Type 2, $e_{ss} = 0$, $k_v = \infty$.

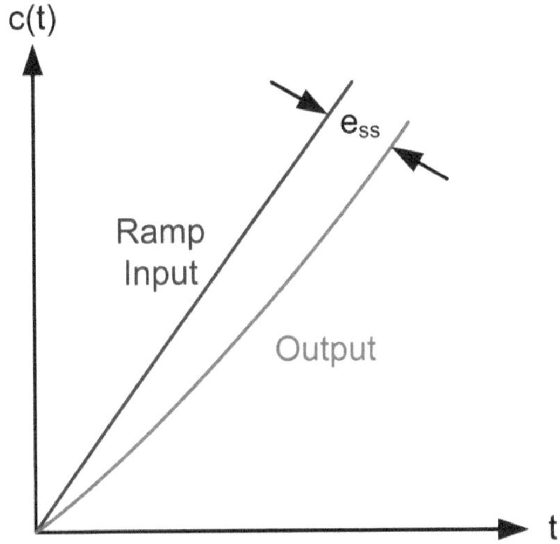


**FIGURE 5.3** Ramp input and steady-state error system of Type 1.

Figure 5.3 shows the ramp input and steady-state error for a Type 1 system.

**c) Steady-state error of systems with acceleration input**

$$R(s) = \frac{R}{s^3}$$

$$e_{ss} = \lim_{s \to 0} \frac{s}{1 + G(s)} \cdot \frac{R}{s^3} = \lim_{s \to 0} \frac{R}{s^2 G(s)} = \frac{R}{k_a} \tag{5.8}$$

where,

$$k_a = \frac{R}{e_{ss}} \tag{5.9}$$

Figure 5.4 shows the parabolic input and the steady-state error for a Type 1 system.

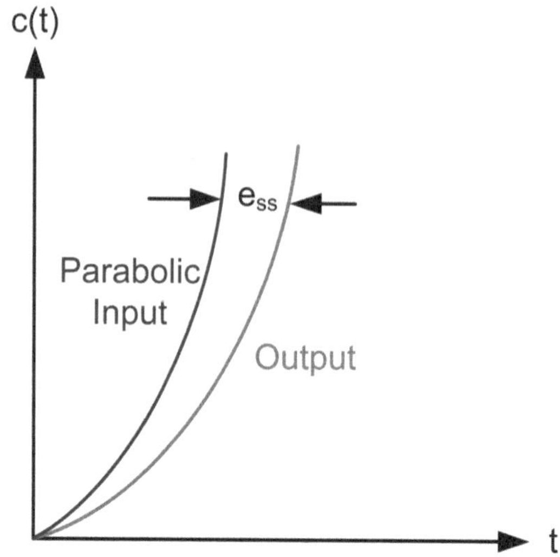


**FIGURE 5.4** Parabolic input and the steady-state error for a system of Type 1.

For Type 0, $e_{ss} = \infty$, $k_a = 0$
For Type 1, $e_{ss} = \infty$, $k_a = 0$
For Type 2, $e_{ss} = \frac{R}{k}$, $k_a = k$
For Type 3 or higher systems, $e_{ss} = 0$, $k_a = \infty$

**Example 5.1:** Consider a feedback control system with $G(s) = \frac{k}{(s+1)(2s+1)}$ and step, ramp, and acceleration inputs. Find $k_p$, $k_v$, and $k_a$ and steady-state errors in each case.

***Solution***

a) $e_{ss} = \frac{1}{1+k}$, $k_p = k$
b) $e_{ss} = \infty$, $k_v = 0$
c) $e_{ss} = \infty$, $k_a = 0$

**Example 5.2:** Consider a feedback control system with $(s) = \frac{k}{s(s+1)}$. Find $k_p$, $k_v$, $k_a$, and steady-state errors for step, ramp, and parabolic inputs.

***Solution***

a) $k_p = \infty$, $e_{ss} = 0$

b) $k_v = k,\ e_{ss} = \frac{R}{k}$

c) $k_a = 0,\ e_{ss} = \infty$

**Example 5.3:** The system transfer function for a unity feedback control system with $(s) = \frac{k}{s^2(s+1)}$. Find $k_p$, $k_v$, and $k_a$ and steady-state errors for step, ramp, and parabolic inputs.

***Solution***

a) $k_p = \infty,\ e_{ss} = 0$

b) $k_v = \infty,\ e_{ss} = 0$

c) $k_a = k,\ e_{ss} = \frac{R}{k}$

Table 5.1 shows the final values of different types of systems along with different inputs.

**Example 5.4:** Find the error coefficient for $G(s)H(s) = \frac{(s+3)}{s(1+0.6s)(1+0.3s)}$.

**TABLE 5.1**

**Final Values of Different Types of Systems and Different Inputs**

| | Step Input | | Ramp Input | | Parabolic Input | |
|---|---|---|---|---|---|---|
| **Type** | $k_p$ | $e_{ss}$ | $k_v$ | $e_{ss}$ | $k_a$ | $e_{ss}$ |
| Type 0 | $k$ | $\frac{A}{1+k}$ | 0 | $\infty$ | 0 | $\infty$ |
| Type 1 | $\infty$ | 0 | $k$ | $\frac{A}{k}$ | 0 | $\infty$ |
| Type 2 | $\infty$ | 0 | $\infty$ | 0 | $k$ | $\frac{A}{k}$ |

***Solution***

$$k_p = \lim_{s\to 0} G(s)H(s) = \lim_{s\to 0} \frac{(s+3)}{s(1+0.6s)(1+0.3s)} = \infty,\ e_{ss} = 0$$

$$k_v = \lim_{s\to 0} sG(s)H(s) = \lim_{s\to 0} \frac{s(s+3)}{s(1+0.6s)(1+0.3s)} = 3,\ e_{ss} = \frac{1}{3}$$

$$k_a = \lim_{s\to 0} s^2 G(s)H(s) = \lim_{s\to 0} \frac{s^2(s+3)}{s(1+0.6s)(1+0.3s)} = 0,\ e_{ss} = \infty$$

**Example 5.5:** For a system having $G(s)H(s) = \frac{k(s+4)}{s(s^3+3s^2+2s)}$, find (i) type of system, (ii) error coefficients, and error due to input $\frac{A}{2}t^2$.

***Solution***

i) $$G(s)H(s) = \frac{k(s+4)}{s^2(s^2+3s+2)} = \frac{k(s+4)}{s^2(s+1)(s+2)}$$

So it is a Type 2 system.

ii) Since Type 2 system's $k_p$ and $k_v$ are $\infty$,

$$k_a = \lim_{s\to 0} s^2 G(s)H(s) = \lim_{s\to 0} \frac{k(s+4)}{(s+1)(s+2)} = 2k$$

Since the input is parabolic, $e_{ss} = \frac{A}{2k}$.

**Example 5.6:** The system transfer function for a unity feedback control system with $G(s)H(s) = \frac{k}{s(Ts+F)}$. Find out the steady-state error for unit ramp input.

***Solution***

$$E(s) = \frac{1}{1+G(s)} = \frac{1}{1+\frac{k}{s(Ts+F)}} R(s) = \frac{1}{1+\frac{k}{s(Ts+F)}} \cdot \frac{1}{s^2}$$

$$e_{ss} = \lim_{s\to 0} \frac{s^2(Ts+F)}{s(Ts+F)+k} = \frac{F}{k}$$

**Example 5.7:** For a unity feedback system having an open loop transfer function, $G(s) = \frac{k(s+2)}{s(s^3+7s^2+12s)}$, determine (i) type of system, (ii) error constants $k_p$, $k_v$, and $k_a$, and (iii) steady-state error for parabolic input.

***Solution***

i) Since there are two poles at the origin of the s-plane, the system is of Type 2.

ii) $$k_p = \lim_{s\to 0} G(s) = \lim_{s\to 0} \frac{k(s+2)}{s^2(s^2+7s+12)} = \infty$$

$$k_v = \lim_{s\to 0} sG(s) = \lim_{s\to 0} \frac{k(s+2)}{s(s^2+7s+12)} = \infty$$

$$k_a = \lim_{s\to 0} s^2G(s) = \lim_{s\to 0} \frac{k(s+2)}{(s^2+7s+12)} = \frac{k}{6}$$

iii) Let the parabolic input be

$$r(t) = t^2$$

$$R(s) = \frac{2}{s^3}$$

$$e_{ss} = \lim_{s\to 0} sE(s) = \lim_{s\to 0} \frac{sR(s)}{1+G(s)H(s)} = \lim_{s\to 0} \frac{sR(s)}{1+G(s)} = \lim_{s\to 0} \frac{s.\frac{2}{s^3}}{1+\frac{k(s+2)}{s^2(s^2+7s+12)}}$$

$$= \lim_{s\to 0} \frac{2(s^2+7s+12)}{s^2(s^2+7s+12)+k(s+3)} = \frac{24}{2k} = \frac{12}{k}$$

**Example 5.8:** For a system having $G(s)H(s) = \frac{k(s+4)}{s(s^3+5s^2+6s)}$, find (i) the type of system, (ii) the error coefficients, (iii) the error due to input $\frac{A}{2}t^2$.

***Solution***

i) $$G(s)H(s) = \frac{k(s+4)}{s(s^3+5s^2+6s)} = \frac{k(s+4)}{s^2(s^2+5s+6)} = \frac{2k(1+0.25s)}{3s^2(1+0.5s)\left(1+\frac{1}{3}s\right)}$$

So it is a Type 2 system.

ii) Since this is a Type 2 system, $k_p$ and $k_v$ are $\infty$.

$$k_a = \lim_{s\to 0} s^2G(s)H(s) = \lim_{s\to 0} \frac{2k(1+0.25s)}{3(1+0.5s)\left(1+\frac{1}{3}s\right)} = \frac{2k}{3}$$

iii) The steady-state error is $\frac{3A}{2k}$.

**Problems**

P5.1. Consider a feedback control system with $G(s)=\frac{2}{(s+1)(2s+1)}$ and step, ramp, and acceleration inputs. Find $k_p$, $k_v$, and $k_a$ and steady-state errors in each case.

P5.2. Find the error coefficient for $G(s)H(s)=\frac{(s+25)}{s(1+0.6s)(1+0.3s)}$.

P5.3. For a system having $G(s)H(s)=\frac{10(s+4)}{s(s^3+3s^2+2s)}$, find (i) type of system, (ii) error coefficients, and error due to input $\frac{1}{2}t^2$.

P5.4. The system transfer function for a unity feedback control system with $(s)H(s)=\frac{3}{s(2s+1)}$. Find out the steady state error for unit ramp input.

P5.5. For a unity feedback system having an open loop transfer function $G(s)=\frac{100(s+2)}{s(s^3+7s^2+12s)}$, determine (i) type of system (ii) error constants $k_p$, $k_v$, and $k_a$, and (iii) steady-state error for parabolic input.

P5.6. For a system having $G(s)H(s)=\frac{0.5(s+4)}{s(s^3+5s^2+6s)}$, find (i) the type of system, (ii) the error coefficients, (iii) the error due to input $2t^2$.

# 6 Time Response Analysis

## 6.1 INTRODUCTION

A control system's output can be analyzed as a function of time, and this is termed a time response analysis. It is the combination of transient response, $c_t(t)$, and steady state, $c_{ss}(t)$.

Thus, the total output response is

$$c(t) = c_t(t) + c_{ss}(t) \tag{6.1}$$

As the names specify, the transient response lasts a short time, but the steady-state response is achieved once the time approaches an infinite value. The parameters of a feedback control system adjust themselves to alter the system's transient and steady-state behavior.

## 6.2 FIRST ORDER SYSTEM—TIME RESPONSE

Various inputs can be given to the first-order system and its response time can be evaluated as follows:

### (a) Impulse Input

Consider a system,

$$\frac{C(s)}{R(s)} = \frac{1}{\tau s + 1}$$

$$R(s) = 1$$

$$C(s) = \frac{1}{\tau s + 1}$$

By applying the Inverse Laplace transform, we get

$$c(t) = \frac{1}{\tau} e^{-\frac{t}{\tau}} \tag{6.2}$$

Steady-state error is

$$e_{ss} = \lim_{s \to 0} sE(s) = \lim_{s \to 0} s\left[R(s) - C(s)\right] = \lim_{s \to 0} s\left[1 - \frac{1}{\tau s + 1}\right]$$

DOI: 10.1201/9781003484202-6

$$= \lim_{s \to 0} s\left[\frac{\tau s + 1 - 1}{\tau s + 1}\right] = \lim_{s \to 0} s\left[\frac{\tau s}{\tau s + 1}\right]$$

$$e_{ss} = 0$$

**(b) Step Input**

Let

$$G(s) = \frac{1}{\tau s + 1}$$

$$R(s) = \frac{1}{s}$$

$$C(s) = \frac{1}{s(\tau s + 1)}$$

$$c(t) = 1 - e^{-\frac{t}{\tau}} \tag{6.3}$$

It can be drawn as shown in Figure 6.1.

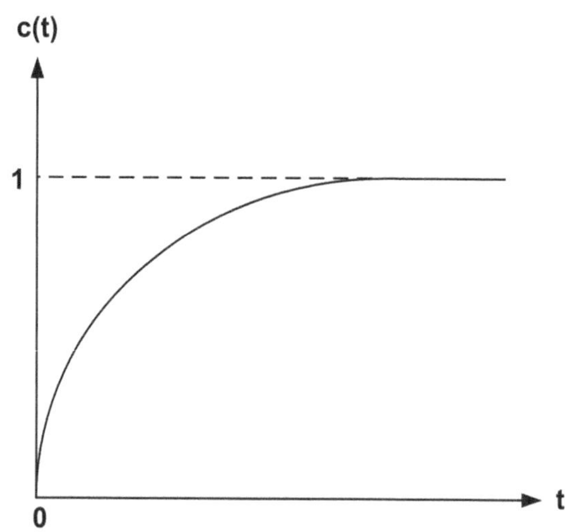


**FIGURE 6.1** Step input for first-order system.

## 6.3 SECOND-ORDER SYSTEM—PERFORMANCE PARAMETERS

The nomenclature of performance parameters for a second-order system's time response can be depicted as given in Figure 6.2 and summarized as follows:

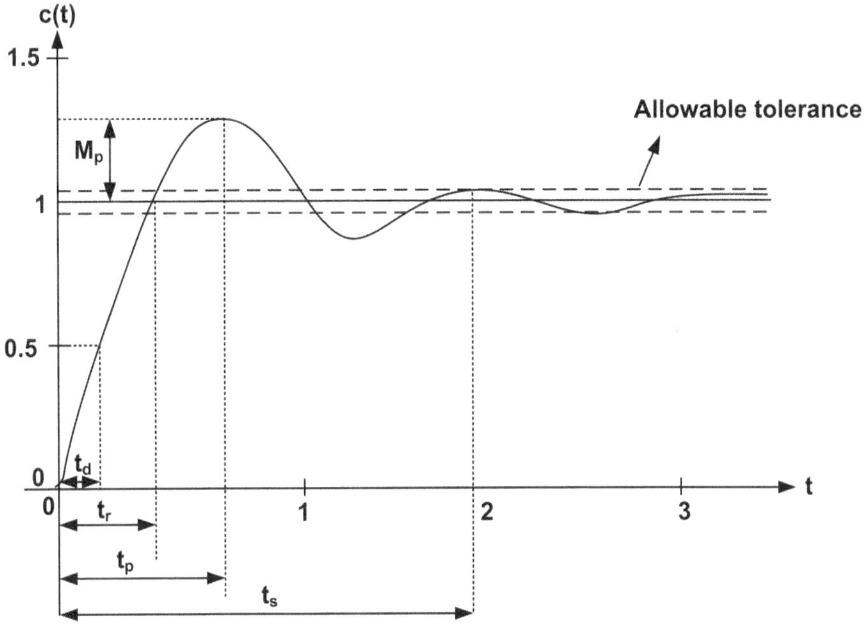


**FIGURE 6.2** Performance parameters for second-order system.

- The natural frequency of oscillations, $\omega_n$
- Damped frequency of oscillations, $\omega_d$
- Damping ratio, $\xi$
- Damping constant, $\alpha = \xi\omega_n$
- Maximum overshoot, $M_p$
- Peak time, $T_p$
- Rise time, $T_r$
- Time constant, $T_c$
- Settling time, $T_s$
- Fifty percent output, $T_d = \frac{1+0.7\xi}{\omega_n}$

Consider a standard second-order system,

$$s^2 + \xi\omega_n s + \omega_n^2 = 0$$

The roots are,

$$s_1, s_2 = -\xi\omega_n \pm j\omega_n\sqrt{1-\xi^2} = -\alpha \pm j\omega_d \tag{6.4}$$

$$\xi = damping \;\; ratio$$

$$\omega_n = underdamped \;\; natural \;\; frequency$$

$$\alpha = \xi\omega_n \, damping \; constant \; (actual \; damping)$$

$$\omega_d = \omega_n\sqrt{1-\xi^2} \; damped \; frequency$$

When,

$$\xi < 1 \; s_1, s_2 = -\xi\omega_n \pm j\omega_n\sqrt{1-\xi^2} \rightarrow underdamped \; case$$

$$\xi = 1 \; s_1, s_2 = -\omega_n \rightarrow critical \; damped \; case$$

$$\xi > 1 \; s_1, s_2 = -\xi\omega_n \pm \omega_n\sqrt{\xi^2-1} \rightarrow overdamped \; case$$

$$\xi = 0 \; s_1, s_2 = \pm j\omega_n \rightarrow undamped \; case$$

If $\omega_n$ is held constant and $\xi$ is varied from $0$ *to* $\infty$, the roots of the characteristic equation will move away from the imaginary axis.

$$cos\theta = \xi$$

$$sin\theta = \sqrt{1-\xi^2}$$

$$tan\theta = \frac{\sqrt{1-\xi^2}}{\xi}$$

The roots of the characteristic equation can be drawn as shown in Figure 6.3:

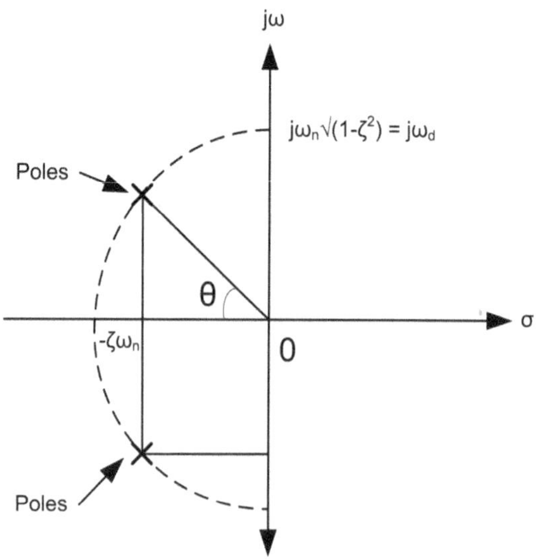


**FIGURE 6.3** Roots of characteristic equation for second-order system.

Next, consider the response of a second-order system to a step input.

$$\frac{C(s)}{R(s)} = \frac{\omega_n^2}{s^2 + 2\xi\omega_n s + \omega_n^2} \tag{6.5}$$

$$C(s) = \frac{\omega_n^2}{s(s^2 + 2\xi\omega_n s + \omega_n^2)}$$

$$c(t) = \mathcal{L}^{-1}\left[\frac{\omega_n^2}{s(s^2 + 2\xi\omega_n s + \omega_n^2)}\right]$$

$$c(t) = 1 - \frac{e^{-\xi\omega_v t}}{\sqrt{1-\xi^2}}\sin\left[\sqrt{1-\xi^2}\,t + tan^{-1}\frac{\sqrt{1-\xi^2}}{\xi}\right] \tag{6.6}$$

For obtaining maxima, differentiate $\omega$ with respect to $t$.

$$\frac{dc}{dt} = -\frac{e^{-\xi\omega_n t}}{\sqrt{1-\xi^2}}.\omega_n\sqrt{1-\xi^2}\cos\left[\omega_n\sqrt{1-\xi^2}\,t + tan^{-1}\frac{\sqrt{1-\xi^2}}{\xi}\right]$$

$$+\frac{e^{-\xi\omega_n t}}{\sqrt{1-\xi^2}}\xi\omega_n\sin\left[\omega_n\sqrt{1-\xi^2}\,t + tan^{-1}\frac{\sqrt{1-\xi^2}}{\xi}\right]$$

Let $\varphi = tan^{-1}\frac{\sqrt{1-\xi^2}}{\xi}$

$$\omega_d = \omega_n\sqrt{1-\xi^2}$$

$$\frac{dc}{dt} = -\frac{e^{-\xi\omega_n t}}{\sqrt{1-\xi^2}}[\omega_n\sqrt{1-\xi^2}\cos[\omega_d t + \varphi] + \xi\omega_n\sin[\omega_d t + \varphi]]$$

$$\tan[\omega_d t + \varphi] = \frac{\omega_n\sqrt{1-\xi^2}}{\xi\omega_n} = \frac{\sqrt{1-\xi^2}}{\xi} \tag{6.7}$$

or

$$\tan\left[\omega_d t + tan^{-1}\frac{\sqrt{1-\xi^2}}{\xi}\right] = \frac{\sqrt{1-\xi^2}}{\xi}$$

can only be satisfied if

$$\omega_d t = n\pi \text{ where, } n = 0,1,2\ldots$$

or

$$t = \frac{n\pi}{\omega_n \sqrt{1-\xi^2}}$$

The first peak will occur for $n = 1$.

$$t_p = \frac{n\pi}{\omega_n \sqrt{1-\xi^2}}$$

It may be noted that for odd $n$, an overdamped response occurs, and for even $n$, an underdamped response occurs.

Substituting the value of time in $c(t)$,

$$c(t)_{max} = 1 - \frac{e^{-\frac{n\pi\xi}{\sqrt{1-\xi^2}}}}{\sqrt{1-\xi^2}} \sin\left[ n\pi + tan^{-1}\frac{\sqrt{1-\xi^2}}{\xi} \right]$$

$$= 1 - \frac{e^{-\frac{n\pi\xi}{\sqrt{1-\xi^2}}}}{\sqrt{1-\xi^2}} \sin\left[ tan^{-1}\frac{\sqrt{1-\xi^2}}{\xi} \right]$$

$$= 1 - \frac{e^{-\frac{n\pi\xi}{\sqrt{1-\xi^2}}}}{\sqrt{1-\xi^2}} \sqrt{1\;\;\xi^2} = 1-(-1)^n e^{-\frac{n\pi\xi}{\sqrt{1-\xi^2}}} \tag{6.8}$$

Percentage overshoot is

$$c_{max} - 1 = 100\frac{e^{-\pi\xi}}{\sqrt{1-\xi^2}}$$

Which is only a function of the damping ratio $\xi$. Since $\xi\omega_n$ is defined as a damping factor, the time constant is defined as

$$T_c = \frac{1}{\xi\omega_n} \tag{6.9}$$

Therefore, for a second order system,

$$\frac{1}{\xi\omega_n} = time\,constant$$

$$\frac{e^{-\pi\xi}}{\sqrt{1-\xi^2}} = Max\,overshoot$$

$$\frac{\pi}{\omega_n\sqrt{1-\xi^2}} = time\ for\ \max\ overshoots$$

Rise time can be obtained by equating

$$c(t) = 1$$

$$c(t) = 1 = 1 - \frac{e^{-\xi\omega_n t}}{\sqrt{1-\xi^2}}\ \sin\left[\omega_d t + tan^{-1}\frac{\sqrt{1-\xi^2}}{\xi}\right]$$

$$0 = -\frac{e^{-\xi\omega_n t}}{\sqrt{1-\xi^2}}\ \sin\left[\omega_d t + tan^{-1}\frac{\sqrt{1-\xi^2}}{\xi}\right]$$

Since, $\frac{e^{-\xi\omega_n t}}{\sqrt{1-\xi^2}} \neq 0$

$$\sin\left[\omega_d t_r + tan^{-1}\frac{\sqrt{1-\xi^2}}{\xi}\right] = 0$$

$$\sin\omega_d t_r \cos\left(tan^{-1}\frac{\sqrt{1-\xi^2}}{\xi}\right) + cos\omega_d t_r \sin(tan^{-1}\frac{\sqrt{1-\xi^2}}{\xi}) = 0$$

$$\sin\omega_d t_r\ \cdot\xi + cos\omega_d t_r\sqrt{1-\xi^2} = 0$$

$$\tan\omega_d t_r = -\frac{\sqrt{1-\xi^2}}{\xi}$$

Thus, rise time $T_r$ is

$$t_r = \frac{1}{\omega_d} tan^{-1}\frac{\omega_d}{-\alpha} = \frac{\pi - \beta}{\omega_d} \tag{6.10}$$

where,

$$\beta = tan^{-1}\frac{\omega_d}{\alpha} = tan^{-1}\frac{\omega_n\sqrt{1-\xi^2}}{\xi\omega_n}$$

$T_s$ settling time is 4 times $T_c$ for 3% $e_{ss}$ and 5 times for 2% $e_{ss}$.

$$T_d = \frac{1 + 0.7\xi}{\omega_n} \tag{6.11}$$

**Example 6.1**

We have a unity feedback control system, as shown in Figure 6.4, whose open-loop transfer function is $G(s) = \frac{10}{s(s+1)}$. If the reference input is a unit step, find $\omega_n$, $\xi$, $\omega_d$, $\alpha$, $T_c$, $T_s$, $T_p$ and $M_p$.

***Solution***

$$\frac{C(s)}{R(s)} = \frac{10}{s^2 + s + 10}$$

$$\frac{C(s)}{R(s)} = \frac{\omega_n{}^2}{s^2 + 2\xi\omega_n s + \omega_n{}^2}$$

$$\omega_n = \sqrt{10} = 3.1622$$

$$2\xi\omega_n = 1$$

$$\xi = 0.158$$

$$\omega_d = \omega_n\sqrt{1 - \xi^2} = 3.122$$

$$\alpha = \xi\omega_n = 0.5$$

$$T_c = \frac{1}{\alpha} = 2sec$$

$$T_p = \frac{\pi}{\omega_d} = 1sec$$

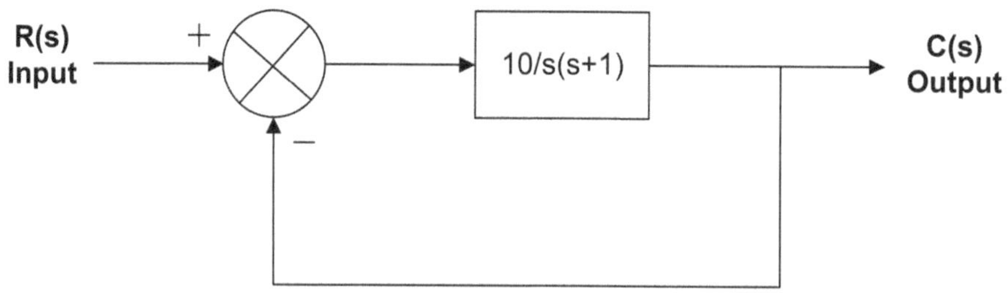


**FIGURE 6.4** Block diagram of a control system.

**Example 6.2**

Consider the mechanical system shown in Figure 6.5(a). The mass oscillates when a 2lb force is applied to the system, as shown in Figure 6.5(b). Determine k, m, and f.

***Solution***

$$\frac{X(s)}{P(s)} = \frac{1}{ms^2 + fs + K}$$

$$P(s) = \frac{2}{s}$$

So that,

$$X(s) = \frac{2}{s(ms^2 + fs + K)}$$

$$X(\infty) = \lim_{s \to 0} sX(s) = \frac{2}{K} = 0.1ft$$

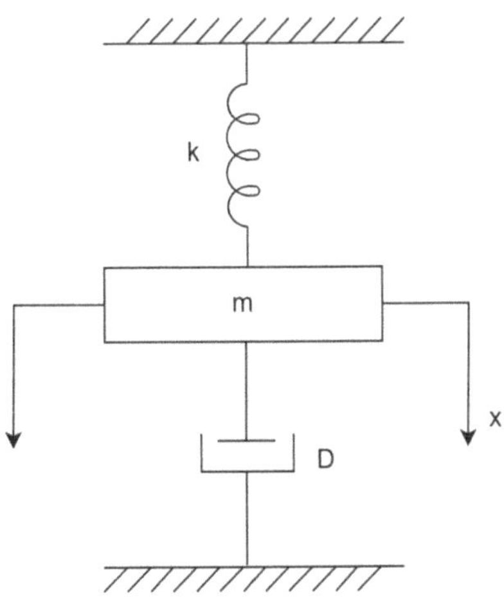


**FIGURE 6.5(A)** A mechanical system.

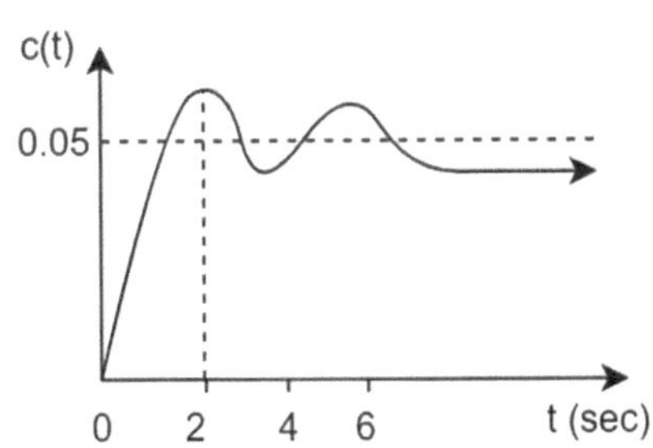


**FIGURE 6.5(B)** Time response analysis for mechanical system.

$$K = 20 lb / ft$$

$$M_p = 9.5\%$$

$$\xi = 0.6$$

$$T_p = \frac{\pi}{\omega_d} = \frac{\pi}{\omega_n\sqrt{1-(0.6)^2}} = \frac{\pi}{0.8\omega_n}$$

$$\omega_n = 1.96 rad$$

Since, $\omega_n^2 = \frac{K}{m} = \frac{20}{m}$

$$m = 5.2\, slugs$$

$$f = 2\xi\omega_n m = 12.2\, lb / ft / sec$$

## 6.4 EFFECTS OF CONTROLLER ELEMENTS

Consider a feedback control system, as shown in Figure 6.6.
The controller modes are defined as the functional behavior of controller elements. The three types of controllers are:

a) Proportional control element K, e
b) Derivative control element D, $\frac{de}{dt}$
c) Integral control element I, $\int e\,dt$

All three combined, the controller is called a proportional integral differential (PID) controller.

The following simple example can demonstrate the effect of each element.

a) Proportional control

$$\text{Let } G(s) = \frac{k}{s(s+1)}$$

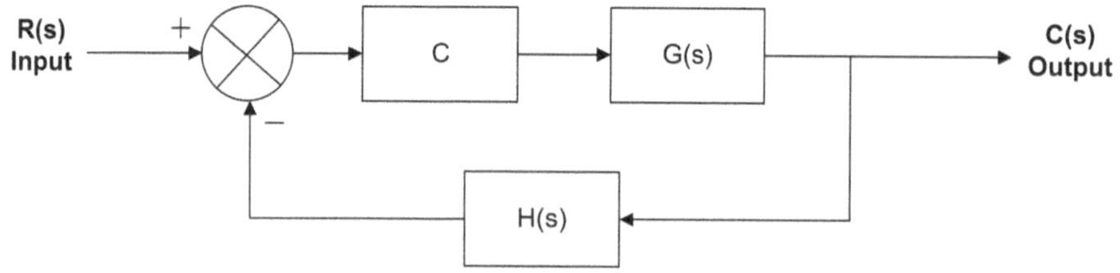


**FIGURE 6.6** Block diagram of a control system.

So that

$$\frac{C(s)}{R(s)} = \frac{G(s)}{1+G(s)} = \frac{k}{s^2+s+k}$$

For k = 1

Which defines $\omega_n = 1$ rad/sec and $\xi = 0.5$

Now let k be the gain of the proportional controller, then

$$\frac{C(s)}{R(s)} = \frac{k}{s^2+s+k}$$

Then, $\omega_n = \sqrt{k}$

$$2\xi\omega_n = 1$$

$$\xi = \frac{0.5}{\sqrt{k}}$$

$$\omega_d = \sqrt{k}\left(1-\frac{0.25}{k}\right)^{\frac{1}{2}}$$

$$\omega_d = \frac{\sqrt{k}}{\sqrt{k}}(k-0.25)^{\frac{1}{2}}$$

$$\omega_d = \sqrt{k-0.25}$$

$$T_p = \frac{\pi}{\sqrt{k-0.25}} < \frac{\pi}{0.866}$$

$$e_{ss} = 0$$

$$M_{p_1} < M_{p_k} \text{ for } k > 1$$

And so on.

b) Derivative control

Figure 6.7 shows the block diagram of a control system.

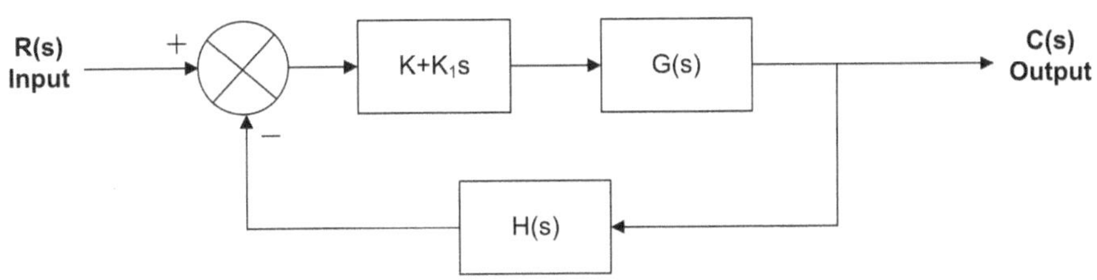


**FIGURE 6.7** Block diagram of a control system.

$$\frac{C(s)}{R(s)}=\frac{\dfrac{k+k_1s}{s(s+1)}}{1+\dfrac{k+k_1s}{s(s+1)}}=\frac{k+k_1s}{s^2+(k_1+1)s+k}$$

$$\omega_n=\sqrt{k}$$

$$2\xi\omega_n=k_1+1$$

$$\xi=\frac{0.5(k_1+1)}{\sqrt{k}}$$

Which is more than the previous for all $k_1>0$

$$\omega_d=\omega_n\sqrt{1-\xi^2}$$

$$=\sqrt{k}\left[1-\frac{0.25(k_1+1)^2}{k}\right]^{\frac{1}{2}}$$

$$=\sqrt{k}\left[\frac{k-0.25(k_1+1)^2}{k}\right]^{\frac{1}{2}}$$

$$=\left[k-0.25(k_1+1)^2\right]^{\frac{1}{2}}$$

$$\omega_{d_2}<\omega_{d_1}$$

$$T_p=\frac{\pi}{\omega_d}=\frac{\pi}{\left(k-0.25(k_1+1)^2\right)^{\frac{1}{2}}}>\frac{\pi}{(k-0.25)^{\frac{1}{2}}}$$

c) Integral Control

$$\frac{C(s)}{R(s)}=\frac{\dfrac{k+k_2/s}{s(s+1)}}{1+\dfrac{k+k_2/s}{s(s+1)}}=\frac{ks+k_2}{s^3+s^2+ks+k_2}$$

The order of the system is increased as one more pole is introduced. The system becomes sluggish, and so on.

$$T_{r_1}=\frac{\pi}{\omega_n}=\frac{2.16\xi+0.60}{\omega_n}$$

**Example 6.3**

Consider liquid level control as given in Figure 6.8. Assume that the controller's set point is fixed, step disturbance of $n_0$ determines the error, $n_0$ is small, and variations in variables from their respective steady-state values are also small. The controller is proportional. What is steady-state error if the controller is not proportional but integral?

***Solution***

When the controller is proportional with gain $k_p$ (assume the transfer function of the valve is unity). Since the set point is fixed, the variation of the set point is zero or $X(s)=0$.

The Laplace transform of *h*(*t*) is

$$H(s)=\frac{k_pR}{RCs+1}E(s)-\frac{R}{RCs+1}N(s)$$

$$E(s)=\frac{R}{RCs+1+k_pR}N(s)$$

Since,

$$N(s)=\frac{n_0}{s}$$

$$E(s)=\frac{R}{RCs+1+k_pR}\cdot\frac{n_0}{s}$$

$$=\frac{Rn_0}{1+k_pR}\left(\frac{1}{s+\dfrac{1+k_pR}{RC}}\right)-\frac{Rn_0}{1+k_pR}\cdot\frac{n_0}{s}$$

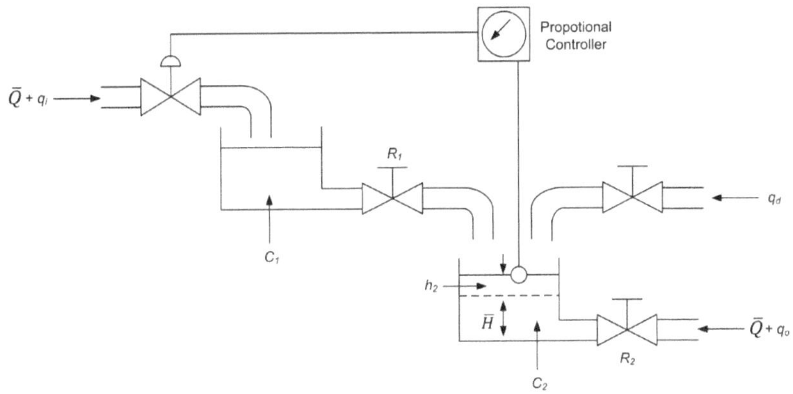


**FIGURE 6.8** Liquid level control system.

$$e(t)=\frac{Rn_0}{1+k_pR}\left(e^{-\frac{1+k_pR}{RC}t}-1\right)$$

$$e(\infty)=-\frac{Rn_0}{1+k_pR}$$

When

$$G_c(s)=\frac{k}{s}$$

$$E(s)=\frac{Rs}{RCs^2+s+kR}N(s)$$

$$e(\infty)=\lim_{s\to 0}\frac{-Rs^2}{RCs^2+s+kR}\cdot\frac{n_0}{s}=0$$

**Example 6.4**

Consider the control system shown in Figure 6.9. Find the value of $k$ and $k_1$ given that $T_p=2sec$ and $\xi=0.4$.

***Solution***

$$\frac{C(s)}{R(s)}=\frac{\frac{k}{s^2}}{1+\frac{k}{s^2}.(1+k_1s)}$$

$$=\frac{k}{s^2+k_1ks+k}$$

$$T_p=\frac{\pi}{\omega_d}=\frac{\pi}{\omega_n\sqrt{1-\xi^2}}$$

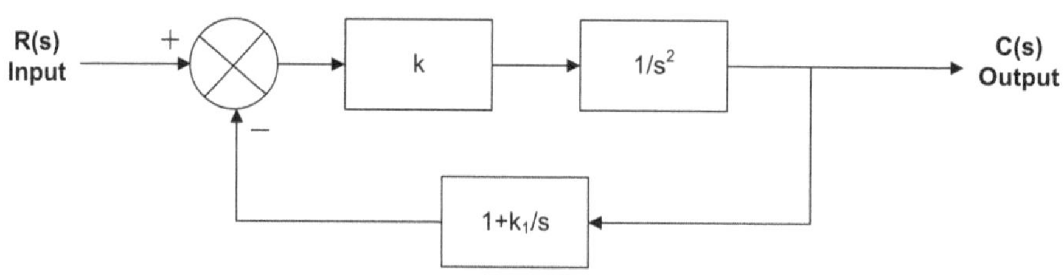


FIGURE 6.9 Block diagram of a control system.

Since $T_p = 2sec$ and $\xi = 0.4$

$$\omega_n = \frac{\pi}{2\sqrt{1-0.16}} = 1.71$$

$$kk_1 = 2\xi\omega_n = 0.8\omega_n$$

$$kk_1 = 0.8\sqrt{k}$$

$$k_1 = \frac{0.8}{\sqrt{k}} = \frac{0.8}{\omega_n} = 0.468$$

$$k = 2.92$$

**Example 6.5**

A system has $G(s) = \frac{k}{s(1+sT)}$ and $H(s) = 1$. Determine the factor by which $k$ should be multiplied to reduce the maximum overshoot from 16.25% to 9.4%.

***Solution***

$$\frac{C(s)}{R(s)} = \frac{\frac{k}{s(1+sT)}}{1+\frac{k}{s(1+sT)}} = \frac{k}{s^2T+s+k} = \frac{\frac{k}{T}}{s^2+\frac{s}{T}+\frac{k}{T}}$$

$$\omega_n = \sqrt{\frac{k}{T}} \text{ rad/sec}$$

$$2\xi\omega_n = \frac{1}{T}$$

$$2\xi\omega_n = \frac{1}{2\omega_n T} = \frac{1}{2\sqrt{\frac{k}{T}}T} = \frac{1}{2\sqrt{kT}}$$

$$\xi_1 = 0.5$$

$$\xi_2 = 0.6$$

$$\frac{\xi_1}{\xi_2} = \frac{0.5}{0.6} = \frac{\frac{1}{2\sqrt{k_1T}}}{\frac{1}{2\sqrt{k_2T}}} = \frac{2\sqrt{k_2T}}{2\sqrt{k_1T}}$$

$$\frac{0.5}{0.6} = \frac{\sqrt{k_2}}{\sqrt{k_1}}$$

$$k_2 = 0.911k_1$$

**Problems**

P6.1. A unity feedback control system with $G(s) = \frac{16}{s(s+4.8)}$ acts as the open loop transfer function. Find (i) $\omega_n$ (ii) $\xi$ (iii) $\omega_d$ (iv) $T_p$ , and (v) $M_p$

P6.2. A control system has $G(s) = \frac{25}{s(s+5)}$ and $H(s) = 1$. Find (i) $\omega_n$ (ii) $\xi$ (iii) $\omega_d$ (iv) $T_d$ (v) $T_r$ (vi) $T_p$ (vii) $M_p$, and (viii) $T_s$

P6.3. A control system is shown in Fig P6.3. Find $k_1$ and $k_2$ so that the peak overshoot of the system $M_p = 9.2\%$ and the $T_p = 0.786\,sec$.

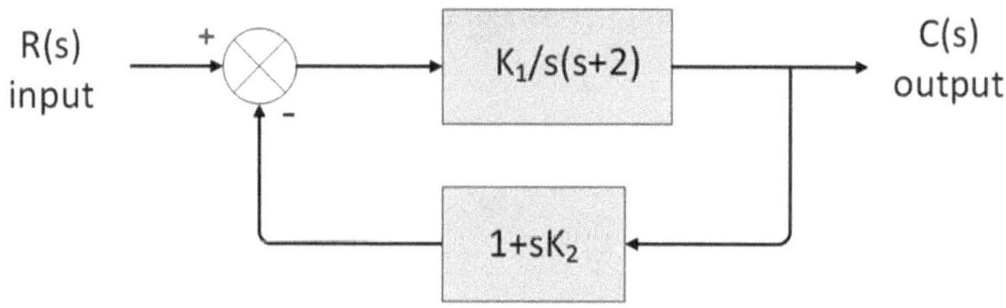


P6.4. A control system is shown in Figure P6.4. Determine the values of $K$, $M$, and $J$ if $M_p$ is 16.2%, $T_p = 2sec$, output at steady state is 5 units, and unit step is the input.

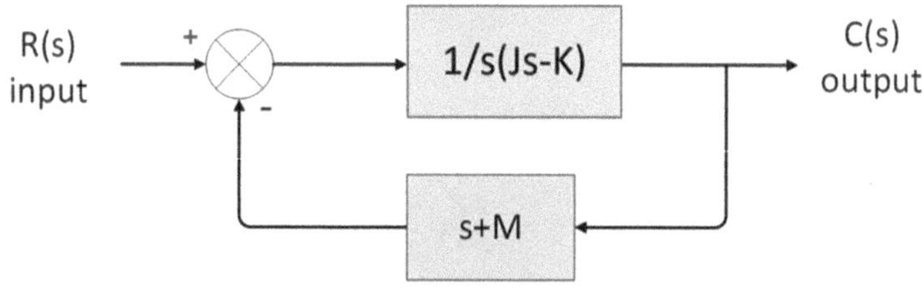


P6.5. Consider Fig. P6.5 where $D_1 = 8\,N.s/m$, $D = 10\,N.s/m$, $F = 10N$, K = 150 N/m and M = 1 kg. Calculate the following:

a) The final displacement of the trolley
b) The maximum displacement of the trolley
c) The time required to reach the maximum displacement.

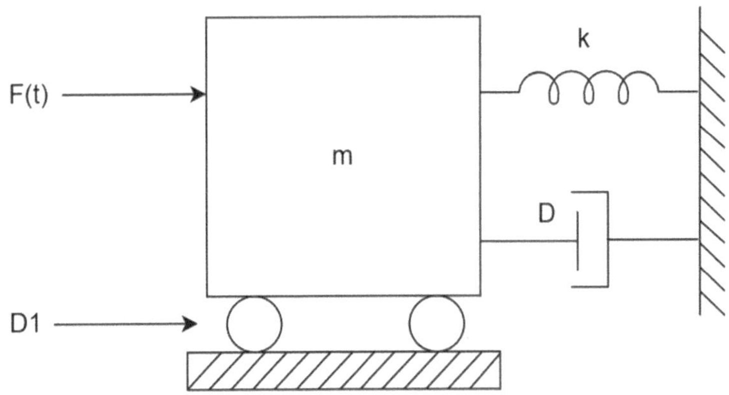


P6.6. Determine *M*, *D*, and *K* for the system shown in Fig P6.6. The system was initially relaxed.

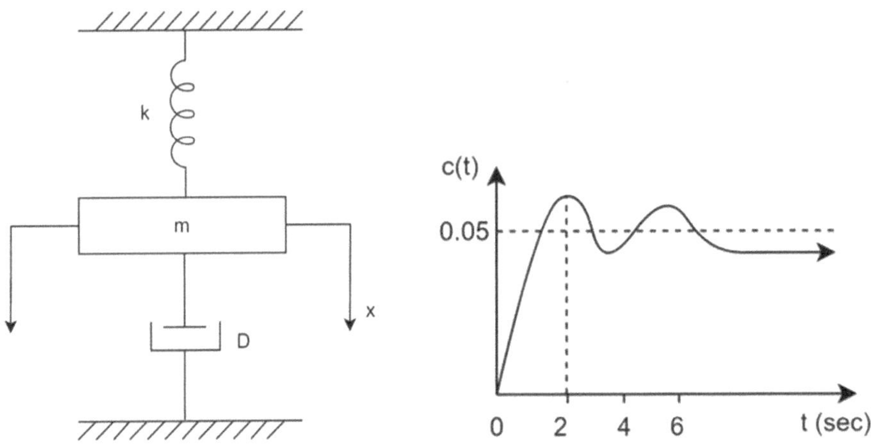


P6.7. An open loop control system transfer function $G(s)=\frac{25}{s(s+6)}$ is given for a unity feedback control system. Determine (i) $\xi$, (ii) peak overshoot, and (iii) $T_p$, time for peak output

P6.8. A flywheel motor delivers a torque of 1600 Nm/rad, and 100 kgm$^2$ is the moment of inertia for this system. The flywheel is controlled by a hand-wheel that controls the flywheel which is given a movement through π/4 rad from rest. Calculate the flywheel's response if the viscous friction coefficient is 400Nm/rad/sec.

P6.9. The block diagram of the control system is shown in Fig P6.9. The input is a unit step function. Determine the characteristic equation of the system and the following features.

a) Undamped frequency of oscillation $\omega_n$
b) Damped frequency of oscillation $\omega_d$
c) Damping ratio $\xi$ and damping factor
d) Maximum overshoot $M_p$
e) Time interval in which maximum and minimum output will occur
f) Settling time $T_s$
g) The number of cycles completed before the output is settled within 2% of the final value.

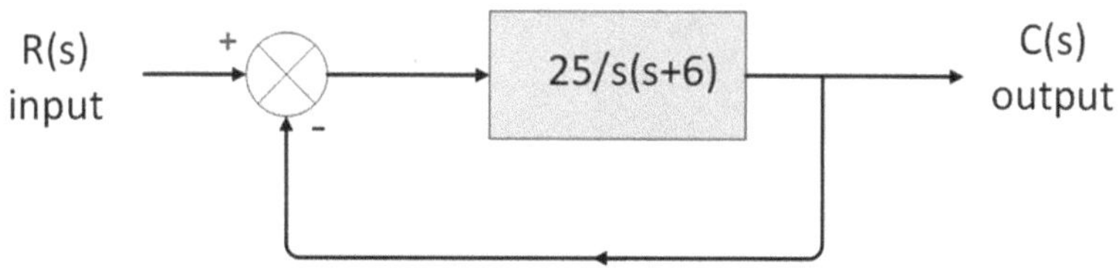


P6.10. Determine the value of $K$ for the feedback control system of Fig P6.10 so that the system will have a damping ratio of 0.6. Also, for this value of $K$, take a unit ramp input and calculate the maximum overshoot settling time and steady-state error.

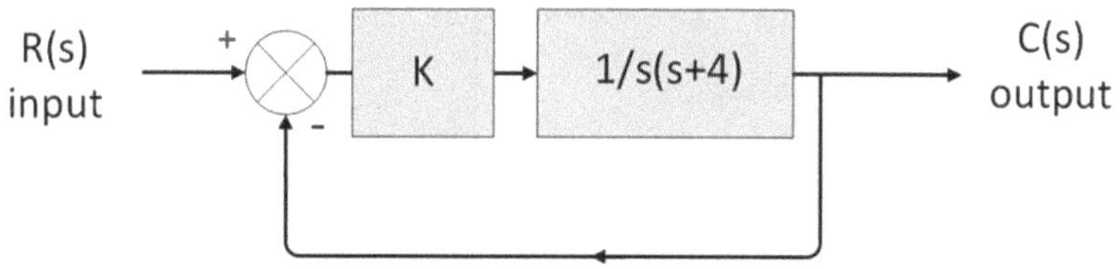


P6.11. Fig P6.11 shows a feedback control system with a derivative feedback control.

a) Find the damping ratio $\xi$ and natural frequency $\omega_n$ for $K_0 = 0$. If a unit ramp input is given, what will the steady-state error be?
b) If the system's damping ratio is 0.6, what will be the value of $K_0$? Also, find the steady-state error for this value of $K_0$ and a unit ramp input.

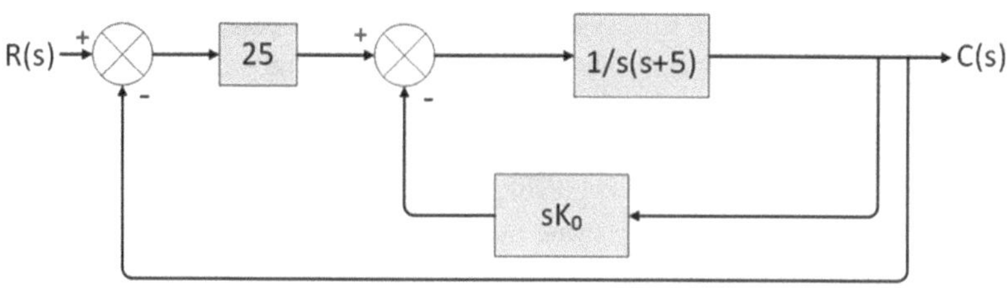

# 7 Stability of Linear Systems

A system is considered stable if its output is bounded for any bounded input. If all the roots of its characteristic equation lie inside the left half of the s-plane, then the system is declared to be stable.

## 7.1 METHODS FOR DETERMINING STABILITY OF LINEAR CONTROL SYSTEMS

a) Routh-Hurwitz Criterion: This method determines the absolute stability of the system and tests whether any root lies in the right half of the s-plane.
b) Root Locus Method: This method uses the graphical location of the roots of the characteristics equation by varying open loop gain or some other parameters of interest. It is used to find the stability of the control system relatively.
c) Bode Plots: This method is used to plot the frequency response in a simplified manner by determining both gain and phase margins. It is also used to find the stability of the control system relatively.
d) Nyquist Plots: This method also indicates gain, and phase margins and figures out the stability in relative terms for a linear feedback control system. It uses Cauchy's theorem and maps contours in the Laplace plane.

## 7.2 THE CRITERION BY ROUTH-HURWITZ

A linear system can be expressed with the characteristic equation as follows:

$$F(s) = a_0 s^n + a_1 s^{n-1} + a_2 s^{n-2} + \ldots + a_{n-1} s + a_n = 0 \tag{7.1}$$

where, all the coefficients are real numbers.

So there should be no roots in the right half plane (RHP), but it is necessary but insufficient.

- All coefficients of $F(s)$ have the same sign.
- All coefficients of $F(s)$ are non-zero.

The necessary and sufficient condition that all the roots of an nth-order polynomial lie in left half plane (LHP) is that Hurwitz determinant $D_k(k = 1, 2, 3...)$ of the $F(s)$ must be all positive.

The following procedure obtains $D_k$.

$$\begin{matrix} a_0 & a_2 & a_4 & a_6 & a_8 & \cdots\cdots \\ a_1 & a_3 & a_5 & a_7 & a_9 & \cdots\cdots \end{matrix} \tag{7.2}$$

DOI: 10.1201/9781003484202-7

The next step is to form the following array of numbers obtained by the indicated operations.

$$
\begin{array}{c|cccc}
s^6 & a_0 & a_2 & a_4 & a_6 \\
s^5 & a_1 & a_3 & a_5 & 0 \\
s^4 & \dfrac{a_1a_2 - a_0a_3}{a_1} = A & \dfrac{a_1a_4 - a_0a_5}{a_1} = B & a_6 & 0 \\
s^3 & \dfrac{Aa_3 - a_1B}{A} = C & \dfrac{Aa_5 \quad a_1a_6}{A} = D & 0 & 0 \\
s^2 & \dfrac{CB - AD}{C} = E & \dfrac{Ca_6}{C} = a_6 & 0 & 0 \\
s^1 & \dfrac{ED - a_6C}{E} = F & 0 & 0 & 0 \\
s^0 & a_6 & 0 & 0 & 0
\end{array}
$$

The roots of the polynomial will be in LHP if the elements of the first column have the same sign.

If there are sign changes in the elements of the first column, the number of sign changes indicates the number of roots in RHP.

$$D_1 = a_1$$

$$D_2 = \begin{vmatrix} a_1 & a_3 \\ a_0 & a_2 \end{vmatrix}$$

$$D_3 = \begin{vmatrix} a_1 & a_3 & a_5 \\ a_0 & a_2 & a_4 \\ 0 & a_1 & a_3 \end{vmatrix}$$

$$D_n = \begin{vmatrix} a_1 & a_3 & a_5 & \cdots\cdots & a_{2n-1} \\ a_0 & a_2 & a_4 & \cdots\cdots & a_{2n-2} \\ 0 & a_1 & a_3 & \cdots\cdots & a_{2n-3} \\ 0 & a_0 & a_2 & \cdots\cdots & a_{2n-4} \\ 0 & 0 & a_1 & \cdots\cdots & a_{2n-5} \\ \vdots & \vdots & \vdots & \vdots & \vdots \\ \vdots & \vdots & \vdots & \vdots & \vdots \\ 0 & 0 & 0 & \cdots\cdots & a_n \end{vmatrix}$$

where, the coefficients with indices larger than n or negative indices are replaced by zeros.

The Routh-Hurwitz criterion states that the necessary and sufficient condition that all the roots of a polynomial, $F(s) = 0$ lie in LHP is that $a_0 > 0$, $D_1 > 0$, $D_2 > 0$ ...... $D_n > 0$.

The previous method is formidable. Therefore, the following technique is used to overcome complexity.

**Example 7.1:** Consider the polynomial F(s) = $s^3 + 2s^2 - 5s - 6 = 0$. Since all the coefficients do not have the same sign, at least one of the roots is in RHP.

The Routh's tabulation is as follows:

$$\begin{array}{lll} s^3 & 1 & 5 \\ s^2 & a_1 & -6 \\ s^1 & \dfrac{-10+6}{2} = -2 & 0 \\ s^0 & -6 & \end{array}$$

Since there is one change of sign, there is one root in the RHP.

**Example 7.2:** Consider the polynomial $(s - 1)^2 (s + 2) = s^3 - 3s + 2 = 0$. The necessary conditions are not fulfilled. The Routhian array is as follows:

$$\begin{array}{lll} s^3 & 1 & -3 \\ s^2 & 0(\epsilon) & 2 \\ s^1 & \dfrac{-3\epsilon - 2}{\epsilon} & . \\ s^0 & 2 & . \end{array}$$

We cannot proceed further.

Replace 0 by $\epsilon > 0$, where $\epsilon$ is infinitesimally small. There are two changes of sign, and therefore, there are two roots in RHP.

Alternatively, replace $s$ *by* $\frac{1}{x}$ and rewrite the polynomial.

$$\text{F(s)} = s^3 - 3s + 2 = 0$$

$$\frac{1}{x^3} - 3\frac{1}{x} + 2 = 0$$

$$1 - 3x^2 + 2x^3 = 0$$

$$\begin{array}{lll} x^3 & 2 & 0 \\ x^2 & -3 & 1 \\ x^1 & \dfrac{2}{3} & . \\ x^0 & 1 & . \end{array}$$

Now, there are two sign changes, so there are two roots in RHP. The first column elements have two changes of sign. Therefore, two roots lie in the RHP, and the system is unstable.

### 7.2.1 Special Cases

The following difficulties may be encountered during applications of the Routh-Hurwitz criterion.

- The first element is that one row of Routh tabulation which is zero, but the others are not.
- The elements in one row of Routh's tabulation are all zero.

If a zero appears in the first position of a row, the elements in the next row become infinite, and Routh's tests break down. To restore the missing power of Routh's test, replace zero by $\epsilon$ or let $s\, by \frac{1}{x}$ and reformulate the polynomial.

*When . . . the elements . . . one row of Routh's table is zero.* Such a condition indicates that these are pairs of real roots with opposite signs, and pairs of complex conjugate roots form a quadratic. In this case, the equation corresponding to the coefficients first above the row is called the auxiliary equation. The order of the auxiliary equation is always even, and it indicates several root pairs that are equal in magnitude but opposite in sign.

To correct such a situation, derivative of the auxiliary equation concerning 's' is taken, zeros are replaced with the coefficients of the resulting equation, and the process is continued.

**Example 7.3:** Consider the polynomial F($s$) = $s^4 + 3s^3 + 5s^2 + 4s + 3$.

$$\begin{array}{cccc} s^4 & 1 & 5 & 3 \\ s^3 & 3 & 4 & 0 \\ s^2 & \frac{11}{3} & 3 & . \\ s^1 & \frac{7}{3} & 0 & . \\ s^0 & 3 & . & . \end{array}$$

There is no change of sign in the first column. Therefore, all roots lie in the LHP, and the system is stable.

**Example 7.4:** Consider the polynomial F($s$) = $s^5 + 2s^4 + 2s^3 + 4s^2 + 11s + 10 = 0$.

$$\begin{array}{cccc} s^5 & 1 & 2 & 11 \\ s^4 & 2 & 4 & 10 \\ s^3 & \epsilon & 6 & 0 \\ s^2 & \frac{4\epsilon - 12}{\epsilon} = \frac{-12}{\epsilon} & 10 & 0 \\ s^1 & 6 & 0 & 0 \\ s^0 & 10 & & \end{array}$$

There are two sign changes, so there are two roots in RHP, and the system is unstable.

**Example 7.5:** Consider the polynomial $F(s) = s^3 + 2s^2 + 4s + k = 0$.

$$\begin{array}{lll} s^3 & 1 & 4 \\ s^2 & 2 & k \\ s^1 & \dfrac{8-k}{2} & 0 \\ s^0 & k & \end{array}$$

$k = 8$ for an element to be zero. For the system to be stable $0 < k < 8$. At $k = 8$, a pair of poles lie at the imaginary axis. So

$$2s^2 = 8, \text{ and } s^2 = 4.$$

$$\text{At } \omega = \tfrac{2\text{rad}}{\text{sec}},\ j\omega = \pm 2$$

**Example 7.6:** Consider the polynomial $F(s) = s^4 + 6s^3 + 21s^2 + 36s + 20 = 0$

$$\begin{array}{llll} s^4 & 1 & 21 & 20 \\ s^3 & 6 & 36 & 0 \\ s^2 & 15 & 20 & \\ s^1 & 28 & 0 & \\ s^0 & 20 & & \end{array}$$

Since there is no change of sign in coefficients in the first column, there are no poles in RHP. Hence, the system is stable.

**Example 7.7:** Consider the polynomial $F(s) = s^5 + 2s^4 + 4s^3 + 8s^2 + 3s + 1 = 0$

$$\begin{array}{llll} s^5 & 1 & 4 & 3 \\ s^4 & 2 & 8 & 1 \\ s^3 & \epsilon & 2.5 & 0 \\ s^2 & \dfrac{8\epsilon - 5}{\epsilon} & 1 & 0 \\ s^1 & \dfrac{2.5\left(\dfrac{8\epsilon - 5}{\epsilon}\right)}{\dfrac{8\epsilon - 5}{\epsilon}} & 0 & 0 \\ s^0 & 1 & . & . \end{array}$$

The coefficients of the first column have two sign changes. Hence, there are two poles in the RHP, and the system is unstable.

**Example 7.8:** Consider the polynomial $F(s) = s^4 + 4s^3 + 4s^2 + 6s + k = 0$

$$\begin{array}{lcc} s^4 & 1 & 7 \quad k \\ s^3 & 4 & 6 \quad 0 \\ s^2 & 5.5 & k \\ s^1 & \dfrac{33-4k}{5.5} & 0 \\ s^0 & k & \end{array}$$

$$k > 0$$
$$33 - 4k$$
$$k < \frac{33}{4}$$
$$k < 8.25$$

Since there is no change of coefficient sign in the first column, there are no poles in RHP. Hence, the system is stable for $0 < k < 8.25$.

**Example 7.9:** Consider the polynomial $F(s) = 2s^4 + 4s^2 + 1 = 0$

$$\begin{array}{lccc} s^4 & 2 & 4 & 1 \\ s^3 & 0(8) & 0(8) & 0 \\ s^2 & 2 & 1 & 0 \\ s^1 & 4 & 0 & 0 \\ s^0 & 1 & & \end{array}$$

Auxiliary equation is:

$$A(s) = 2s^4 + 4s^2 + 1$$
$$\frac{dA(s)}{ds} = 8s^3 + 8s = 0$$

As there is no change of sign, no roots are in RHP. Hence, the system is stable.

**Problems**

P7.1 Use the Routh-Hurwitz criterion to find the roots for the polynomial $F(s) = s^3 + s^2 - 5s - 6 = 0$

P7.2 Discuss the stability of the following polynomial, $F(s) = s - 2s + 1 = 0$

P7.3 Discuss the stability of the following polynomial equation of a control system.

$$F(s) = s^4 + 3s^3 + 5s^2 + 4s + 3 = 0$$

P7.4 Discuss the stability of the following polynomial equation of a control system.

$$F(s) = 2s^5 + 2s^4 + 2s^3 + 2s^2 + 2s + 1 = 0$$

P7.5 Analyze the polynomial $F(s) = s^3 + s^2 + s + k = 0$

P7.6 Discuss the stability of the following polynomial equation of a control system.

$$F(s) = 2s^4 + 6s^3 + 21s^2 + 6s + 2 = 0$$

P7.7 Discuss the stability of the following polynomial equation of a control system.

$$F(s) = 8s^5 + 4s^4 + 2s^3 + s^2 + 3s + 1 = 0$$

P7.8 Discuss the stability of the following polynomial equation of a control system.

$$F(s) = 3s^4 + 4s^3 + 7s^2 + 4s + k = 0$$

P7.9 Discuss the stability of the following polynomial equation of a control system.

$$F(s) = 3s^4 + 6s^2 + 1 = 0$$

# 8 Control System Analysis and Design by Root Locus Analysis

## 8.1 INTRODUCTION

A closed-loop control system's relative stability and transient performance are directly related to the locations of closed-loop roots of the characteristics equation in the s-plane. Adjusting one or more system parameters to obtain suitable root locations is frequently necessary. Therefore, it is worthwhile to study the travelling of closed loop poles as one of the parameters is varied, which is usually the controller's gain. The root locus method was introduced by Evans in 1948 and has been developed and utilized extensively in control system practice. It is a graphical method for sketching the locus of the roots in the s-plane as a parameter is varied. The root locus method provides a graphical representation; therefore, an appropriate sketch can be used to obtain qualitative information concerning the stability and performance of the system. This technique is also applicable to multi-loop systems. In summary, root locus indicates how the closed loop poles and zeros should be modelled so that system performance is according to design specifications.

## 8.2 ROOT LOCUS PLOTS

### 8.2.1 Angle and Magnitude Conditions

Consider the characteristics equations of a feedback control system given in Figure 8.1.

$$\frac{C(s)}{R(s)}=\frac{G(s)}{1+G(s)H(s)} \tag{8.1}$$

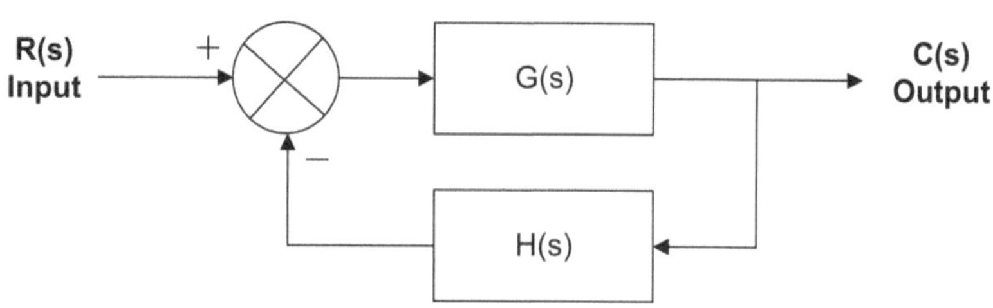


**FIGURE 8.1** Block diagram of a feedback control system.



DOI: 10.1201/9781003484202-8

The characteristic equation of the system is:

$$1 + G(s)\,H(s) = 0$$

$$G(s)\,H(s) = -1 \tag{8.2}$$

Since $G(s)\,H(s)$ is a complex quantity, it can be written as follows:

a. Angle Condition

$$\angle G(s)H(s) = \pm 180^{o}\,(2k+1) \qquad k = 0,1,2,\ldots,n \tag{8.3}$$

b. Magnitude Condition

$$|G(s)\,H(s)| = 1 \tag{8.4}$$

The value of s, which fulfils the angle and magnitude conditions, is the root of the characteristic equation or the closed loop poles.

### 8.2.2 Root Locus Plot of a Second-Order System

Consider the feedback control system shown in Figure 8.2.

$$\frac{C(s)}{R(s)} = \frac{K(s)}{s^2 + s + K} \tag{8.5}$$

The characteristic equation is:

$$s^2 + s + K = 0$$

The objective is to find the roots of this equation as $K$ is varied from 0 to ∞.

$$s_1, s_2 = -\frac{1 + \sqrt{1-4K}}{2}$$

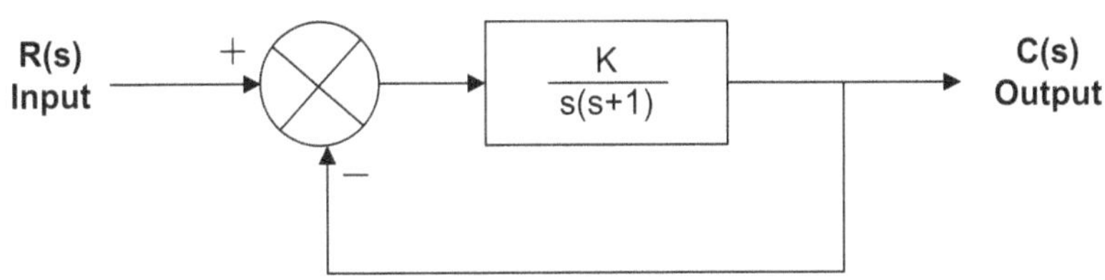


**FIGURE 8.2** Block diagram of second-order feedback system.

$$s_1 = -\frac{1}{2} + \frac{1}{2}\sqrt{1-4K}, s_2 = -\frac{1}{2} - \frac{1}{2}\sqrt{1-4K} \tag{8.6}$$

The roots are real for $K \leq \frac{1}{4}$ and complex for $K > \frac{1}{4}$.

Figure 8.3(a) shows the root locus of a second order system, and Figure 8.3(b) shows the angle condition at any point of root locus. It can be shown that any point on the root locus satisfies the angle condition.

$$\angle \frac{k}{s(s+1)} = -\angle s - (\angle s + 1) = \pm 180(2K+1), \text{ k=0, 1,2, . . .} \tag{8.7}$$

Consider the point $P$ in the root locus in Figure 8.3(b). The angles of $s$ and $s$+1 are $\theta_1$ and $\theta_2$, and the magnitudes of s and s+1 are $|s|$ and $|s+1|$. The angles are positive; when measured counterclockwise, the sum of $\theta_1$ and $\theta_2$ is clearly 180°. If $P$ is located on the real axis, then $\theta_1 = 180$ and $\theta_2 = 0$.

If the closed loop poles are specified on the root locus, the corresponding value of $k$ is determined. If $s = -\frac{1}{2} \pm j2$. Then the corresponding value of $K$ is;

$$|G(s)H(s)| = \left|\frac{K}{s(s+1)}\right|_{s=-\frac{1}{2}+j2} = 1$$

$$K = \left(-\frac{1}{2} + j2\right)\left(-\frac{1}{2} + j2 + 1\right)$$

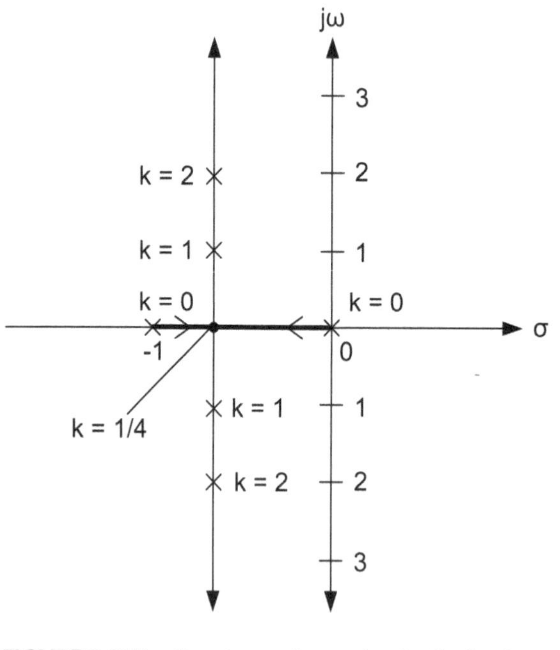


**FIGURE 8.3(A)** Root locus of second-order feedback system.

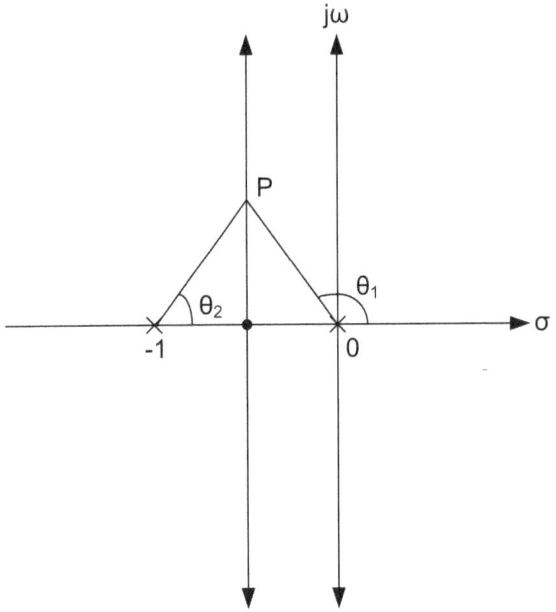


FIGURE 8.3(B) Angle condition at any point of root locus.

$$= \left| \left( -\frac{1}{2} + j2 \right) \left( \frac{1}{2} + j2 \right) \right| = \frac{17}{4} \tag{8.8}$$

### 8.2.3 General Rules for the Construction of Root Locus

**a. Obtain the characteristic equation.**

$$1 + G(s)\,\mathrm{H}(s) = 0$$

Arrange the previous characteristic equation in the form so that the parameter of interest appears as the multiplying factor in the form

$$1 + \frac{K(s+z_1)(s+z_2)\ldots(s+z_m)}{(s+p_1)(s+p_2)\ldots(s+p_n)} = 0 \tag{8.9}$$

$K$, in this case, is positive. However, in the case of positive feedback, $K < 0$, the angle condition is modified. From the factored form of the open loop transfer function, poles and zeros in the s-plane are located.

For example, if $G(s) = \frac{s+3}{s(s+2)}$, Figure 8.4 shows a pole-zero plot in s-plane.

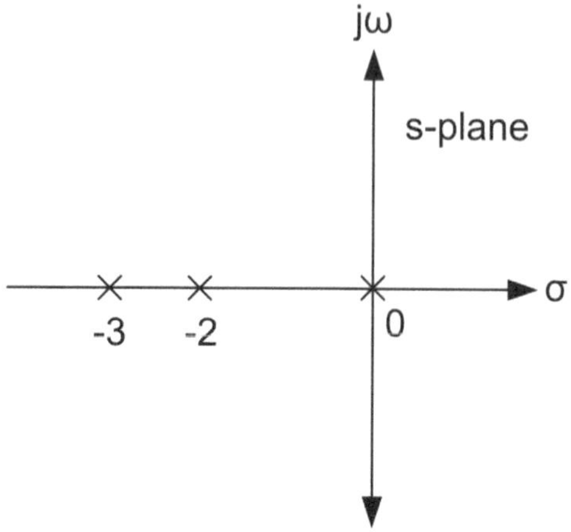


**FIGURE 8.4** Pole-zero plot in s-plane.

**b. Find the starting point and termination points of the root locus.**

The points on the root locus corresponding to $K = 0$ are the open loop poles. This can be seen as follows:

$$\lim_{K=0}\left|\frac{K(s+z_1)(s+z_2)\ldots(s+z_m)}{(s+p_1)(s+p_2)\ldots(s+p_n)}\right| = \lim_{K\to 0}\frac{1}{K}$$

It implies that the value of $s$ must approach the open loop poles. Thus, each root loci originates as a pole of the open loop transfer function $G(s)$ H($s$). As $K$ is increased to infinity, each root locus approaches either a zero of the open loop transfer function $G(s)$ H($s$) or infinity in the complex plane.

$$\lim_{K\to\infty}\left|\frac{K(s+z_1)(s+z_2)\ldots(s+z_m)}{(s+p_1)(s+p_2)\ldots(s+p_n)}\right| = \lim_{K\to 0}\frac{1}{K} = 0 \tag{8.10}$$

Hence, the value of $s$ must reach one of the open loop zeros or an open loop zero at infinity so that $G(s)$ H($s$) would have the same number of poles and zeros if zero at infinity is counted.

c. **A root locus plot will have just as many branches as there are roots of the characteristic equation,** that is, the number of branches is equal to the number of open-loop poles.
d. **The root loci are symmetric to the real axis.**
e. **Determination of the asymptotes of the root locus.**

If the test point $s$ is located far from the origin, then the angle of each complex quantity may be considered the same. One open-loop pole and one open-loop zero cancel the effects of the other. Therefore, the root loci for a very large value must be asymptotic to straight lines whose angles are given by:

$$\text{Angle of asymptotic} = \frac{\pm 180(2K+1)}{n-m}, \qquad k = 0,1,2\ldots \tag{8.11}$$

n = number of finite poles of $G(s)$ H($s$)
m = number of finite zeros of $G(s)$ H($s$)

For $K = 0$, the corresponding intersects on the real axis. The points of intersection is determined as:

$$\sigma = \frac{\sum poles - zeros}{p-z} \tag{8.12}$$

where, σ is called centroid and is always real as complex poles cancel each other. The point of intersection is determined as follows:

$$G(s)H(s) = \frac{K\left[s^m + (z_1 + z_1 + \ldots + z_m)s^{m-1} + \ldots + z_1 z_2 z_m\right]}{s^n + (p_1 + p_2 + \ldots + p_n)s^{n-1} + \ldots + p_1 p_2 p_n}$$

If a test point is located very far from the origin, the $G(s)$ H($s$) may be written as:

$$G(s)H(s) = \frac{K}{s^{n-m} + \left[(p_1 + p_2 + \ldots + p_n) - (z_1 + z_1 + \ldots + z_m)\right]s^{n-m-1}} \tag{8.13}$$

Since the characteristic equation is:

$$G(s)H(s) = -1$$

$$s^{n-m} + \left[(p_1 + p_2 + \ldots + p_n) - (z_1 + z_1 + \ldots + z_m)\right]s^{n-m-1} + \ldots = -K \tag{8.14}$$

For a large value of $s$, it may be approximated as:

$$[s + \frac{\left[(p_1 + p_2 + \ldots + p_n) - (z_1 + z_1 + \ldots + z_m)^{n-m}\right]}{n-m} = 0 \tag{8.15}$$

If the abscissa of the intersection of asymptotes and the real axis is denoted by $-\sigma_a$, the

$$-\sigma_a = -\frac{(p_1 + p_2 + \ldots + p_n) - (z_1 + z_1 + \ldots + z_m)}{n - m} \tag{8.16}$$

Because all the poles and zeros occur in conjugate points $-\sigma_a$ is always a real quantity.

**f. Determination of the root loci on the real axis.**

Root loci on the real axis are determined by open loop poles and zeros lying on it. Thus, complex poles and zeros do not affect the location of root loci on the real axis because the angle contribution of a pair of complex conjugate poles and zeros is 360° on the real axis. Each portion of the root locus on the real axis extends over a range from a pole or zero to another pole and zero. For instance, choose a test point on the real axis to ascertain the existence of the root locus on the real axis. If the total number of real poles and real zeros to the right of this test point is added, then this point lies on the root locus. The root locus and its complement are from alternate segments along the real axis.

**g. Breakaway points on the real axis.**

The breakpoints on the complete root loci of $1 + G(s)\,\mathrm{H}(s) = 0$ must satisfy

$$d\frac{G(s)H(s)}{ds} = 0 \tag{8.17}$$

Let,

$$G(s)H(s) = \frac{KB(s)}{A(s)}$$

$$1 + \frac{KB(s)}{A(s)} = 0 \tag{8.18}$$

***Proof***

$$A(s) + KB(s) = 0 \tag{8.19}$$

If $K$ is varied, $\Delta K$ then is determined as:

$$A(s) + (K + \Delta K)B(s) = 0$$

Dividing by $A(s) + KB(s)$,

$$1+\frac{\Delta KB(s)}{A(s)+KB(s)}=0 \tag{8.20}$$

Or $1 + \Delta KF(s) = 0$

Where

$$F(s)=\frac{A(s)}{A(s)+KB(s)} \tag{A}$$

Since the denominator of $F(s)$ is the same as the left-hand side of the characteristic equation, at a point very close to the nth order root $s = si$, which corresponds to a breakaway point of n loci, F(s) can be approximated by:

$$F(s)=\frac{Ai}{(s-si)}=\frac{Ai}{(\Delta s)^n} \tag{8.21}$$

where, $Ai$ is a constant.

Substituting the value of $F(s)$ in (A),

$$1+\frac{\Delta KAi}{(\Delta s)^n}=0 \tag{8.22}$$

From which,

$$\frac{\Delta K}{\Delta s}=-\frac{(\Delta s)^{n-1}}{Ai} \tag{8.23}$$

Taking the limits on both sides:

$$\lim_{\Delta k\to 0}\frac{\Delta K}{\Delta s}=\frac{dk}{ds}=0 \tag{8.24}$$

It implies that a breakaway point has

$$\frac{dk}{ds}=0 \tag{8.25}$$

Since their roots should also satisfy the characteristic equation,

$$1 + KG(s)\,H(s) = 0 \tag{8.26}$$

Differentiating with respect to $s$ yields:

$$\frac{dk}{ds}\big(G(s)H(s)\big) + K\frac{d(G(s)H(s)}{ds} = 0 \tag{8.27}$$

Since, $\frac{dk}{ds} = 0$

$$K\frac{d\left[G(s)H(s)\right]}{ds} = 0 \tag{8.28}$$

As $K$ is not zero.

$$\frac{d(G(s)H(s)}{ds} = 0 \tag{8.29}$$

It is important to note that the condition for the breakaway point given by the previous equation is necessary but is not sufficient. In other words, all breakaway points must satisfy the previous equation, but not all solutions are breakaway points. In general, all the real solutions of the previous equation breakaway points on the root loci. Because of the conjugate symmetry of the root loci, the breakaway point and breaking points either lie on the real axis or occur in complex conjugate pairs. If the root locus lies between two open loop poles on the real axis, then at least one breakaway point exists between the two poles. Similarly, if the root locus lies between the adjacent zeros (one zero may be located at $-\infty$) on the real axis, one breaking point always exists between two zeros.

If the root locus lies between an open pole and a zero (finite or infinite) on the real axis, there may exist a breakaway point (break-in point), or there may exist both breakaway or break-in points. For example,

$$f(s) = s^3 + 3s^2 + 2s + k = 0 \tag{8.30}$$

$$k = -\left(s^3 - 3s^2 - 2s\right) \tag{8.31}$$

$$\frac{dk}{ds} = -3s^2 - 6s - 2 = 0$$

$$3s^2 + 6s + 2 = 0$$

$$s_{1,2} = -0.423, -1.577$$

So the breakaway point will exist on −0.423. In general, a breakaway point may involve more than two root loci.

The root loci must approach and leave a breakaway point at an angle of $\frac{180^\circ}{n}$ apart, where $n$ is the number of roots to approach and leave the point. The same is true for complex breakaway points.

**Example 8.1:** Consider the following open loop transfer function.

$$G(s) = s(s+4)(s+2-j4)(s+2+j4)$$

$$F(s) = s^4 + 8s^3 + 36s^2 + 80s + k = 0$$

$$\frac{dk}{ds} = \frac{d}{ds}\left(s^4 + 8s^3 + 36s^2 + 80s\right) = 0$$

$$= 4s^3 + 24s^2 + 72s + 80 = 0$$

The roots are

$$s_1 = -2 \qquad s_{2,3} = -2 \pm j\sqrt{6}$$

Figure 8.5 shows the breakaway point of two root loci.

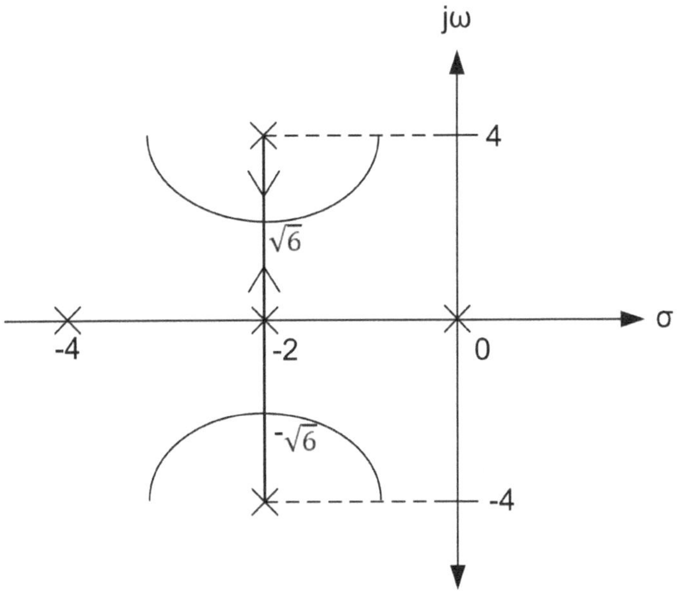


**FIGURE 8.5** Breakaway point of two root loci.

The breakaway points can also be a complete method of moments. Let $(-a)$ be a breakaway point of the root loci on the real axis. This can be determined by the following equation:

$$\sum_{\substack{\text{real zeros to the}\\ \text{left of } -a}} \left(\frac{1}{z_i - a}\right) - \sum_{\substack{\text{real poles to the}\\ \text{left of } -a}} \left(\frac{1}{p_j - a}\right) + \sum_{\substack{\text{complex zeros}\\ \text{to the left of } -a}} \left(\frac{2(\propto_k - a)}{(\propto_k - a)^2 + B_k^2}\right)$$
$$- \sum_{\substack{\text{complex poles}\\ \text{to the left of } -a}} \left(\frac{2(\propto_m - a)}{(\propto_m - a)^2 + B_m^2}\right) = \sum_{\substack{\text{real zeros to}\\ \text{the right of } -a}} \left(\frac{1}{a - z_i}\right) - \sum_{\substack{\text{real poles to}\\ \text{the right of } -a}} \left(\frac{1}{a - p_j}\right)$$
$$+ \sum_{\substack{\text{complex zeros to}\\ \text{the right of } -a}} \left(\frac{2(a - \propto_k)}{(a - \propto_k)^2 + B_k^2}\right) - \sum_{\substack{\text{complex poles}\\ \text{to the right of } -a}} \left(\frac{2(a - \propto_m)}{(a - \propto_m)^2 + B_m^2}\right) \quad (8.32)$$

**h. Computation of angle of departure (or angle of arrival) of root loci from the complex poles (or at the complex zeros).**

To sketch root loci, one must find the directions of the root loci near the complex pole or zeros. If a test point is chosen and moved in the vicinity of the complex pole or zero, the sum of angular combinations from all other poles and zeros can be considered to remain the same. Therefore, the angle of departure (or angle of arrival) of the root locus from the complex pole (or at the complex zero) can be found by subtracting from 180° the sum of all other poles and zeros in questions with appropriate signs. The angle of departure, as calculated, is shown in Figure 8.6.

$$\text{Angle of departure} = 180 - (\theta_1 + \theta_2 - \theta_3) \quad (8.33)$$

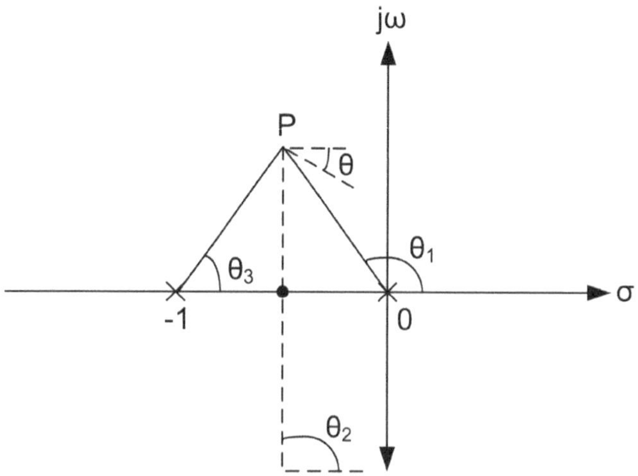


**FIGURE 8.6** Calculation of angle of departure from a complex pole.

**i. Point of intersection on the imaginary axis.**

The points where the root loci intersect the imaginary axis can be found by Routh-Harwirtz Criterion or by putting $s = j\omega$ in the characteristic equation, equating real and imaginary points to zero and solving the $\omega$ and $K$.

Consider

$$G(s) = \frac{k}{s(s+1)(s+2)} \tag{8.34}$$

The characteristic equation is $1 + \frac{k}{s(s+1)(s+2)}$

$$s(s+1)(s+2) + k = 0$$

$$s^3 + 3s^2 + 2s + k = 0$$

$$\begin{array}{c|cc} s^3 & 1 & 2 \\ s^2 & 3 & k \\ s^1 & \frac{6-k}{3} & \\ s^0 & k & \end{array} \tag{8.35}$$

$$k_c = 63s^2 + 6 = 0$$

$$s^2 = -2$$

$$\omega = \sqrt{2}$$

Or

$$-j\omega^3 - 3\omega^2 + 2j\omega + k = 0$$

$$+j\omega^2 = 2j\omega$$

$$\omega^2 = 2$$

$$\omega = \sqrt{2}$$

$$k = 2\omega^2 = 3.2 = 6$$

**j. Determination of parameter *k* at a specific root.**

$$k_s = \frac{\prod_{j=1}^{np} \left|(s + p_j)\right|}{\prod_{i=1}^{nz} \left|(s + z_i)\right|} \tag{8.36}$$

It can be either computed analytically or measured by measuring the lengths of poles and zeros from the point and computing k.

### 8.2.4 Representative Examples

**Example 8.2:** Draw the root locus of a second-order system.

$$D(s) = \frac{k}{s(s+1)}$$

***Solution***

$$k = s^2 + s$$

$$\frac{dk}{ds} = 2s + 1 = 0, \qquad s = -\frac{1}{2}$$

Figure 8.7 shows the root locus of a second order system of Example 8.2.

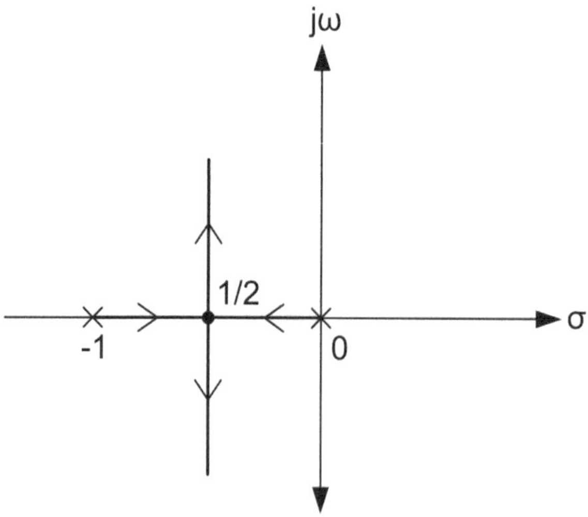


**FIGURE 8.7** Root locus of a second-order system.

**Example 8.3:** Draw the root locus of a third-order system.

$$G(s) = \frac{k}{s(s+1)(s+100)}$$

***Solution***

$$CG = \frac{0+1+100}{3} = 33.3$$

Angles ±60°, 180°

Breakaway point,

$$k = \left(s^3 + 101s^2 + 100s\right)$$

$$\frac{dk}{ds} = 3s^2 + 202s + 100, \qquad s_1, s_2 = -0.5, -66$$

$$\begin{array}{c|cc} s^3 & 1 & 100 \\ s^2 & 100 & k \\ s^1 & \dfrac{10100-k}{101} & \\ s^0 & k & \end{array}$$

$$k_c = 10100$$

$$\omega_c = 10\,rad\,/\,sec$$

Figure 8.8 shows the root locus of a third order system of Example 8.3.

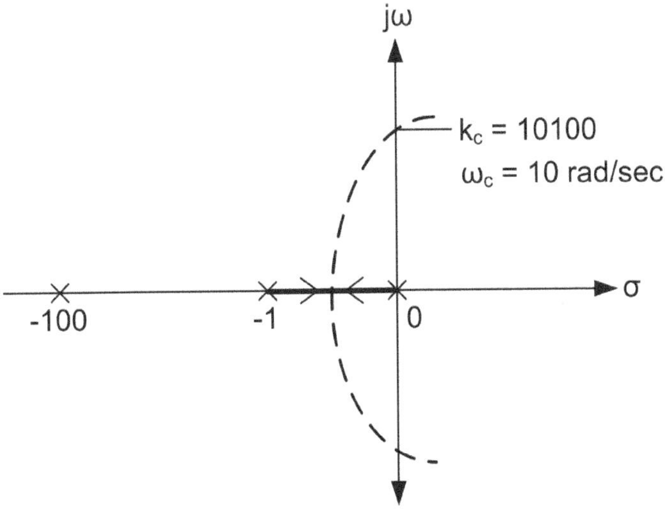


**FIGURE 8.8** Root locus of a third-order system.

**Example 8.4:** Draw the root locus of a third-order system.

$$G(s)=\frac{k}{s(s+1)^2}$$

***Solution***

$$CG=\frac{2}{3}=-0.67$$

$$k=s\left(s^2+2s+1\right)=s^3+2s^2+s$$

$$\frac{dk}{ds}=3s^2+4s+1,\ s_1,s_2=\frac{-4\pm\sqrt{16-12}}{6}=\frac{-4\pm 2}{6}=-1,-0.33$$

$$\begin{array}{c|cc} s^3 & 1 & 2 \\ s^2 & 2 & k \\ s^1 & 4-k & \\ s^0 & k & \end{array}$$

$$k_c=4$$

$$\omega_c=\sqrt{2}=1.44\,rad/sec$$

Figure 8.9 shows the root locus of a third order system of Example 8.4.

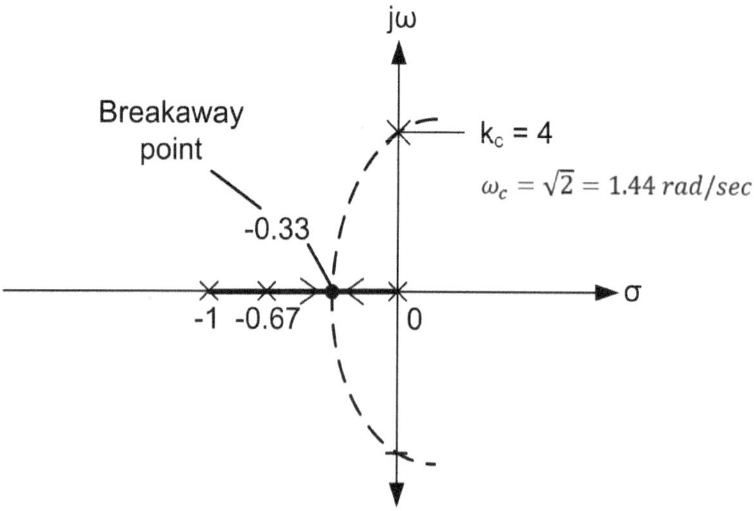


**FIGURE 8.9** Root locus of a third-order system.

**Example 8.5:** Draw the root locus of a third order system.

$$G(s) = \frac{k}{s(s+0.5)(s+1)}$$

***Solution***

$$CG = \frac{1.5}{3} = -0.5$$

$$k = s\left(s^2 + 1.5s + 0.5\right)$$

$$k = s^3 + 1.5s^2 + 0.5s$$

$$\frac{dk}{ds} = 3s^2 + 3s + 0.5$$

$$\frac{-3 \pm \sqrt{9-6}}{6} = \frac{-3 \pm \sqrt{3}}{6} = -0.5 + 0.3 = -0.2$$

$$s_1, s_2 = -0.2, -0.8$$

$$s^3 + 1.5s^2 + 0.5s + k$$

$$\begin{array}{c|ll} s^3 & 1 & 0.5 \\ s^2 & 1.5 & k \\ s^1 & 0.75 - k & \\ s^0 & k & \end{array}$$

$$k_c = 0.75$$

$$\omega_c = 1.417\, rad/sec$$

Figure 8.10 shows the root locus of a third order system of Example 8.5.

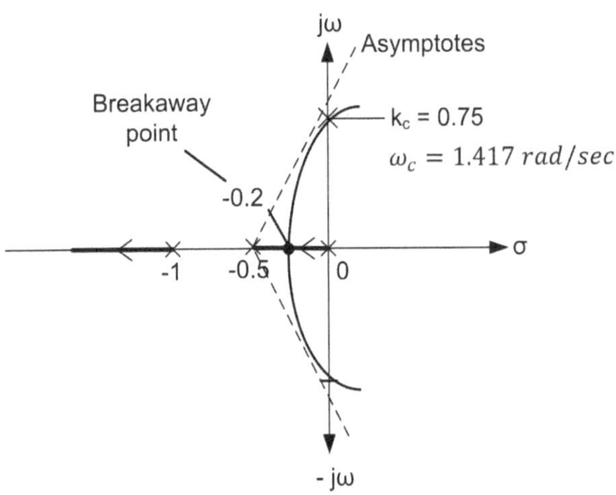


**FIGURE 8.10** Root locus of a third-order system.

## 8.3 EFFECT OF ADDING OPEN LOOP POLES AND ZEROS

Figure 8.11 shows the respective root locus of the equations given here:

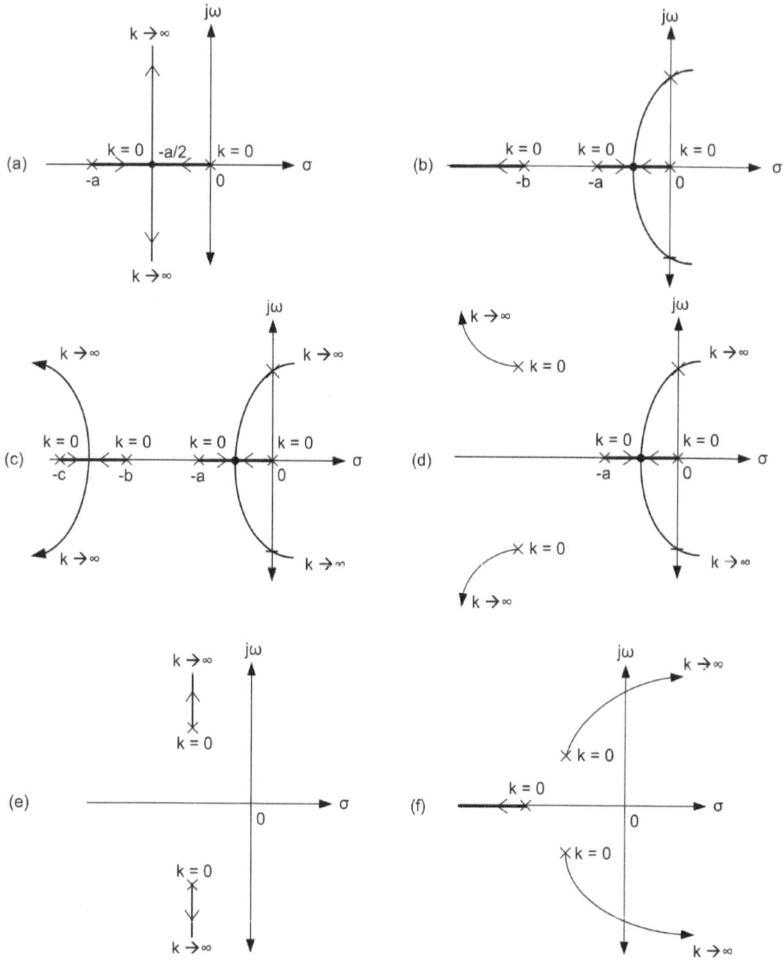


**FIGURE 8.11** Effects of adding open loop poles and zeros.

(a) $s + \dfrac{k}{s(s+a)}$

(b) $\dfrac{K}{s(s+a)(s+b)}$

(c) $\dfrac{k}{s(s+a)(s+b)(s+c)}$

(d) $\dfrac{k}{s(s+a)(s^2+2as+b)}$

(e) $\dfrac{k}{s^2+2as+b}$ $(a^2<4b)$

(f) $\dfrac{k}{(s+c)(s^2+2as+b)}$ $(a^2<4b)$

Some more situations can be drawn as shown in Figure 8.12 wherein the effect of adding the open-loop pole and zeros can be observed.

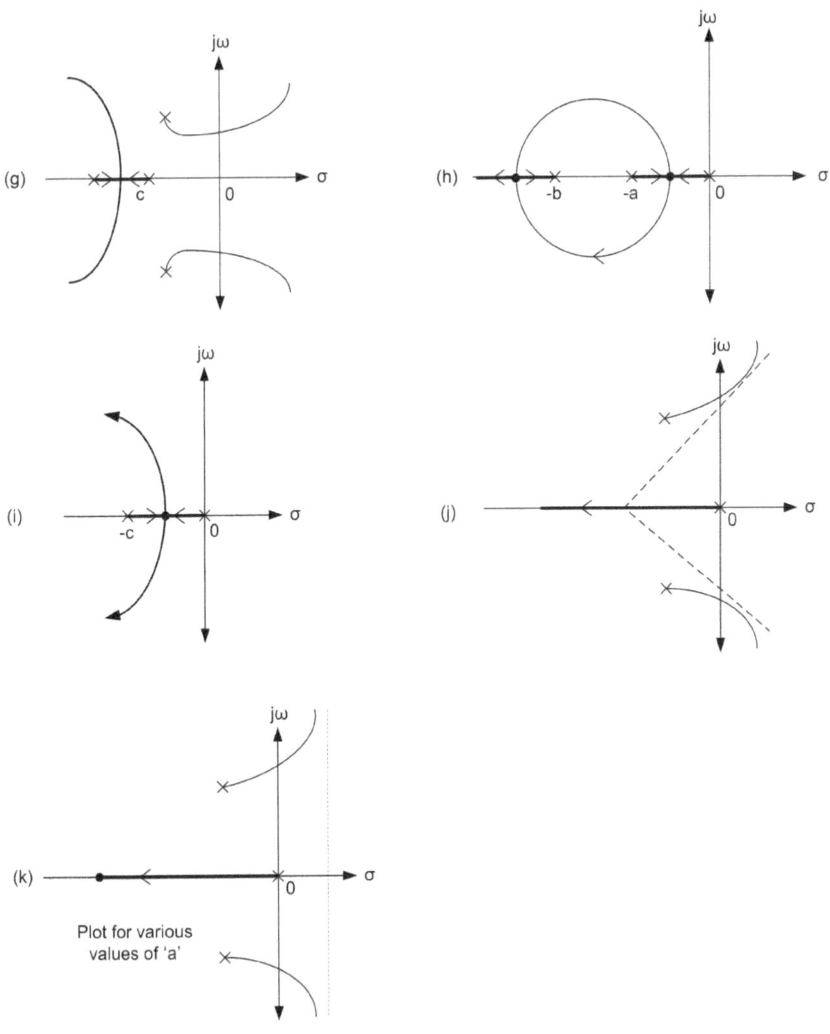


**FIGURE 8.12** Effects of adding open loop poles and zeros.

(g) $s+\dfrac{k}{(s+c)(s+d)(s^2+as+b)}$

(h) $\dfrac{s+b}{s(s+a)}$

(i) $\dfrac{s^2+as+b}{s(s+c)}$

(j) $\dfrac{k}{s(s^2+2s+2)}$

(k) $\dfrac{k(s+a)}{s(s^2+2s+2)}$

## 8.4 SYSTEM WITH TRANSPORTATION LAG OR DEAD TIME

Consider Figure 8.13 which shows a block diagram of a control system with transportation lag.

$$1+G(s)=1+\frac{k_c}{s+1}=0 \tag{8.37}$$

$$s+1+ke^{-Ts}=0 \tag{8.38}$$

$$\frac{ke^{-Ts}}{s+1}=-1 \tag{8.39}$$

$$\angle\frac{ke^{-Ts}}{s+1}=e^{-Ts}-\angle s+1=\pm180(2k+1),\ k=0,1 \tag{8.40}$$

To find the angle, $e^{-Ts}$, write $s=\sigma+j\omega$. Then,

$$e^{-Ts}=e^{-T\sigma-j\omega} \tag{8.41}$$

$$e^{-j\omega T}=\angle cos\omega T-jsin\omega T,\ e^{-t\sigma}=0 \tag{8.42}$$

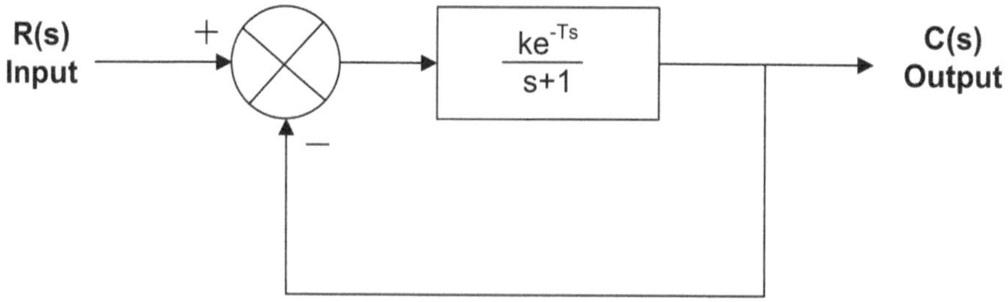


**FIGURE 8.13** Block diagram of control system with transportation lag.

$$= -\omega T\ rad = 57.3\,\omega T\ degree$$

$$57.3\,\omega T - \angle s+1 = \pm 180(2k+1)$$

$$\angle s+1 = \pm 180(2k+1) - 57.3\,\omega T$$

The angle contribution of $e^{-Ts}$ is zero for $\omega = 0$. The real axis from $-1$ to $\infty$ form a part of root loci. The root locus plot is given in Figure 8.14.

$$\lim_{s\to-\infty} \frac{ke^{-Ts}}{s+1} = \left.\frac{\frac{d}{ds}\left(ke^{-Ts}\right)}{\frac{d}{ds}(s+1)}\right|_{s=-\infty}$$

$$= \frac{-Tke^{-Ts}}{1} = -kTe^{-Ts}\Big|_{s\to-\infty} \tag{8.43}$$

Therefore, $s \to -\infty$ is a pole of the open loop transfer function. Thus, loci start at $s = -1$ and $s = -\infty$, and terminate at $s = \infty$. Figure 8.15 shows the root locus of control system for Figure 8.13.

Since there are an infinite number of values from $e^{-Ts} - \angle s+1 = \pm 180(2k+1)$, there are infinite numbers of root loci as the value of k (k = 0, 1, 2, ...) goes from zero to infinity. For example, if $k = 1$, the angle condition becomes:

$$\angle s+1 = \pm 540 - 57.3\,\omega T$$

$$= -3\pi - \omega T\ rad$$

The construction of root loci for $k = 1$ is the same as for $k = 0$.

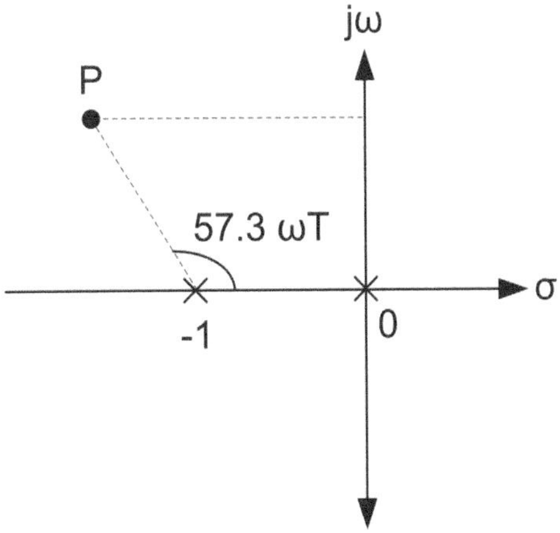


**FIGURE 8.14** Root locus plot.

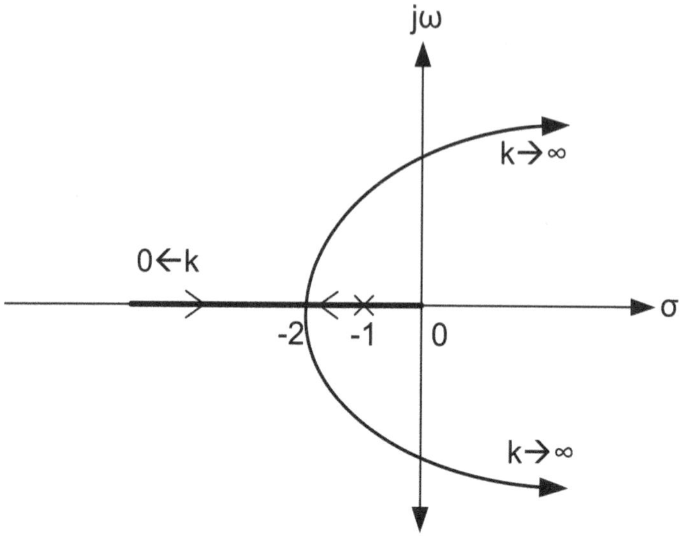


FIGURE 8.15 Root locus of the control system of Figure 8.13.

## 8.5 ROOT LOCI FOR PARAMETERS OTHER THAN *K*

Consider the block diagram of a second-order system shown in Figure 8.16. The characteristic equation can be written as: $1+\frac{k}{s(s+a)}=0$

$$s^2+as+k=0 \tag{8.44}$$

Dividing by $s^2+k$,

$$1+\frac{as}{s^2+k}=0$$

Then, from magnitude condition, $\frac{a|s_1|}{|s_1^2+k|}=1$

And

$$\angle s_1-\left(\angle s_1+j\sqrt{k}\right)+\angle s-j\sqrt{k}=\pm 180(2k+1) \tag{8.45}$$

To draw the root loci:

- Number of separate root loci = 2.
- Angle of asymptote, $\frac{(2k+1)180}{1}=180^{\circ}$
- Breakaway point

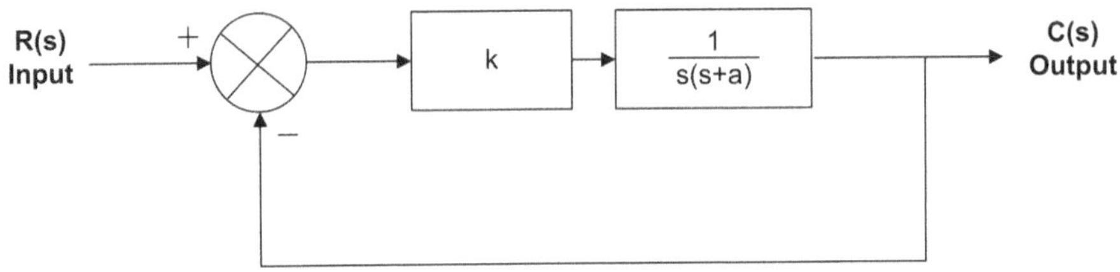


**FIGURE 8.16** Block diagram of second-order system.

$$s^2 + as + k = 0$$

$$as = -\frac{\left(s^2 + k\right)}{s}$$

$$\frac{da}{ds}\left(\frac{2s.s - \left(s^2 + k\right)}{s^2}\right) = 0$$

$$s^2 - k = 0,\ s^2 = k,\ s = \sqrt{k}$$

$$S^2 + as + k = 0$$

$$\begin{array}{c|cc} s^2 & 1 & k \\ s^1 & a & \\ s^0 & k & \end{array}$$

Figure 8.17 shows the root locus for parameters other than k.

**Example 8.6:** The open loop transfer function of a unity feedback system is:

$$G(s) = \frac{k}{s(s+1)(s+2)}$$

Draw its root locus after computing all features. Find the design gain, margin and phase margin for $16.25\%(\xi = 0.5)$ overshoot in the time response $(16.25\%)$ and $\cos 0.5 = 60^{\circ}$.

***Solution***

$$G(s)H(s) = \frac{k}{s(s+1)(s+2)}$$

The characteristic equation is:

$$-s(s+1)(s+2) = k = -\left(s^3 + 2s^2 + 2s\right)$$

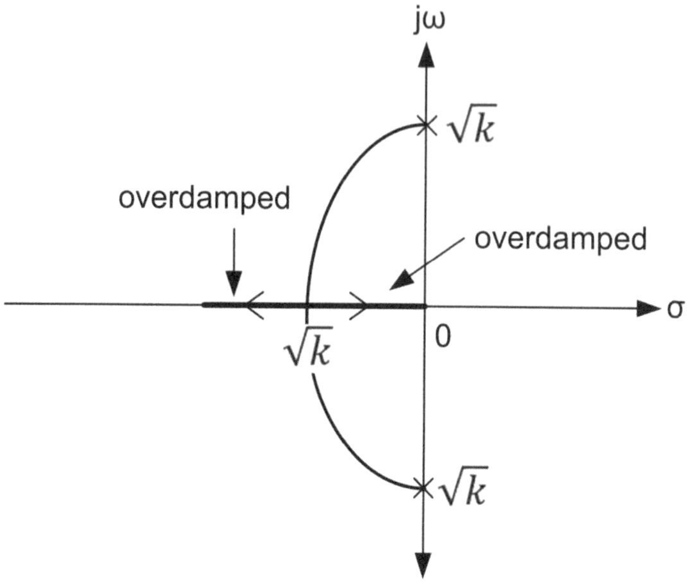


**FIGURE 8.17** Root locus for parameter other than k.

The angle of asymptotes is 60°, 180°, 360°.

Centre of gravity of asymptotes, $\sigma = \dfrac{-1-2}{3} = -1$

Breakaway points:

$$\frac{dk}{ds} = 3s^2 + 6s + 2 = 0$$

$$s_{1,2} = -0.42, -1.58$$

The intersection of $j\omega$ axis:

$$\begin{array}{c|cc} s^3 & 1 & 2 \\ s^2 & 3 & k \\ s^1 & 6-k & \\ s^0 & k & \end{array}$$

$$k_c = 6$$

$$\omega_c = 1.414\, rad/sec$$

*kd* for 16.25% overshoot is:

$$kd = \left|(-0.337 + j0.58)(-0.337 + j0.58 + 1)(-0.337 + j0.58 + 2)\right| = 1.0383 \approx 1.04$$

Third pole for $kd = 1.04$

$$x(-0.337 + j0.58)(-0.337 - j0.58) = 1.04 = -2.3326$$

$$\text{Gain margin} = \frac{k_c}{k_d} = \frac{6}{1.04} = 5.8$$

Phase margin:

$$|GH(jw)| = 1$$

$$\left|\frac{1}{jw(jw+1)(jw+2)}\right| = 1$$

$$\left|jw(2-w^2) - 3w^2\right| = 1$$

$$\left(4 + \omega^4 - 4\omega^2 + 9\omega^2\right)^{\frac{1}{2}} = \frac{1}{\omega}$$

$$\omega^6 + 5\omega^4 + \omega^2 = 1$$

$$\omega_1 = 0.6$$

$$|j0.6(2 - 0.6 + 1)(j0.6 + 2|$$

$$-1.9\,rad \cong -110^{\circ}$$

Phase margin = 180 − 110 = 70°. Figure 8.18 shows the root locus.

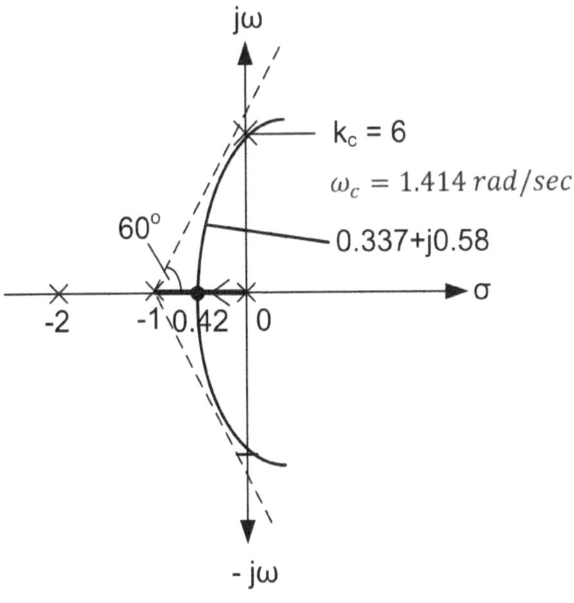


**FIGURE 8.18** Root locus for Example 8.6.

**Example 8.7:** Prove that the root locus is a circle of a given transfer function

$$G(s) = \frac{k(s+2)(s+3)}{s(s+1)}$$

***Solution***

$$\angle G(s) = \tan^{-1}\frac{\omega}{\sigma+2} + \tan^{-1}\frac{\omega}{\sigma+3} = 100 + \tan^{-1}\frac{\omega}{\sigma} + \tan^{-1}\frac{\omega}{\sigma+1}$$

Taking tangents on both sides,

$$\frac{\dfrac{\omega}{\sigma+2}+\dfrac{\omega}{\sigma+3}}{1-\dfrac{\omega}{\sigma+2}.\dfrac{\omega}{\sigma+3}} = \frac{\dfrac{\omega}{\sigma}+\dfrac{\omega}{\sigma+1}}{1-\dfrac{\omega}{\sigma}.\dfrac{\omega}{\sigma+1}}$$

$$\frac{2\sigma+5}{\sigma^2+\sigma5+6-\omega^2} = \frac{2\sigma+1}{\sigma^2+\sigma-\omega^2}$$

$$\sigma^2+3\sigma+\omega^2 = \frac{-3}{2}$$

$$\left(\sigma+\frac{3}{2}\right)^2+\omega^2 = \frac{-3}{2}+\frac{9}{4}$$

$$\left(\sigma+\frac{3}{2}\right)^2+\omega^2 = \left(\sqrt{\frac{3}{4}}\right)^2$$

$$r = \sqrt{\frac{3}{4}} = \sqrt{\frac{3}{2}} = 0.866$$

$$\text{Center } \frac{-3}{2} = 0.866.$$

**Example 8.8:** Draw the root loci for feedback control system shown in Figure 8.19 and governed by following equation.

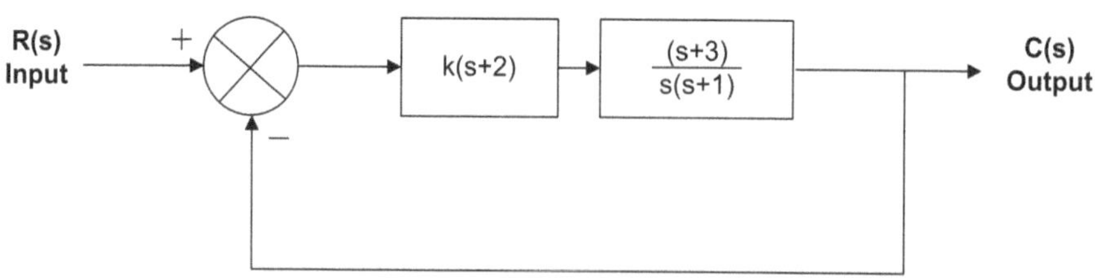


FIGURE 8.19 Block diagram of a feedback control system.

$$G(s) = \frac{k(s+2)(s+3)}{s(s+1)}$$

***Solution***

- Number of separate root loci = 2
- No asymptotes
- Breakaway and break-in points

$$\frac{k(s+2)(s+3)}{s(s+1)} + 1 = 0$$

$$\frac{k(s^2+5s+6)}{s^2+s} = -1$$

$$k = -\left(\frac{s^2+s}{s^2+5s+6}\right)$$

$$\frac{dk}{ds} = (2s+1)(s^2+5s+6) - (s^2+s)(2s+5)$$

$$2s^2 + 6s + 3$$

$$s_1 = -0.634,\ s_2 = -2.366$$

Figure 8.20 shows the root locus.

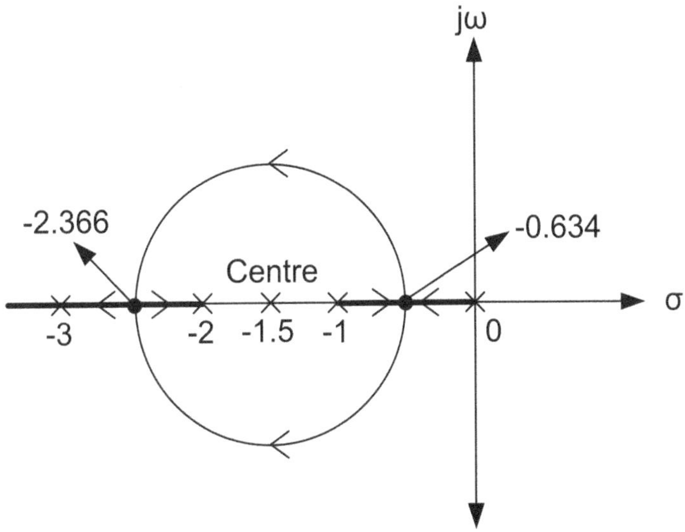


**FIGURE 8.20** Root locus of control system in Example 8.8.

**Example 8.9:** A control system's open loop transfer function with unity feedback is $G(s)=\frac{k(s+1)}{s(s-3)}$. Draw its root locus and range of $k$ for which the system is stable.

***Solution***

$$G(s)=\frac{k(s+1)}{s(s-3)}$$

The characteristic equation is:

$$1+\frac{k(s+1)}{s(s-3)}$$

$$s^2-s+ks+k=0$$

$$s^2+s(k-3)+k=0$$

$$\begin{array}{c|ll} s^2 & 1 & k \\ s^1 & k-3 & \\ s^0 & k & \end{array}$$

$$k_c=3$$

$$\omega_c=1.732\,rad\,/\,sec$$

The system is stable for $k\geq 3$

Further, one can find the radius and center of the circle for this root locus by:

$$\frac{\frac{\omega}{\sigma+1}-\frac{\omega}{\sigma}}{1+\frac{\omega}{\sigma+1}.\frac{\omega}{\sigma}}=\frac{\omega}{\sigma+3}+180^\circ$$

$$\frac{-1}{\sigma^2+\sigma+\omega^2}=\frac{1}{\sigma-3}$$

$$-\sigma+3=\sigma^2+\sigma+\omega^2$$

$$\sigma^2+2\sigma+\omega^2=3$$

Radius = 2

$$(\sigma+1)^2+\omega^2=4$$

Center of circle = (–1,0)

Figure 8.21 shows the root locus.

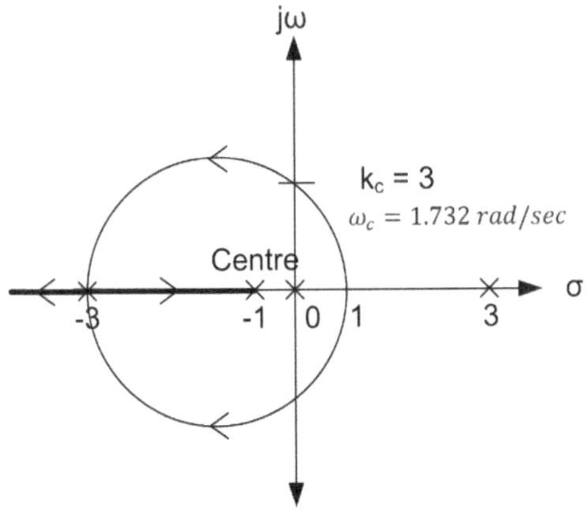


FIGURE 8.21 Root locus of control system in Example 8.9.

**Example 8.10:** Draw root locus for the control system whose open loop transfer function is

$$G(s) = \frac{k}{s(s+1)(s^2+4s+13)}$$

***Solution***

$$G(s) = \frac{k}{s(s+1)(s^2+4s+13)} = \frac{k}{s(s+1)(s+2+j3)(s+2-j3)}$$

Angle of asymptote $= \pm 45^\circ \pm 135^\circ$

$$\partial = \frac{0+1+4}{4} = \frac{5}{4} = -1.25$$

Breakaway point $-0.467$
Angle of departure $= -142.3^\circ$
Critical gain $k$ and critical frequency.

$$s^4 + 5s^3 + 17s^2 + 13s + k = 0$$

| | | | |
|---|---|---|---|
| $s^4$ | 1 | 17 | $k$ |
| $s^3$ | 5 | 13 | |
| $s^2$ | $\frac{72}{5}$ | $k$ | |
| $s^1$ | $\frac{72*13}{5} - 5k$ | | |
| $s^0$ | $\frac{72}{5}$ | | |

$$k_c = 37.44$$

$$\omega_c = \pm 1.6125\, rad/sec$$

Figure 8.22 shows the root locus.

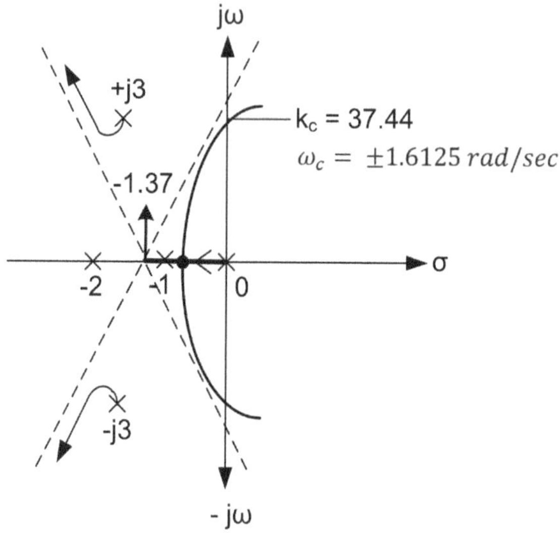


**FIGURE 8.22** Root locus of control system in Example 8.10.

**Example 8.11:** Use angle condition and check whether points $-1 + j0$ and $-3 + j$ lie on a root locus of a system given by

$$G(s) = \frac{k}{(s+1)(s+2)}$$

***Solution***

$$G(s)H(s)\big|_{s=-1+j0} = \frac{k}{(-1+j+1)(-1+j+2)} = \frac{k}{(1\angle 90^\circ)(\sqrt{2}\angle 90^\circ)} = \frac{k}{\sqrt{2}}\angle -180^\circ$$

Since the angle condition is fulfilled, $(-1 + j)$ lies on the root locus.

$$G(s)H(s)\big|_{s=-3+j} = \frac{k}{(s+1)(s+2)}\bigg|_{s=-3+j} = \frac{k}{(-3+j+1)(-3+j+1)}$$

$$\frac{k}{(-2+j)(-1+j)} = \frac{1}{1-3j} = \frac{k}{3.162} = \angle -71.56$$

Since the angle condition is not fulfilled, (3 + *j*) does not lie on the root locus.

**Example 8.12:** If $s = -0.85$ lies on the root locus of a system having $G(s)H(s) = \frac{k}{s(s+1)(s+3)}$, find the value of k using the magnitude condition.

***Solution***

$$G(s)H(s)\big|_{-0.85} = \left|\frac{k}{-0.85(-0.85+1)(-0.85+3)}\right| = \frac{k}{0.274}$$

Since $G(s)H(s)\big|_{s=-0.85} = 1$, $\frac{k}{0.274} = 1$, $k = 0.274$

The characteristic equation for $G(s)H(s) = \frac{k}{s(s+1)(s+3)}$

$$= s^3 + 4s^2 + 3s + k = 0$$

$$\begin{array}{c|cc} s^3 & 1 & 3 \\ s^2 & 4 & k \\ s^1 & 12-k & \\ s^0 & k & \end{array}$$

$$k_c = 12$$

If $s = -0.85$, the other two roots, $-0.75 \pm j1.12$, can be determined.

**Example 8.13:** The open loop transfer function of a system is given as

$$G(s)H(s) = \frac{k}{s(s+1)(s+2)}$$

Draw the root locus diagram and establish the condition for stability concerning the value of gain *k*. Calculate the phase margin and gain margin when the value of gain is reduced equal to 4.

***Solution***

$$\sigma = \frac{-1-2-0}{3} = -1$$

The angles of asymptotes are 60°,180°, 300°

The characteristics equation is:

$$s(s+1)(s+2) + k = 0$$

$$\frac{dk}{ds} = \frac{d}{ds}\left(s^3 + 2s^2 + 2s\right)$$
$$= 3s^2 + 6s + 2 = 0$$

Breakaway point is −0.426.

Routharian array for the characteristic equation is:

$$s^3 + 2s^2 + 2s + k = 0$$

$$\begin{array}{c|cc} s^3 & 1 & 2 \\ s^2 & 3 & k \\ s^1 & \frac{6-k}{3} & 0 \\ s^0 & k & \end{array}$$

The condition for stability is $0 < k < 6$.

$$k_c = 6$$
$$\omega_c = \pm j\sqrt{2}\ rad/sec$$

Figure 8.23 shows the root locus.

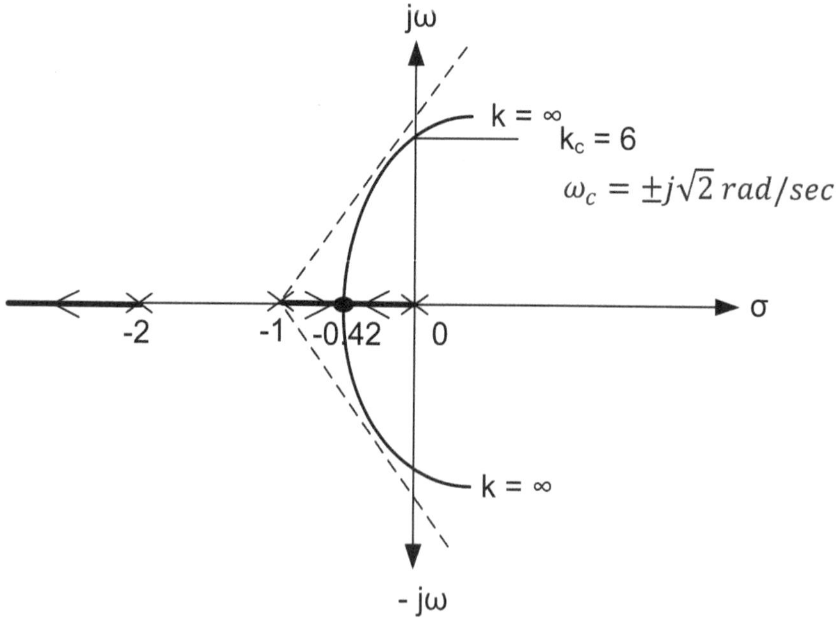


**FIGURE 8.23** Root locus of control system in Example 8.13.

The distance from P for gain, $k = 4$ is

$$\text{Gain margin} = \frac{value\ of\ critical\ k_c}{value\ of\ desired\ k_d} = \frac{6}{4} = 1.5$$

Phase margin is: $|G(j\omega)H(j\omega)| = 1$ for $j\omega = 1.2$

$$\text{Phase margin} = 18 + (-\theta_{p1} - \theta_{p2} - \theta_{p3}) = 180^{\circ} - [90^{\circ} + 45^{\circ} + 20^{\circ}] \cong 25^{0}$$

**Example 8.14:** The transfer function of a unity feedback system is $G(s) = \frac{k}{s(s+4)}$. Draw the root locus plot and determine the value of $k$ for a damping ratio of 0.5.

***Solution***

$$\sigma = \frac{-4}{2} = -2$$

The angle of asymptotes: $\frac{(2n+1)*180}{2}$, $n = 0,1$

Breakaway point = $90^{\circ}$, $270^{\circ}$

The characteristic equation is:

$$s(s+4) + k = 0$$

$$s^2 + 4s + k = 0$$

$$k = -(s^2 + 4s)$$

$$\frac{dk}{ds} = -(2s+4) = 0$$

$$s = -2$$

$$\xi = 0.5 = \cos\theta \text{ or } \theta = 60^{\circ}$$

Figure 8.24 shows the root locus.

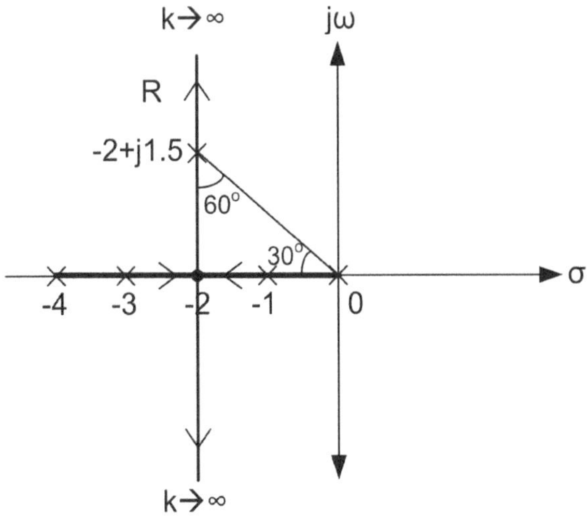


**FIGURE 8.24** Root locus of control system in Example 8.14.

At this point $R$, the value of $s = -2 - j1.5$

The magnitude criterion is $|G(s)H(s)| = 1$

$$\left|\frac{k}{s(s+4)}\right| = 1$$

$$\left|\frac{k}{(-2+j1.14)(-2+j1.14+4)}\right| = 1$$

$$\frac{k}{\left(\sqrt{5.3}*\sqrt{5.3}\right)^2} = \frac{k}{5.3} = 1$$

$$k = 5.3$$

## 8.6 SUMMARY OF RULES FOR THE CONSTRUCTION OF ROOT LOCI

1. Locate poles and zeros on the s-plane.
2. The $k = 0$ points of the root loci are poles of $G(s)H(s)$.
3. The $k = ¥$ points of the root loci are the zeros of $G(s)H(s)$.
4. Number of separate Loci.
   $z = z$ if $z > p$, $p = z$ if $p > z$
5. The root loci are symmetric with respect to real axis.
6. Asymptotes of root loci
   For large values of $s$, the root loci are asymptotic to straight lines with angles given by

$$\theta_k = \tfrac{(2k+1)\pi}{p-z},\ k = 0,1,\ldots,p-z \tag{8.46}$$

7. Intersection of asymptotes
   a) The intersection of the $n = p - z$ number of asymptotes lies on the real axis only.
   b) The intersection of asymptotes on the real axis is given by:

$$\sigma = \frac{-\Sigma(b_1 - a_1)}{n} = \frac{\Sigma\, poles\, of\, G(s)H(s) - \Sigma\, zeros\, of\, G(s)H(s)}{p-z} \tag{8.47}$$

8. Root loci on the real axis
   Positive $k$ root loci will exist on the real axis only if the total number of poles and zeros of $G(s)H(s)$ to the right of the section is odd.
9. The angle of departure from poles and angle of arrival at zeros of root loci.

$$-\left(\theta_{p1} + \theta_{p2} + \ldots\right) + \left(\theta_{z1} + \theta_{z2} + \ldots\right) = 280(2k+1) \tag{8.48}$$

10. The intersection of the Root Loci with the imaginary axis is determined by the Routh-Harwitz criterion.
11. Breakaway points on the root loci.
12. Calculation of $k$ on the root loci.

$$|k| = \frac{\prod_{j=1}^{n} |s + p_j|}{\prod_{i=1}^{m} |s + z_i|}$$

or

$$k = \frac{Product\ of\ lengths\ of\ vectors\ drawn\ from\ the\ poles\ of\ G(s)H(s)}{Product\ of\ lengths\ of\ vectors\ drawn\ from\ the\ zeros\ of\ G(s)H(s)} \quad (8.49)$$

**Example 8.15:** Consider the control system shown in Figure 8.25.

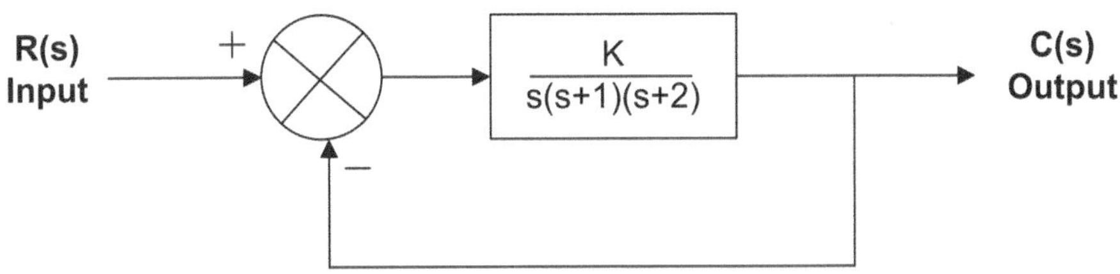


**FIGURE 8.25** Block diagram of a control system.

***Solution***

$$\angle G(s) = \angle \frac{k}{s(s+1)(s+2)}$$

$$= -\angle s - \angle(s+1) - \angle(s+2) \pm 180(2k+1)$$

$$|G(s)| = \left| \frac{k}{s(s+1)(s+2)} \right| = 1$$

Asymptotes = $60^o, 180^o, 300^o$

Intersection: $-\frac{0+1+2}{3} = -1$

$$f(s) = s^3 + 2s^2 + 2s + k$$

$$k = \left(-s^3 + 2s^2 + 2s\right)$$

$$\frac{dk}{ds} = -3s^2 + 6s + 2, \frac{-6 \pm \sqrt{36+24}}{6}$$

$$s_1 = -0.4226, -1.5774$$

$$k_c = s^3 + 2s^2 + 2s + k$$

$$\begin{array}{|ll} 1 & 2 \\ 3 & k \\ \dfrac{6-k}{3} & \\ k & \end{array}$$

$$k_c = 6$$
$$\omega_c = \sqrt{2}\ rad/sec$$

Figure 8.26 shows the root locus.

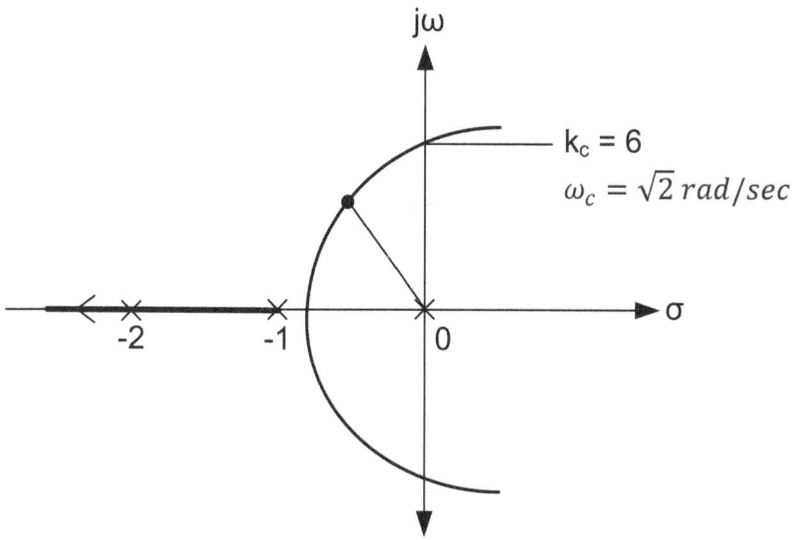


**FIGURE 8.26** Root locus plot for Example 8.15.

Moreover, to determine the dominant complex conjugate poles such that the damping ratio is 0.5,

$$\pm\cos^{-1}\xi = 0.5,\ \theta = 60^o$$
$$s_1 = -0.3337 + j0.580,\ s_2 = 0.3337 - j0.5780$$

The value of $k$ that yields such poles is

$$k = \left|s(s+1)(s+2)\right|_{s=-0.3337+J0.5780} = 1.0380$$

Using the value of $k$, the third pole is found at $s = -2.3326$.

## 8.7 GAIN AND PHASE MARGIN

- The gain margin is defined as the ratio of critical gain to the design gain, $\frac{k_c}{k_d}$.
- The phase margin
- The phase margin of a feedback control system can be determined from the root locus. It is necessary to find the point $j\omega_1$ on the $j\omega$ axis for which $|GH(j\omega)| = 1$.
- For the design value of $k$,

$$\left|\frac{D(j\omega_1)}{N(j\omega_1)}\right| = k \text{ design} \tag{8.50}$$

The phase margin is then computed from $\arg(GH(j\omega_1))$:

$$\varphi_{PM} = 180^0 + \arg GH(j\omega_1) \tag{8.51}$$

## 8.8 POLE PLACEMENT

Instead of moving closed loop poles into the region of a complex plane, it is possible to place one or more closed loop poles at the desired places. There are two methods to perform:

- An algebraic compensator.
- A fixed structure compensator like PID.

### 8.8.1 Algebraic Compensator

Consider the block diagram shown in Figure 8.27. The closed-loop transfer function is $C(s) = \frac{G_c(s)G_p(s)}{1+G_c(s)G_p(s)}$, where $G_p(s)$ is known and cannot be changed.

Suppose it is desired that a closed loop transfer is $T(s)$. Then,

$$T(s) = \frac{G_c(s)G_p(s)}{1+G_c(s)G_p(s)}$$

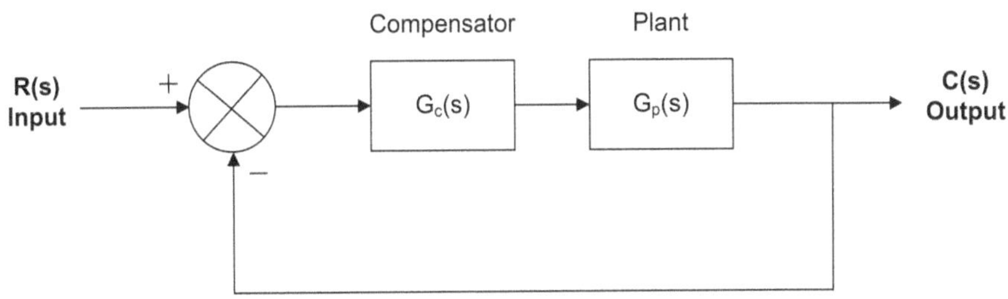


**FIGURE 8.27** Block diagram of controller and plant.

$$T(s)\big(1+G_c(s)G_p(s)\big)=G_c(s)G_p(s)$$

$$G_c(s)G_p(s)\big(1-T(s)\big)=T(s)$$

$$G_c(s)=\frac{T(s)}{G_p(s)(1-T(s))}$$

Define $T(s)=\frac{N_t(s)}{D_t(s)}$, $\quad G_p(s)=\frac{N_p(s)}{D_p(s)}$

Then,

$$G_c(s)=\frac{N_t(s)D_p(s)}{N_p(s)\big[D_t(s)-N_t(s)\big]}$$

A gain $k$ can be included with $G_c(s)$ where $k=1$ for the correct pole placement. The open loop transfer function will be:

$$G_c(s)G_p(s)=\frac{kT(s)}{1-T(s)}$$

The root locus depends only on $T(s)$ and not on the plant.

The closed loop pole that generates the denominator of $T(s)$ should be selected for desired transients such as time constant, settling time, etc.

Suppose $T(s)$ is given by

$$T(s)=\frac{b_2s^2+b_1s+b_0}{s^3+a_2s^3+a_1s+a_0}$$

Then, for a steady-state response, we usually examine

$$T_E(s)=1-T(s)$$
$$=\frac{s^3+(a_2-b_2)s^2+(a_1-b_1)s+(a_0-b_0)}{s^3+a_2s^2+a_1s+a_0}$$

In order to become system type one, $a_0=b_0$ and for type two, $a_0=b_0, a_1=b_1$.

Suppose the closed loop poles are at −1 and −2, then,

$$T(s)=\frac{2}{(s+1)(s+2)}=\frac{2}{s^2+3s+2}, \text{ type 0}$$

**Example 8.16:** Suppose the second-order system

$$G_p(s)=\frac{1}{(s+1)(s+2)}=\frac{N_p(s)}{D_p(s)}$$

is to be compensated, so that the closed-loop poles become −2 and −4 and the system becomes type 1.

$$T(s)=\frac{8}{(s+2)(s+4)}=\frac{8}{s^2+6s+8}=\frac{N_t(s)}{D_t(s)}$$

The system becomes type one because the $s^0$ terms in the numerator and denominator are the same.

$$G_c(s)=\frac{N_t(s)*D_p(s)}{N_p(s)\left[D_t(s)-N_t(s)\right]}=\frac{8(s+1)(s+3)}{1\left[(s+2)(s+4)-8\right]}$$

$$=\frac{8(s+1)(s+3)}{s(s+3)}$$

$$G_c(s)G_p(s)=\frac{8k}{s(s+6)}$$

**Example 8.17:** Consider the system shown here.

$$\boldsymbol{G_p(s)}=\frac{1}{\boldsymbol{s}\left(\boldsymbol{s}^2+4\boldsymbol{s}+5\right)}=\frac{\boldsymbol{N_p(s)}}{\boldsymbol{D_p(s)}}$$

Suppose that the closed loop poles are $-1\pm j1$ and −1. To obtain the type 1 option, the denominator and numerator coefficient must be equal.

$$T(s)=\frac{2}{(s+1)(s+1+j1)(s+1-j1)}=\frac{2}{s^3+2s^2+4s+2}=\frac{N_t(s)}{D_t(s)}$$

$$G_c(s)=\frac{N_t(s)D_p(s)}{N_p(s)\left[D_t(s)-N_t(s)\right]}$$

$$=\frac{2s\left(s^2+4s+5\right)}{1\left[s^3+3s^2+4s+2-2\right]}$$

$$G_c(s)=\frac{2\left(s^2+4s+5\right)}{s^2+3s+4}$$

$$G_c(s)G_p(s)=\frac{2\left(s^2+4s+5\right)}{s^2+3s+4}*\frac{1}{2\left(s^2+4s+5\right)}$$

$$=\frac{2}{s\left(s^2+3s+4\right)}$$

### 8.8.2 Fixed Structure Compensation

$$G_c(s) = \frac{N_c(s)}{D_c(s)},\ G_p(s) = \frac{N_p(s)}{D_p(s)} \tag{8.52}$$

$$T(s) = \frac{N_c(s) * N_p(s)}{D_c(s)D_p(s) + N_c(s)N_p(s)} \tag{8.53}$$

The characteristic polynomial is:

$$D_c(s)D_p(s) + N_c(s)N_p(s) \tag{8.54}$$

The order of the closed-loop system is the order of the plant plus the order of the compensator. The number of close loop poles that can be placed is the number of multiplying factors in the compensator.

If the closed loop system is fifth order and the compensator has three multiplying factors, then three closed loop poles can be placed anywhere desired. The remaining two closed-loop poles and the three multiplying factors of the compensator become dependent variables that must be computed from equations for coefficients for each of the five powers of *s* in the characteristic polynomial, except for the highest power of *s*, which always has a unity coefficient.

In general, if the closed-loop system is of order "*n*" and the compensator has *m* multiplying factors, then the designer can place *m* CLPS as desired. The remaining $n-m$ CLPS and the *m* compensator multiplying factors are calculated from *n* equations, are always linear, and may be simultaneous. PID is used as:

$$G_c(s) = k_d s + k_p + \frac{k_i}{s} = \frac{k_d s^2 + k_p s + k_i}{s} = \frac{N_c(s)}{D_c(s)} \tag{8.55}$$

where $k_d$ causes the compensator output to depend.

**Example 8.18:** Consider a PID pole placement, where

$$G_p(s) = \frac{1}{(s+1)(s+2)} = \frac{N_p(s)}{D_p(s)}$$

***Solution***

The plant is of second order, $G_c(s) = \frac{k_D s^2 + k_p s}{s} = \frac{N_c(s)}{D_c(s)}$, so the closed loop system is third order. $(D_c(s) = s)$, thus, raising the order by 1 and raising the type to 1. A PID has three multiplying factors, so $m = 3$ and all three closed loop poles can be placed.

Suppose the two previous close loop poles are selected as −3 and −4 while the remaining pole is placed at −24 to be non-dominant.

$$P(s) = s(s+1)(s+2) + k_d s^2 + k_p s + k_c = (s+3)(s+4)$$

$$= s^3 + 31s^2 + 180s + 288$$
$$= s^3 + (k_d + 3)s^2 + (k_p + 2)s + k_i$$
$$3 + k_d = 31,\ k_d = 28$$
$$2 + k_p = 180,\ k_p = 178$$
$$k_i = 288$$

The PID compensator is:

$$\frac{28s^2 + 178s + 288}{s} = \frac{28}{s}\left(s^2 + 6.36s + 10.3\right)$$
$$= \frac{28}{s}(s + 3.18 + 10.418)(s + 3.18 - j0.418)$$
$$G_c(s)G_p(s) = \frac{28k(s + 3.18 + j0.418)(s + 3.18 - j0.418)}{s(s+1)(s+2)}$$

**Example 8.19:** Consider a PID pole placement, where

$$G_p(s) = \frac{1}{s\left(s^2 + 4s + 5\right)}$$

***Solution***

The system becomes 4th order $(n = 4)$. Since $m = 3$ in the PID compensator, only three of the four closed loop poles can be placed. Here, $n - m = 1$ closed-loop pole and three PID multiplying factors are the four unknowns to be determined from four coefficients for four powers of $s$. Suppose the same three closed-loop poles are to be placed as $(s + 1 \pm j1), (s + 1)$. It is convenient to call the other close loop pole as $-a$.

$$P(s) = s^2\left(s^2 + 4s + 5\right) + k_d s^2 + k_p s + k_i$$
$$= (s + a)(s + 1 + j)(s + 1 - j)(s + 1)$$
$$= s^4 + (3 + a)s^3 + (3a + 4)s^2 + (4a + 2)s + 2a$$
$$= s^4 + 4s^3 + (5 + k_d)s^2 + k_p s + k_i$$

$$3 + a = 4,\ a = 1$$
$$3a + 4 = 5 + k_d,\ k_d = 2$$
$$4a + 2 = k_p = 6$$
$$2a = k_i,\ k_i = 2$$
$$G_c(s) = \frac{2s^2 + 6s + 2}{s} = \frac{2}{s}\left(s^2 + 3s + 1\right)$$

$$2(s+2.62)(s+0.38)\,G_c(s)G_p(s)=\frac{2k(s+2.62)(s+0.38)}{s^2(s^2+4s+5)}$$

**Example 8.20:** The open loop transfer function of a system is $G(s)=\frac{25(s+2)}{s(s+1)(s+5)}$. The system is required to be stable with a damping ratio of $\xi=0.7$ and $k=25$.

***Solution***

To reach the value of $\xi=0.7$, the value of $k=0.7$. Therefore, we must compensate.

$$\frac{\theta_c}{\theta}=\frac{k'(s+z)}{(s+p)},\ p>z$$

Let $z=1,\ p=10$

So that $k=25$ as desired. The resulting transfer function is:

$$G(s)=\frac{k(s+2)}{s(s+1)(s+5)(s+10)}$$

Figure 8.28 shows the root locus.

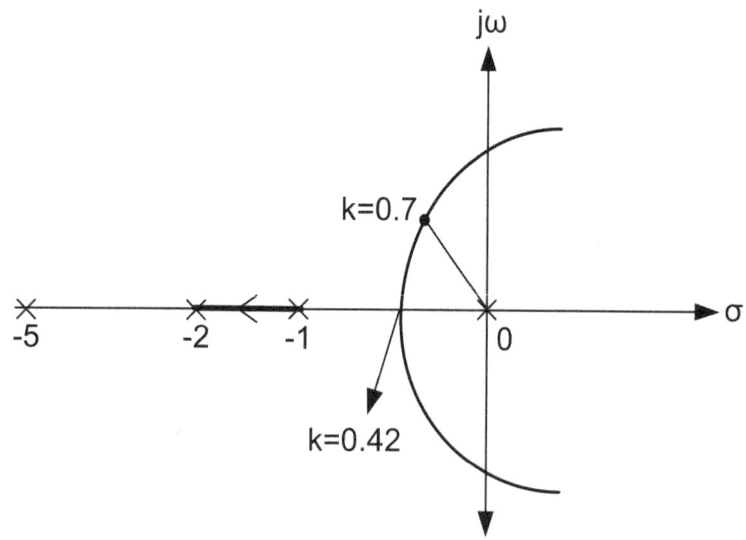


**FIGURE 8.28** Root locus of Example 8.20.

**Problems**

P8.1 Draw the root locus of a third-order system.

$$G(s)=\frac{k}{s(s+1)(s+10)}$$

P8.2 Draw the root locus of a third-order system.

$$G(s)=\frac{k}{s(s+1)(s+2)}$$

P8.3 The open loop transfer function of a unity feedback control system is $G(s)=\frac{k(s+2)}{s(s+1)}$. Draw its root locus.

# 9 Frequency Domain Analysis of Control Systems

## 9.1 FREQUENCY DOMAIN SPECIFICATION

Figure 9.1 shows the plot of resonance peak vs. resonance frequency. Following key terminologies show the frequency domain specifications.

- Resonance Peak, $\mu_r$: It is the maximum magnitude value as the frequency, $\omega$, is varied.
- Resonance Frequency, $\omega$: It is the frequency at which the peak resonance, $\mu_r$, occurs.
- Bandwidth (BW): The frequency at which the magnitude of $\mu_{(j\omega)}$ drops by 3 dB. If the bandwidth is large, higher frequencies will pass through the system. This indicates a large rise in time. Therefore, BW gives an idea about transient response. Small BW means lower frequencies are passed, so the response is sluggish. BW also indicates the sensitivity to parameter variations.
- Cut-off rate: The slope of the magnitude curve near the cut-off frequency, $\omega_c$, is known as the cut-off rate.

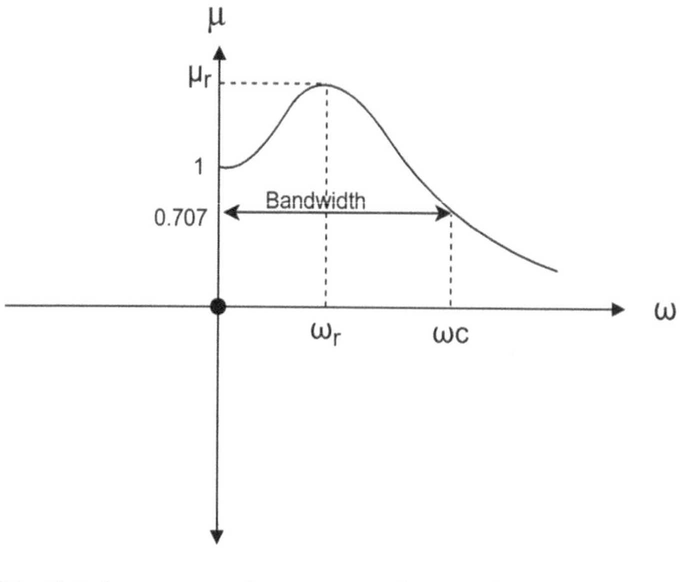


**FIGURE 9.1** Plot of resonance peak vs. resonance frequency ($\mu_r$ vs. $\omega$).



DOI: 10.1201/9781003484202-9

- Phase Margin: If $\varphi$ is the phase angle of a system at unity gain, the phase margin is given by 180 + $\varphi$. The frequency at which gain is unity is known as gain crossover frequency. Phase Margin is a measure of relative stability.
- Gain Margin: The frequency at which the phase angle of the transfer function is –180° is known as phase crossover frequency. The gain at this frequency is $|G(j\omega)\, H(j\omega)|$, which gives the gain margin.

## 9.2 THE CORRELATION BETWEEN TIME AND FREQUENCY DOMAIN SPECIFICATIONS FOR A SECOND-ORDER SYSTEM

The relationship between time and frequency domain specification is important before designing a control system. Consider a standard second-order system as shown in Figure 9.2.

$$G(s) = \frac{\omega n^2}{s(s+2\xi\omega n)} \text{ and } H(s) = 1 \tag{9.1}$$

$$T(s) = \frac{C(s)}{R(s)} = \frac{\omega_n^2}{s^2 + 2\xi\omega_n + \omega_n^2} \tag{9.2}$$

Let $s = j\omega$,

$$T(j\omega) = \frac{C(s)}{R(s)} = \frac{\omega_n^2}{(j\omega)^2 + 2\xi\omega_n(j\omega) + \omega_n^2}$$

$$T(j\omega) = \frac{1}{1 + 2j\xi\left(\frac{\omega}{\omega_n}\right) - \left(\frac{\omega}{\omega_n}\right)^2} \tag{9.3}$$

Let $= \frac{\omega}{\omega_n}$, we have

$$T(j\omega) = \frac{1}{1 + 2j\omega\xi - u^2} \tag{9.4}$$

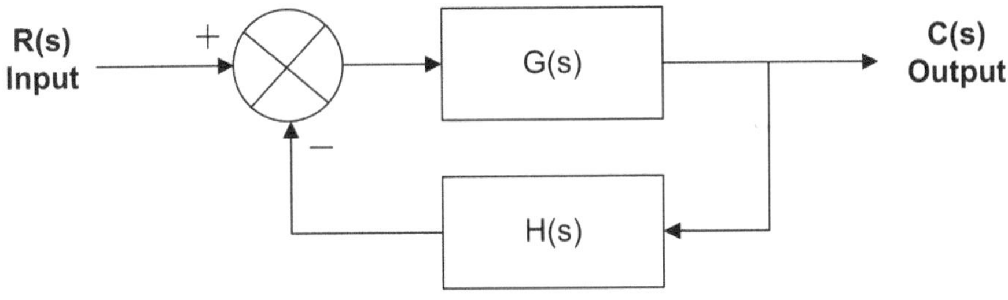


**FIGURE 9.2** Block diagram of a feedback control system.

$$|T(j\omega)| = \mu = \frac{1}{\left(1-u^2\right)+\left(2u\xi\right)^2} \tag{9.5}$$

$$\angle T(j\omega) = \tan^{-1}\frac{2u\xi}{1-u^2} \tag{9.6}$$

Putting $\left.\frac{d\mu}{du}\right|_{u=u_r} = 0$, the resonant frequency, $\omega_r$, can be obtained as:

$$\left.\frac{d\mu}{du}\right|_{u=u_r} = \frac{-1}{2}\frac{\left[2\left(1-u_r^2\right)\left(-2u_r\right)+8\xi^2 u_r\right]}{\left[\left(1-u_r^2\right)+\left(2u_r\xi\right)^2\right]^{\frac{3}{2}}} = 0 \tag{9.7}$$

i.e., $4u_r\xi^2 - 4u_r^3 - 8u_r\xi^2 = 0$

$$u_r\left(1-2\xi^2-u_r^2\right) = 0 \tag{9.8}$$

$$u_r = 0 \text{ gives } \frac{\omega_r}{\omega_n} = 0 \text{ or } \omega_r = 0.$$

Therefore, $1-2\xi^2-u_r^2 = 0$

$$\text{Or } u_r = \sqrt{1-2\xi^2}$$

$$\text{Or } \frac{\omega_r}{\omega_n} = \sqrt{1-2\xi^2}$$

$$\omega_r = \omega_n\sqrt{1-2\xi^2} \tag{9.9}$$

$$\mu_r = \left.G(j\omega)\right|_{u=u_r} = \frac{1}{\sqrt{\left(1-u_r^2\right)^2+\left(2u_r\xi\right)^2}}$$

$$\mu_r = \frac{1}{2\xi\sqrt{1-\xi^2}} \tag{9.10}$$

If $\sqrt{1-\xi^2} \geq 0, \quad \xi \leq 0.707$

$$\varphi_r = \left.G(j\omega)\right|_{u=u_r} = \tan^{-1}\frac{\sqrt{1-2\xi^2}}{\xi} \tag{9.11}$$

The BW of the system can be calculated as:

$$0.707 = \frac{1}{\left[\left(1-u_c^2\right)^2+2\left(\xi u_c\right)^2\right]^{\frac{1}{2}}} \tag{9.12}$$

Where $u_c = \dfrac{\omega_c}{\omega_n}$

$$\left(1-u_c^2\right)^2 + \left(2u_c\xi\right)^2 = 2$$

$$u_c^2 = \left(1-2\xi^2\right) \pm \sqrt{4\xi^4 - 4\xi^2 + 2}$$

$$\omega_c = \omega_n \left\{\left(1-2\xi^2\right) + 4\xi^2 - 4\xi^2 + 2\right\}^{\frac{1}{2}} \tag{9.13}$$

$$BW = \omega_c = \omega_n \left\{\left(1-2\xi^2\right) + \sqrt{4\xi^4 - 4\xi^2 + 2}\right\}^{\frac{1}{2}} \tag{9.14}$$

It may be observed from previous equations and expressions that

- BW decreases as $\xi$ increases.
- For $0 < \xi < 0.707$, BW and $\mu_r$ are directly proportional.
- BW decreases due to an increase in rise time.

**Example 9.1:** Determine the frequency domain specifications for a second-order system with unity feedback and $G(s) = \frac{100}{s(s+4)}$.

***Solution***

$$\mu(s) = \frac{C(s)}{R(s)} = \frac{G(s)}{1+G(s)H(s)} = \frac{100}{s^2 + 4s + 100}$$

$$= \frac{\omega_n^2}{s^2 + 2\xi\omega_n s + \omega_n^2}$$

Since, $\omega_n^2 = 100$, $\omega_n = 10\,rad/sec$

$$2\xi\omega_n = 4$$

$$\xi = \frac{4}{2*10} = 0.2$$

$$\mu_r = \frac{1}{2\xi\sqrt{1-\xi^2}} = \frac{1}{2*0.2\sqrt{1-(0.2)^2}} = 2.60$$

$$\omega_r = \omega_n\sqrt{1-2\xi^2} = 10\sqrt{1-2(0.2)^2} = 9.6$$

$$\omega_c = \omega_n \left\{\left(1-2\xi^2\right) + \sqrt{4\xi^4 - 4\xi^2 + 2}\right\}^{\frac{1}{2}} = 15.1\,rad/sec$$

**Example 9.2:** A unity feedback system having $G(s)=\frac{k}{s(s+Ts)}$ is subjected to a unit step input.

The overshoot is 23%, and the resonant frequency is 9 *rad/sec* in the frequency domain. Calculate *k* and *T*. Also, determine (i) gain crossover frequency, (ii) phase margin, and (iii) resonant peak.

***Solution***

$$G(s)=\frac{k}{s(s+Ts)} \text{ and } H(s)=1$$

$$\frac{G(s)}{R(s)}=\frac{k}{Ts^2+s+k}=\frac{k/T}{s^2+\frac{1}{T}s+\frac{k}{T}}$$

$$\omega_n^2=\frac{k}{T} \text{ or } \omega_n=\sqrt{\frac{k}{T}}$$

$$2\xi\omega_n=\frac{1}{T} \text{ or } \xi=\frac{1}{2\sqrt{kT}}$$

For unit step input, $\mu_p = 23\%$. So,

$$23 = e^{-\frac{\pi\xi}{\sqrt{1-\xi}}}$$

$$\xi = 0.4238$$

In frequency domain,

$$\omega_r = 9 = \omega_n\sqrt{1-2\xi^2}$$

$$9=\omega_n\sqrt{1-(2*0.42382)}$$

$$\omega_n = 11.243\, rad/sec$$

Also $k = T\omega^2$,

$$\xi^2=\frac{1}{4kT}=\frac{1}{4T^2\omega_n^2}$$

$$\xi^2=\frac{1}{4*T^2*11.243^2}$$

$$T^2=\frac{1}{4\xi^2*11.243^2},\quad T=0.1048$$

$$k = 0.1048 * 11.243^2 = 13.247$$

$$\omega_{gc} = \omega_n \sqrt{\sqrt{4\xi^2 + 1} - 2\xi^2} = 9.429\, rad / sec$$

$$P_\mu = \tan^{-1} \frac{2\xi}{\sqrt{-2\xi^2 + \sqrt{4\xi^2 + 1}}}$$

$$\tan^{-1} \frac{2 * 0.4 * 0.2438}{\sqrt{-2 * 0.4238 + 1.0625}} = 45.300$$

$$\mu_r = \frac{1}{2\xi\sqrt{1-\xi^2}} = \frac{1}{2 * 0.4238 * \sqrt{1 - 0.4238^2}} = 1.3025$$

**Example 9.3:** The forward transfer function of a unity feedback control system is $G(s) = \frac{64}{s(s+5)}$. Find $\omega_n$, $\omega_r$, $\mu_r$ and $\omega_b$.

***Solution***

$$G(s) = \frac{64}{s(s+5)} = \frac{\omega_n^2}{s^2 + 2\xi\omega_n s + \omega_n^2}$$

Comparing the two expressions:

$$\omega_n = 8$$

$$2\xi\omega_n = 5,\ \xi = 0.3125$$

$$\omega_r = \omega_n \sqrt{1 - 2\xi^2} = 7.1\, rad / sec$$

$$\mu_r = \frac{1}{2\xi\sqrt{1-\xi^2}} = 1.78$$

$$\omega_b = \sqrt{1 - 2\xi^2 + \left(2 - 4\xi^2 + 4\xi^4\right)^{\frac{1}{2}}}$$

$$= 11.25\, rad / sec$$

**Example 9.4:** A feedback control system is shown in Figure 9.3. The peak overshoot to a step input is 12%. Determine the resonant peak $\mu_r$ and the value of $k$, normalized resonant frequency, and bandwidth (BW).

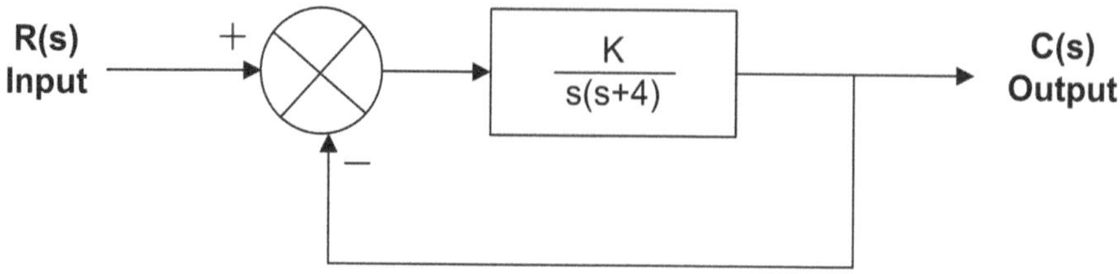


**FIGURE 9.3** A feedback control system.

***Solution***

$$\frac{C(s)}{G(s)}=\frac{\frac{k}{S(s+4)}}{1+\frac{k}{S(s4)}}=\frac{k}{s^2+4s+k}$$

Comparing with the standard equation,

$$\frac{C(s)}{R(s)}=\frac{\omega_n^2}{s^2+\xi\omega_n^2+\omega_n^2}$$

$$\omega_n^2=k,\ \omega_n=\sqrt{k}$$

$$2\xi\omega_n=k,\ \xi=\frac{4}{2\sqrt{k}}$$

$$\mu_p=12\%=e^{-\frac{\pi\xi}{\sqrt{1-\xi^2}}}$$

$$\xi=0.559$$

$$k=\frac{16}{4\xi^2}=\frac{4}{(0.559)^2}=12.8$$

$$\omega_n=\sqrt{k}=3.578$$

Then,

$$\mu_r=\frac{1}{2\xi\sqrt{1-\xi^2}}=\frac{1}{2*0.559*\sqrt{1-(0.559)^2}}=1.078$$

$$\omega_r=\omega_n\sqrt{1-\xi^2}=3.578*\sqrt{1-2*0.559^2}$$

$$\omega_r = 2.191$$

$$BW = \omega_n\sqrt{1-2\xi^2+2-\sqrt{4\xi^2+4\xi^4}}$$

$$= 4.298$$

$$= 20\log_{10} 4.298 = 12.66\,dB$$

## 9.3 POLAR PLOTS

### 9.3.1 Introduction

Polar plots of $G(s)\,H(s)$ open loop transfer function of a control system are necessary for the study of relative stability by the Nyquist method. In this process, frequency ($\omega$) is the independent variable for $\omega = 0$ to $\omega = \infty$ and magnitude with angle is the dependent variable. The open loop function, $G(s)\,H(s)$ plot, determines the relative stability as it passes the left $x - axis$ of the s-plane. The system is stable if the plot is to the left side of $-1 + j0$ when it passes through the $x - axis$, critically stable when it passes through on the $-1 + j0$ and unstable when it is right to $-1 + j0$ of the $x - axis$. Unless stated otherwise, the open loop transfer function is $G(s)\,H(s)$ instead of $G(s)H(s) = 1$.

### 9.3.2 Polar Plots of Typical Functions

**a.** $G(s) = \frac{1}{s}$

Let $s = j\omega$, then

$$|G(j\omega)| = \left(\frac{1'}{j\omega}\right) = \frac{1}{\omega} \tag{9.15}$$

And $\angle G(j\omega) = -90^0$

Table 9.1 shows a few values of $\omega$ corresponding to $|G(j\omega)|$ and $\angle G(j\omega)$.

The polar plot of $G(s) = \frac{1}{s}$ is shown in Figure 9.4.

**b.** $G(s) = \frac{1}{1+as}$

Putting $s = j\omega$ $G(s) = \frac{1}{1+as}$

$$G(j\omega) = \frac{1}{1+ja\omega} = \frac{1}{\sqrt{1+\omega^2 a^2}}\angle -\tan^{-1} a\omega \tag{9.16}$$

Table 9.2 shows a few values of $\omega$ corresponding to $|G(j\omega)|$ and $\angle G(j\omega)$.

The polar plot of $G(s) = \frac{1}{1+as}$ is as shown in Figure 9.5.

**TABLE 9.1**
**Values of $\omega$ $|G(j\omega)|$ and $\angle G(j\omega)$**

| $\omega$ | $\lvert G(j\omega)\rvert$ | $\angle G(j\omega)$ |
|---|---|---|
| 0 | $\infty$ | –90° |
| 1 | 1 | –90° |
| 10 | 0.1 | –90° |
| 100 | 0.01 | –90° |
| . | . | . |
| . | . | . |
| . | . | . |
| $\infty$ | 0 | –90° |

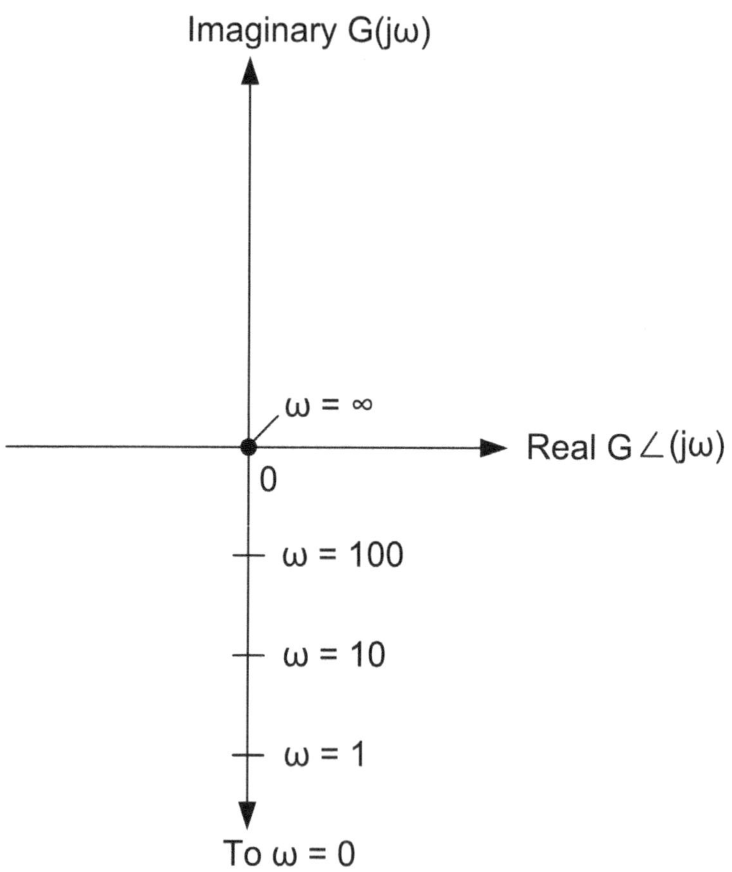


**FIGURE 9.4** Plot of G(s) = 1/s.

**TABLE 9.2**
**Values of $\omega$, $|G(j\omega)|$ and $\angle G(j\omega)$**

| $\omega$ | $\lvert G(j\omega)\rvert$ | $\angle G(j\omega)$ |
|---|---|---|
| 0 | 1 | 0° |
| 1/a | 0.707 | –45° |
| 10/a | 0.1 | –84.28° |
| 100/a | 0.01 | –89.42° |
| . | . | . |
| . | . | . |
| . | . | . |
| ∞ | 0 | –90° |

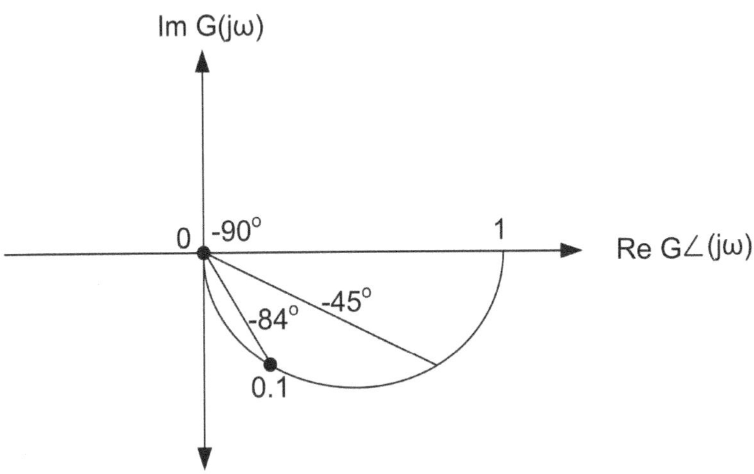


**FIGURE 9.5** Plot of G(s) = 1/(1+as).

c. $G(s)=\dfrac{1}{(s+1)(s+2)(s+3)}$

$$G(s)=\frac{1}{(j\omega+1)(j\omega+2)(j\omega+3)}$$

$$\left|G(j\omega)\right|=\frac{1}{\sqrt{(\omega^2+1)}\sqrt{(\omega^2+4)}\sqrt{(\omega^2+9)}} \tag{9.17}$$

$$\angle G(j\omega)=-\tan^{-1}\omega-\tan^{-1}\frac{\omega}{2}-\tan^{-1}\frac{\omega}{3} \tag{9.18}$$

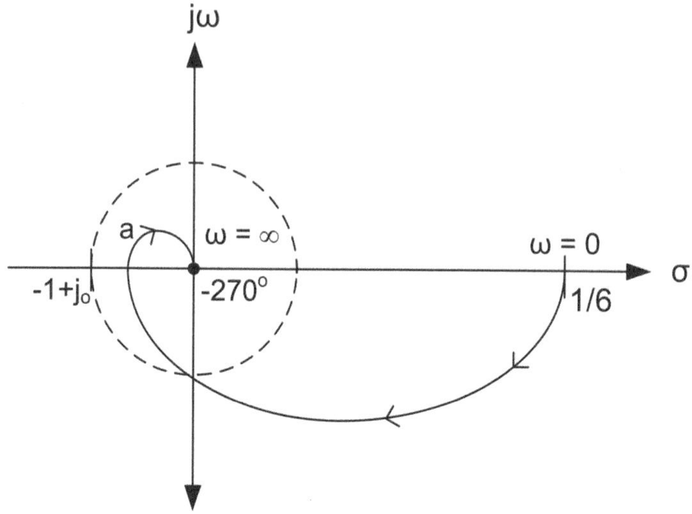


**FIGURE 9.6** A typical plot of a third-order system in Type 0.

When $\omega = 0\left|G(j\omega)\right| = \frac{1}{6}$ and $\angle G\,(j\omega) - 0$, Figure 9.6 shows the typical plot for the previous $G(s)$. The plot may pass through the $x-axis$ depending on the gain of G(s). Gain margin $= \frac{1}{a}$ and phase margin = θ°.

**Example 9.5:** A unity feedback control system has an open-loop transfer function. $G(s) = \frac{8}{(s+1)3}$. Draw the polar plot. Find the gain margin and phase margin.

***Solution***

$$G(s) = \frac{8}{(s+1)3}$$

$$G(s) = \frac{8}{(j\omega+1)3}$$

$$\mu = \frac{8}{\sqrt{\omega^2+1}}\,,\ \varphi = -3\tan^{-1}\omega$$

Table 9.3 shows the comparative values of $\omega\,\mu$ and $\varphi$. Figure 9.7 shows the plot of a third-order system that is critically stable.
The Gain margin $= \frac{1}{1} = 1$ and phase margin $= -180 + 180 = 0$.

**d.** $G(s) = \dfrac{1}{s(s+1)}$

**TABLE 9.3**

**Comparative values of ω μ and φ**

| $\omega$ | Magnitude (μ) | Angle (φ) |
|---|---|---|
| 0 | 8 | 0° |
| 0.5 | 5.71 | −88.5° |
| 1 | 3.66 | −135° |
| 1.732 | 1.00 | −180° |
| 5.0 | 0.06 | −252° |
| 10 | 0.00 | −262° |
| 100 | 0.00 | −265° |
| ∞ | 0.00 | −270° |

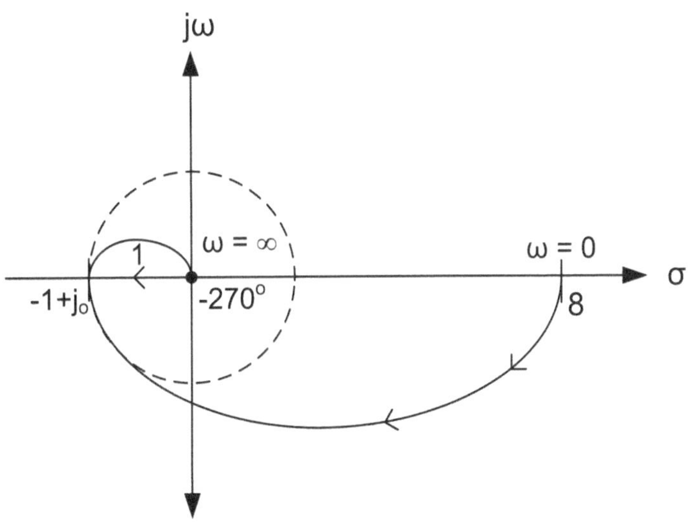


**FIGURE 9.7** Plot of a third-order system that is critically stable.

$$G(j\omega) = \frac{1}{j\omega(j\omega+1)}$$

$$|G(j\omega)| = \frac{1}{\omega + \sqrt{\omega^2+1}}$$

$$\angle G(j\omega) = -90^0 - \tan^{-1}\omega$$

Table 9.4 shows the values of $\omega$ $|G(j\omega)|$, $\angle G(j\omega)$. Figure 9.8 shows the graph of second order and type control system.

**TABLE 9.4**
**Values of $\omega$ $|G(j\omega)|$, $\angle G(j\omega)$**

| $\omega$ | $\lvert G(j\omega)\rvert$ | $\angle G(j\omega)$ |
|---|---|---|
| 0 | $\infty$ | $-90°$ |
| 1 | 0.707 | $-135°$ |
| 10 | 0.01 | . |
| 100 | 0.001 | . |
| $\infty$ | 0 | $-180°$ |

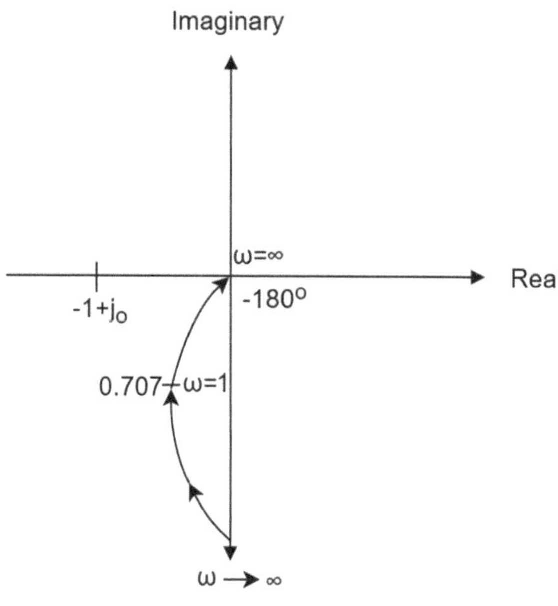


**FIGURE 9.8** Graph of second order control system.

**Example 9.6:** A unity feedback control system has an open loop transfer function $G(s) = \frac{k}{s(s+1)(s+2)}$. Draw its typical plot and discuss stability.

$$G(s) = \frac{k}{s(s+1)(s+2)}$$

$$G(j\omega) = \frac{k}{j\omega(j\omega+1)(j\omega+2)}$$

$$|G(j\omega)| = \frac{k}{\omega * \sqrt{\omega^2+1} * \sqrt{\omega^2+4}}$$

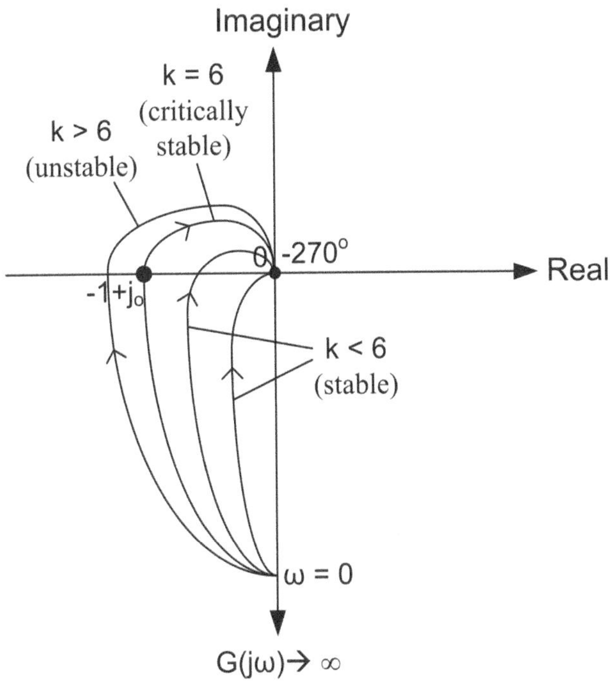


**FIGURE 9.9** A qualitative study of a third-order, type 1 control system.

$$\angle G(j\omega) = -90^0 - \tan^{-1}\omega - \tan^{-1}\frac{\omega}{2}$$

When

$$\omega = 0 \qquad |G(j\omega)| = \infty, \qquad \angle G(j\omega) = -90^0$$
$$\omega = \infty \qquad |G(j\omega)| = 0, \qquad \angle G(j\omega) = -270^0$$

Figure 9.9 depicts the typical plot of a qualitative study of a third-order, type 1 control system.

## 9.4 STABILITY OF FEEDBACK CONTROL SYSTEM USING FREQUENCY DOMAIN ANALYSIS

The phenomena can be explained with some examples as follows.

**Example 9.7:** The open loop transfer function of a feedback control system is given as:

$$G(s)H(s) = \frac{k}{s(s^2 + 2s + 2)}$$

Determine the maximum k value for the close loop system to be stable.

***Solution***

The characteristics equation is

$$s^3 + 2s^2 + 2s + k = 0$$

Let $s = j\omega$

$$-j\omega^3 - 2\omega^2 + 2j\omega + k = 0$$

$$-j\omega^3 + 2j\omega = 0, \ \omega^2 = 2$$

$$-2\omega^2 + k = 0, \ \omega = \sqrt{2}\ rad/sec$$

$$-2\omega^2 = k, \ k = 4$$

Therefore, at $k < 4$, the system will be stable.

**Example 9.8:** Examine the closed loop stability of a system whose open loop transfer function is given by:

$$G(s)H(s) = \frac{40}{(s+1)(s+2)}$$

***Solution***

$$G(s)H(s) = \frac{40}{(s+1)(s+2)}$$

Let $s = j\omega$

$$G(s)H(s) = \frac{40}{(1+j\omega)(2+j\omega)}$$

$$= \frac{40(1-j\omega)(2-j\omega)}{(1+j\omega)(2+j\omega)(1-j\omega)(2-j\omega)}$$

$$= \frac{40\left(2-3j\omega-\omega^2\right)}{\left(1+\omega^2\right)(4+j\omega)}$$

$$= \frac{40\left(2-\omega^2\right)}{\left(1+\omega^2\right)\left(4+\omega^2\right)} + j\frac{-120\omega}{\left(1+\omega^2\right)\left(4+\omega^2\right)}$$

$$Re\left[G(j\omega)H(j\omega)\right] = \frac{40\left(2-\omega^2\right)}{\left(1+\omega^2\right)\left(4+\omega^2\right)}$$

$$Im\left[G(j\omega)H(j\omega)\right] = \frac{-120\omega}{\left(1+\omega^2\right)\left(4+\omega^2\right)}$$

The intersection of the polar plot with the real axis can be found by:

$$Im\left[G(j\omega)H(j\omega)\right] = 0 i.e \frac{-120\omega}{\left(1+\omega^2\right)\left(4+\omega^2\right)} = 0$$

At $\omega = 0$ or $\omega = \infty$, the intersection with the imaginary axis can be found by:

$$Re\left[G(j\omega)H(j\omega)\right] = 0$$

$$\frac{40\left(2-\omega^2\right)}{\left(1+\omega^2\right)\left(4+\omega^2\right)} = 0$$

$$2-\omega^2 = 0$$

$$\omega = \pm\sqrt{2} = \pm 1.414$$

Figure 9.10 shows the polar plot for $G(s)H(s) = \frac{40}{(s+1)(s+2)}$. Table 9.5 shows the values of real and imaginary part of $G(j\omega)H(j\omega)$ for $\omega$ varying from 0 to $\infty$.

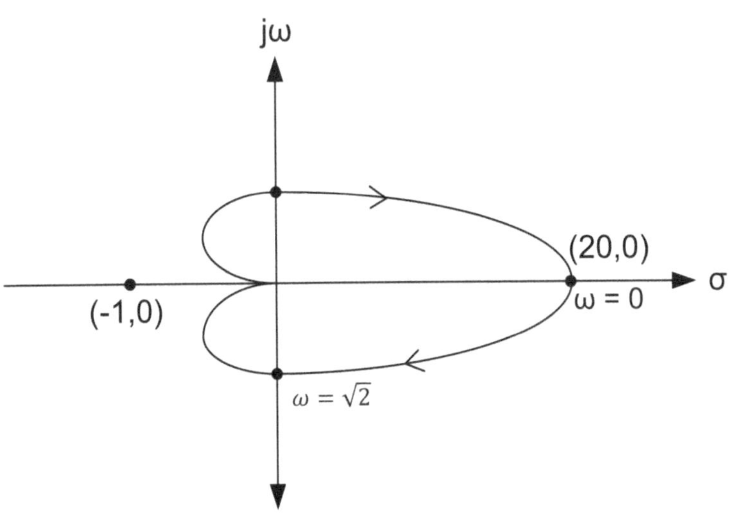


**FIGURE 9.10** Polar plot for G(s)H(s) = 40/(s + 1)(s + 2).

**TABLE 9.5**

**Values of Real and Imaginary Part of $G(j\omega)\ H\ (j\omega)$ for $\omega$ Varying from 0 to ∞**

| $\omega$ | 0 | 1 | $\sqrt{2}$ | ∞ |
|---|---|---|---|---|
| Real $[G(j\omega)\ H\ (j\omega)]$ | 20 | 4 | 0 | 0 |
| Im $[G(j\omega)\ H\ (j\omega)]$ | 0 | j12 | –j9.6 | 0 |

**Example 9.9:** The open loop transfer function of a system is

$$G(s)H(s) = \frac{k}{s(s+2)(s+3)}$$

Determine the value of $k$ for which the system will remain stable with a phase margin of 60° and a gain margin of 6 dB.

***Solution***

The frequency domain transfer function is

$$G(j\omega)H(j\omega) = \frac{1}{j\omega(2+j\omega)(3+j\omega)}$$

Phase margin $= \angle G(j\omega)H(j\omega) + 180^0$

$$\angle G(j\omega)H(j\omega) = -90^0 - \tan^{-1}\frac{\omega}{2} - \tan^{-1}\frac{\omega}{3}$$

For the required phase margin

$$60^0 = -90^0 - \tan^{-1}\frac{\omega}{2} - \tan^{-1}\frac{\omega}{3} + 180^0$$

$$30^0 = \tan^{-1}\frac{\omega}{2} + \tan^{-1}\frac{\omega}{3}$$

$$\tan 30 = \frac{\left[\frac{\omega}{2} + \frac{\omega}{3}\right]}{1 - \frac{\omega}{2} * \frac{\omega}{3}}$$

$$0.57 = \frac{5\omega}{6-\omega^2}$$

$$\omega = 0.5$$

Magnitude $G(j\omega)H(j\omega)$ at $\omega = 0.5$ must be equated to determine the maximum value of gain $k$.

$$G(j\omega)H(j\omega) = \frac{k}{j\omega(2+j\omega)(3+j\omega)} = \frac{k(2-j\omega)(3-j\omega)}{j\omega(4+\omega^2)(9+\omega^2)}$$

$$= \frac{k(6-5j\omega-\omega^2)}{j\omega(4+\omega^2)(9+\omega^2)}$$

$$\frac{j5\omega}{j\omega(4+\omega^2)(9+\omega^2)} + \frac{k(6-\omega^2)}{j\omega(4+\omega^2)(9+\omega^2)}$$

$$= \frac{5k}{(4+\omega^2)(9+\omega^2)} - j\frac{k(6-\omega^2)}{\omega(4+\omega^2)(9+\omega^2)}$$

$$\left|G(j\omega)H(j\omega)\right|_{\omega=0.5} = \frac{5k}{(4+0.25)(9+0.25)} = 1$$

$$k = 7.86$$

Let the polar plot intersect the negative real axis at a distance from the origin.
Then gain margin $6 = 20\log_{10}\frac{1}{a}$

$$a = 0.5$$

At the point of intersection, the imaginary part is equal to zero.

$$\frac{k(6-\omega^2)}{\omega(4+\omega^2)(9+\omega^2)} = 0$$

So that, $\omega^2 = 6$ or $\omega = \sqrt{6} = 2.45\,rad/sec$

Thus $\left|\frac{5k}{(4+\omega^2)(9+\omega^2)}\right|_{at\,\omega^2=6} = \frac{5k}{150} = 0.5$ or $k = 15$

So the system will remain stable for a gain margin of 6 dB.

## 9.5 BODE ANALYSIS

A Bode plot consists of two plots: (*i*) a Magnitude plot of a sinusoidal transfer function, and (ii) a plot of phase angle, both plotted against frequency in a logarithmic scale. The standard representation of the magnitude plot is $20\log|G(j\omega)|$ to the base of 10, and phase is drawn in degrees. The main advantage of using the logarithmic plot is that the multiplication of magnitudes can be converted into addition.

It is simpler than the polar plots to sketch an appropriate log-magnitude curve using a straight-line asymptotic method. Logarithmic representation is useful as it shows the transfer function's low-frequency and high-frequency characteristics in one diagram.

### 9.5.1 Basic Factor of $G(j\omega)\ H(j\omega)$

a) Gain k
b) Integral and derivative $j\omega^{\pm 1}$
c) First order factor $(1+j\omega_T)^{\pm 1}$
d) Quadratic form $\left[1+2\xi\left(\frac{j\omega}{\omega_n}\right)+\left(\frac{j\omega}{\omega_n}\right)^2\right]$

A little explanation of the factors mentioned previously is provided as under:

a) Gain $k$
For positive $k$

$$20\log_{10} k \text{ dB}$$

b) Integral and derivative factor $j\omega^{\pm 1}$

$$20\log\left(\frac{1}{j\omega}\right) = -20\log\omega \tag{9.19}$$

The shape of the line is $-20\ dB/decade$.
The phase angle is $-90^0$
Similarly,

$$20\log(j\omega) = \pm 20\log\omega\, dB \tag{9.20}$$

Phase angle $+90^0$
The slope of the line is $+20\ dB/decade$. If the factor is $(j\omega)^{\pm n}$, then it is $-20n\log\omega\, dB$, $+20n\log\omega\, dB$.
Slope: $-20n\ dB/decade$, $+20n\ dB/decade$.
Phase: $-90n^0$, $+90n^0$.

c) First order Factor $(1+j\omega T)^{\pm 1}$

$$20\log\left(\frac{1}{1+j\omega T}\right) = -20\log\sqrt{1+\omega^2+T^2}\,dB \tag{9.21}$$

For $\omega \ll \frac{1}{T} \cong -20\log 1 = 0\,dB.$

For $\omega \gg \frac{1}{T} \cong -20\log \omega T dB.$

In a straight-line approximation, at $\omega = \frac{1}{T}$, the log magnitude equals zero. At $\omega = \frac{\omega}{T}$, the log magnitude equals $-20\,dB$ and so on.

Actually, at $= \frac{1}{T}$, $log - magnitude = -3\,dB$. Thus, the value of $-20\log \omega T\ dB$ decreases by 20 $dB$ for every decade of $\omega$. For $\omega \gg \frac{1}{T}$ the log magnitude is a straight line with a slope of $-20\ dB/decade.$

The frequency at which two asymptotes meet is called corner frequency $\left(\omega = \frac{1}{T}\right)$. The corner frequency divides the frequency curve into two regions.

The phase angle, $\phi = \tan^{-1}\omega T$.

At $\omega = 0$, $\phi = 0$.

At corner frequency $\phi = -\tan^{-1}\frac{T}{T} = -45^0$.

At $\omega = \infty$, $\phi = -90^0$.

The error in the magnitude curve is $-20\log\sqrt{2} = -3.03\,dB$ at the corner frequency. The error at one decade above and below the corner frequency is $\pm 0.04\,dB$. A similar treatment is valid for $(1+j\omega T)$. The phase angles for $(1+j\omega T)^{\pm 1}$ are:

$$\pm 45^0 \text{ at } \omega = \frac{1}{T}$$

$$\pm 26^0 \text{ at } \omega = \frac{1}{2T}$$

$$\pm 5.7 \text{ at } \omega = \frac{1}{10T}$$

$$\pm 63.4^0 \text{ at } \omega = \frac{2}{T}$$

$$\pm 84.3^0 \text{ at } \omega = \frac{10}{T}$$

d) Quadratic Factor $\left[1+2\xi\left(\frac{j\omega}{\omega_n}\right)+\left(\frac{j\omega}{\omega_n}\right)^2\right]$

$$G(j\omega) = \left[1+2\xi\frac{j\omega}{\omega_n}+\left(\frac{j\omega}{\omega_n}\right)^2\right]^{\mp 1} = \frac{1}{1+2\xi\frac{j\omega}{\omega_n}+\left(\frac{j\omega}{\omega_n}\right)^2} \tag{9.22}$$

If $\xi > 1$, then its real poles. However, $0 < \xi < 1$ results in complex poles.

The asymptotic response is:

$$20\log\left|\frac{1}{1+2\xi\frac{j\omega}{\omega_n}+\left(\frac{j\omega}{\omega_n}\right)^2}\right|$$

$$= -20\log\sqrt{\left\{1-\left(\frac{\omega}{\omega_n}\right)\right\}^2+\left(2\xi\frac{\omega}{\omega_n}\right)^2}$$

When $\omega \ll \omega_n$, the log-magnitude is $-20\log 1 \cong 0\,dB$. For $\omega \ll \omega_n$, the log-magnitude is $-20\log\frac{\omega^2}{\omega_n^2} = -40\log\frac{\omega}{\omega_n}\,dB$ which is a straight line *of* $40\,dB\,/\,decade$. The high-frequency asymptote intersects the low-frequency asymptote at $\omega = \omega_n$. The two asymptotes are independent of $\xi$. Near $\omega = \omega_n$, a resonant peak occurs, which is a function of $\xi$. The magnitude of error near $\omega = \omega_n$ depends on $\xi$. The phase angle of the quadratic factor $\left[1+2\xi\frac{j\omega}{\omega_n}+\left(\frac{j\omega}{\omega_n}\right)^2\right]^{-1}$ is:

$$\phi = \angle\frac{1}{1+2\xi\frac{j\omega}{\omega_n}+\left(\frac{j\omega}{\omega_n}\right)^2} \tag{9.23}$$

$$= -\tan^{-1}\left[\frac{2\xi\frac{\omega}{\omega_n}}{1+\left(\frac{\omega}{\omega_n}\right)^2}\right]$$

The phase angle is a function of both $\omega$ and $\xi$. At $\omega = 0$, the phase angle equals to zero. At $\omega = \omega_n$ $\phi = -90^0$ regardless of $\xi$. At $\omega = \infty$, the phase angle becomes $-180^0$. The phase angle curve is skew-symmetric about the inflexion point (the point where $\phi = -90^0$).

### 9.5.2 General Procedure for Plotting Logarithmic Frequency Response Curves

i. Rewrite the transfer function $G(j\omega)\,H(j\omega)$ as a product of the basic factor.
ii. Identify the corner frequencies associated with these basic factors.

iii. Draw asymptotic *log – magnitude* curves with proper slopes between the corner frequencies.
iv. The phase angle curve can be drawn by adding the angle curves of individual frequencies.

**Example 9.10:** For the following transfer function, find the corner frequencies.

$$G(s)=\frac{10(s+3)}{s(s+3)(s^2+s+2)}$$

***Solution***

$$G(j\omega)=\frac{10(j\omega+3)}{j\omega(j\omega+2)\left((j\omega)^2+j\omega+2\right)}$$

$$=\frac{7.5\left(\frac{j\omega}{3}+1\right)}{j\omega\left(\frac{j\omega}{2}+1\right)\left(\frac{(j\omega)^2}{2}+\frac{j\omega}{2}+1\right)}$$

This function is composed of the following factors:

$$7.5,\ (j\omega)^{-1},\ \left(1+\frac{j\omega}{3}\right)\left(1+\frac{j\omega}{2}\right)^{-1},\ \left(1+\frac{j\omega}{2}+\frac{(j\omega)^2}{2}\right)^{-1}$$

The third, fourth, and fifth corner frequencies are $3\,and\,2\sqrt{2}$, respectively.

### 9.5.3 Minimum Phase System and Non-Minimum Phase System

$$G_1(j\omega)=\frac{1+j\omega T}{1+j\omega T_1},\ G_2(j\omega)=\frac{1-j\omega T}{1+j\omega T_1}\ (0<T<T_1)$$

The previous two equations differ from each other by factor $G_2(j\omega)=\frac{1-j\omega T}{1+j\omega T}$. The magnitude of the factor is always unity, but the phase angle equals $-2\tan^{-1}\omega T$ and varies from 0 to $-180^0$ as $\omega$ is increased from zero to $\infty$. The magnitude and angle characteristics are directly related to a minimum phase system. If the magnitude curve of the system is specified over a frequency range, then the phase angle curve is uniquely determined and vice versa. This, however, does not hold for a non-minimum phase system. Figure 9.11 shows the graph of minimum and non-minimum systems.

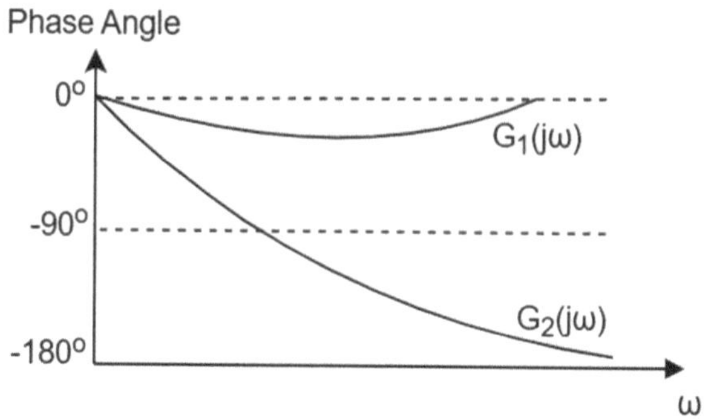


**FIGURE 9.11** Graph of minimum and non-minimum systems.

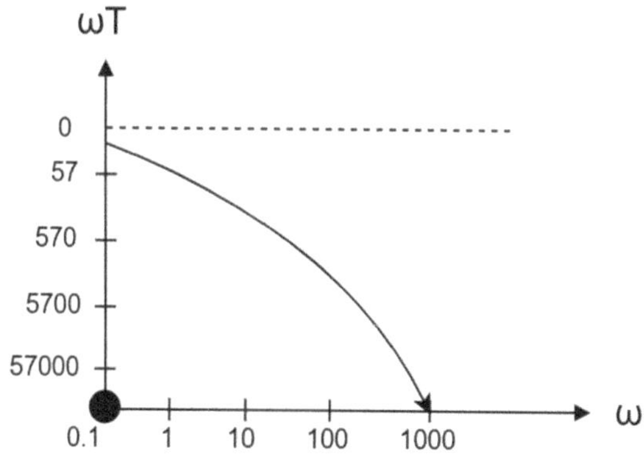


**FIGURE 9.12** Graph of the phase of transport delay.

### 9.5.3.1 Transportation Lag

Transportation lag is a non-minimum phase behavior and has an excessive phase lag with no attenuation at high frequencies. Consider

$$G(j\omega) = e^{-j\omega T}$$

The magnitude is always equal to unity.

$$\left|G(j\omega)\right| = \left|\cos \omega T - \sin \omega T\right| = 1$$

Therefore, the magnitude curve is at 0 *dB*. Figure 9.12 shows graph of the phase of transport delay.

$$G\,(j\omega) = -\omega T\ rad = -57.3\ \omega T\ degrees$$

The phase angle varies linearly with $\omega$.

**Example 9.11:** Draw the Bode Diagram for $G(j\omega) = \dfrac{e^{j\omega L}}{1 + j\omega T}$.

***Solution***

$$20\log\left|G(j\omega)\right| = 20\log\left|e^{-j\omega L}\right| + 20\log\left|\frac{1}{1 + j\omega T}\right|$$

$$= 0 + 20\log\left|\frac{1}{1 + j\omega T}\right|$$

The phase angle is

$$\angle G(j\omega) = \angle e^{-j\omega L} + \angle\frac{1}{1 + j\omega T}$$

$$= -\omega L - \tan^{-1}\omega T$$

This may be seen as follows:

$$G(j\omega)H(j\omega) = \frac{k_v}{j\omega} \text{ for } \omega \ll 1$$

Thus,

$$20\log\left|\frac{k_v}{j\omega}\right|_{\omega=1} = 20\log k_v$$

The intersection of the initial $-20dB\,/\,decade$ segment with the $0dB$ line has a frequency numerically equal to $k_v$. To see this, define the frequency at this intersection to be $\omega_1$, then

$$\left|\frac{k_v}{j\omega_1}\right| = 1,\ k_v = \omega_1$$

As an example, $G(s) = \frac{k}{s(J+F)}$. If we define the corner frequency to be $\omega_2$ and the frequency at the intersection of the $-40dB\,/\,decade$ segment with $0dB$ line to be $\omega_3$, then

$$\omega_2 = \frac{F}{J},\ \omega_3^2 = \frac{k}{J}$$

Since $\omega_1 = k_v = \frac{k}{F}$, it follows $\omega_1\omega_2 = \omega_3^2$ or $\frac{\omega_1}{\omega_3} = \frac{\omega_3}{\omega_2}$.

On the Bode diagram,

$$\log\omega_1 - \log\omega_3 = \log\omega_3 - \log\omega_2$$

Thus, $\omega_3$ is pushed midway between $\omega_1$ and $\omega_2$.

### 9.5.4 Relationship Between System Type and Log Magnitude Curve

The static position, velocity, and acceleration error constants define the low-frequency behavior of the type 0, 1, 2 system. For a given system, only one static error constant is finite and significant. The type of the system determines the slope of *log – magnitude* at low frequencies.

#### 9.5.4.1 Determination of Static Position Error Constants

Figure 9.13 shows an example of the log-magnitude plot of the type 0 system. In such a system, the magnitude of $G(j\omega)\,H(j\omega)$ equals $k_p$ at low frequencies.

Or

$$\lim_{\omega\to 0} G(j\omega)H(j\omega) = k_p \tag{9.24}$$

#### 9.5.4.2 Determination of Static Velocity Error Constants

Figure 9.14 shows an example of a log-magnitude curve of a type 1 system. Subsequently, for higher frequencies, the initial $-20\,dB\,/\,decade$ intersection drops to $-40\,dB\,/\,decade$. Then substituting $\omega_m = \frac{1}{\tau\sqrt{\propto}}$,

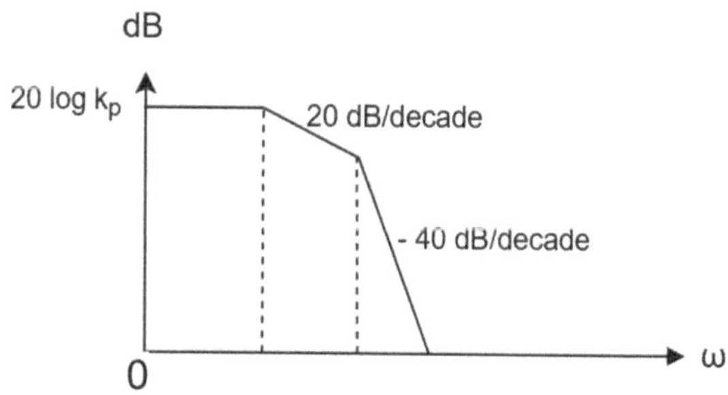


**FIGURE 9.13** Log-magnitude curve of type 0.

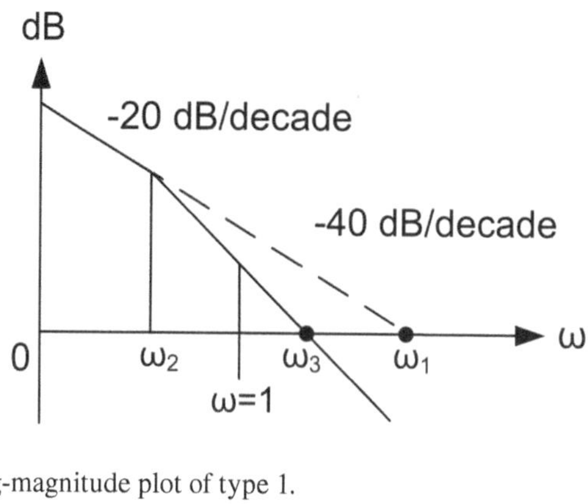


**FIGURE 9.14** Log-magnitude plot of type 1.

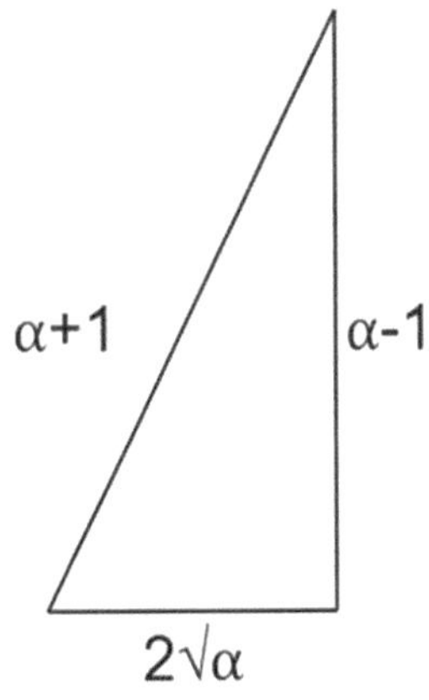


**FIGURE 9.15** Description of trigonometric identities.

$$\tan\varphi_m = \frac{\left(\propto\sqrt{\propto}\right)-\left(\frac{1}{\sqrt{\propto}}\right)}{1+1}$$

$$\tan\varphi_m = \frac{\propto -1}{2\sqrt{\propto}} \tag{9.25}$$

$$\sin\varphi_m = \frac{\propto -1}{\propto +1} \tag{9.26}$$

Figure 9.15 shows the description of trigonometric identities. This is very useful for calculating the necessary $\propto$ ratio between the compensator's pole and zero to provide the required maximum phase lead.

**Example 9.12:** A feedback control system has a loop transfer function:

$$GH(s) = \frac{k}{s(s+2)}$$

A steady-state error for a ramp input of less than 5% of the magnitude ramp is desired.

***Solution***

Therefore, we require

$$k_v = \frac{A}{ess} = \frac{A}{0.05A} = 20$$

### 9.5.5 Phase Lead Compensation

Consider the circuit diagram in Figure 9.16. To write down its transfer function to understand the phenomena of phase lead compensation, we can write

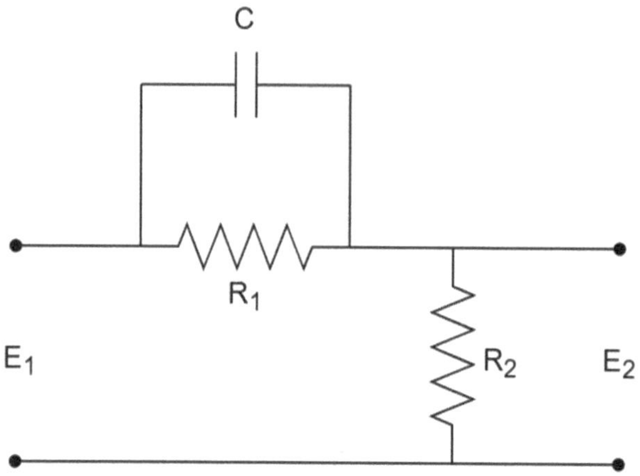


**FIGURE 9.16** Circuit diagram.

$$\frac{E_2}{E_1} = G(s) = \frac{R_2}{R_2 + \dfrac{R_1}{1 + R_1 cs}} \tag{9.27}$$

$$= \frac{R_2(1 + R_1 cs)}{R_2(1 + R_1 cs) + R_1}$$

$$= \frac{R_2}{R_1 R_2}\left[\frac{1 + R_1 cs}{1 + \dfrac{R_1 R_2}{R_1 + R_2} cs}\right]$$

$$\tau = \frac{R_1 R_2 c}{R_1 + R_2}, \quad \propto = \frac{R_1 + R_2}{R_2}$$

$$G_c(s) = \frac{1 + \propto \tau s}{\propto (1 + \tau s)}, \quad G(j\omega) = \frac{1 + R\tau\omega}{\propto (1 + j\tau\omega)} \tag{9.28}$$

The maximum value of the phase lead occurs at $\omega_m$ where $\omega_m$ is the geometric mean, $p = \frac{1}{\tau}$, $z = \frac{1}{\propto\tau}$ that is, maximum phase lead occurs halfway between the pole and zero frequencies on the logarithmic frequency scale.

$$\omega_m = \sqrt{zp} = \frac{1}{\tau\sqrt{\propto}} \tag{9.29}$$

To obtain the equation of max phase lead angle, the phase angle

$$\varphi = \tan^{-1}\left\{\frac{\propto \omega\tau - \omega\tau}{1+(\omega\tau)^2 \propto}\right\} \tag{9.30}$$

Calculating $\propto$, we obtain

$$\frac{\propto -1}{\propto +1} = \sin 30^0 = 0.5,\ \propto = 3$$

The $\phi_m$ occurs at $\omega_m$, and this frequency will be selected so that the new crossover frequency and $\omega_m$ coincide. The magnitude of the lead network at $\omega_m$ $10\log \propto (\propto + 3)$ is 4.8$d$B. The compensated crossover frequency is then evaluated where the magnitude of $GH(j\omega)$ is $-4.8\,d$B and thus,

$$\omega_m = \omega_c = 8.4\, rad/sec$$

Drawing the compensated magnitude line so that it intersects $0\,dB$ at $\omega = \omega_c = 8.4$, we can find $z = 4.8$.

$$p = \propto,\ z = 14.4$$

Therefore, compensation network is:

$$G_c(s) = \frac{1}{3}\frac{\left(1+\frac{s}{4.8}\right)}{\left(1+\frac{s}{14.4}\right)}$$

The total DC gain will be raised by 3 to account for $\propto$. Then compensated loop transfer function:

$$G_c(s)GH(s) = \frac{20\left(\frac{s}{4.8}+1\right)}{s(0.5s+1)\left(\frac{s}{14.4}+1\right)}$$

The crossover phase angle is now $-136.3^0$, and the total magnitude gain is $20\log \propto$.

**Example 9.13:** Assume the lead compensator $G_c(s)$ to be $G_c(s) = k_c \frac{s+\frac{1}{T}}{s+\frac{1}{\propto T}}$ $0 < \propto < 1$.

***Solution***

$$\frac{C(s)}{R(s)} = \frac{G_c(s)\frac{1}{s^2(s+1)}}{1+G_c(s)\frac{1}{s^2(s+1)}} = \frac{G_c(s)(s+1)}{s^2(s+1)+G_c(s)}$$

From the specifications given, as $\xi = 0.5$, and $\omega_n = 0.2$, the dominant CLP's are:

$$s = -0.1 \pm j0.1732$$

The angle deficiency is found to be $-70.89^0$ from the graphical approach. The poles and zeros of the lead compensator are found to be –0.4791 and –0.08349, respectively. Thus,

$$G_c(s) = k_c \frac{s+0.08349}{s+0.4791}$$

The value of $k_c$ can be determined from

$$k_c = \left| \frac{(s+0.4891)s^2(s+1)}{s+0.08349} \right|_{s=-0-1-j0.1732}$$

$$k_c = 0.08782$$

$$G_c(s) = 0.08732\frac{s+0.08349}{s+0.4791}$$

**Example 9.14:** A transfer function is $G(s) = \frac{10}{s(s+2)(s+8)}$. Given $k_c = \frac{10}{16}$ and $k_v = 80$ is required, find $G_c(s)$.

***Solution***

We need to increase the open loop gain by 128. This implies that we need a lag compensator. The root locus of the uncompensated system reveals that it is not possible to bring the dominant CLP's to $-2 \pm j2\sqrt{3}$ by just a gain adjustment (also needs a lead compensator). So assume that the transfer function of a lead-lag compensator is:

$$G_c(s) = k_c \frac{\left(s+\frac{1}{T_1}\right)\left(s+\frac{1}{T_2}\right)}{\left(s+\frac{\beta}{T_1}\right)\left(s+\frac{1}{\beta T_2}\right)}$$

The angle deficiency at the desired CLP's $s=-2+j2\sqrt{3}$ is:

$$\text{Angle } = 120^0 + 90^0 + 30^0 - 180^0 = 60^0$$

The lead portion of the lag-lead compensator must contribute to this angle. Since the choice of $T_1$ is not unique. Let $\frac{1}{T_1} = 3.65$. The pole $\frac{\beta}{T_1}$ is determined from the requirement that

$$\angle \left. \frac{s+3.65}{s+\dfrac{\beta}{T_1}} \right|_{s=-2+j2\sqrt{3}} = \angle \frac{-2+j2\sqrt{3}+3.65}{-2+j2\sqrt{3}+\dfrac{\beta}{T_1}} = 60$$

$$\frac{\beta}{T_1} = 45.7, \ \beta = 12.52$$

For the lag portion of the compensator, we may choose $\dfrac{1}{\beta T_2} = 0.01$

$$\frac{1}{T_2} = 0.1252$$

Noting that $\left. \left| \dfrac{s+0.1252}{s+0.01} \right| \right|_{s=-2+j2\sqrt{3}} = 0.986$

$$\angle \left. \frac{s+0.1252}{s+0.01} \right|_{s=-2+j2\sqrt{3}} = -1.450$$

So that,

$$G_c(s) = 128\left(\frac{s+3.65}{s+45.7}\right)\left(\frac{s+0.1252}{s+0.01}\right)$$

## 9.6 NYQUIST STABILITY CRITERION

It is a frequency domain method and has the following desirable features:

a) It provides the same amount of information on a control system's absolute stability as the Routh-Hurwitz Criterion.
b) In addition to the absolute system stability, the Nyquist criterion indicates a stable system's degree of stability and how the system stability may be improved.
c) It gives information on the frequency domain response of the system.
d) It can be used to stabilize the system during time delays.

e) It can be modified for nonlinear systems.
f) In short, the Nyquist criterion is a graphical method for determining the stability of a closed-loop system by investigating the properties of the frequency-domain plots of $G(s)\,H(s)$. The following is relevant:

   i. Identification of poles and zeros.
      Loop transfer function zeros = zeros of $G(s)\,H(s)$
      Loop transfer function poles = poles of $G(s)\,H(s)$
      Closed loop transfer function poles = zeros of $1+G(s)\,H(s)$
   ii. The poles of $1+G(s)\,H(s)$ are the same as the poles of $G(s)\,H(s)$.
   iii. For a closed loop system to be stable, there is no restriction on the location of poles and zeros of the loop transfer function $G(s)\,H(s)$, but the poles of the closed-loop transfer function or the roots of a characteristic equation must all be in the left-hand plane (LHP).

Consider a typical closed-loop control system with feedback as given in Figure 9.17.

$$G(s)H(s)=\frac{k(s+a)}{s(s+b)(s+c)(s+d)} \tag{9.31}$$

Poles of $G(s)H(s)=s(s+b)(s+c)(s+d)$.
Zeros of $G(s)H(s)=k(s+a)$.

$$\frac{G(s)}{1+G(s)H(s)}=\frac{\dfrac{k(s+a)}{s(s+b)(s+c)}}{1+\dfrac{s(s+a)}{s(s+b)(s+c)(s+d)}}=\frac{k(s+a)}{s(s+b)(s+c)(s+d)+k(s+a)}$$

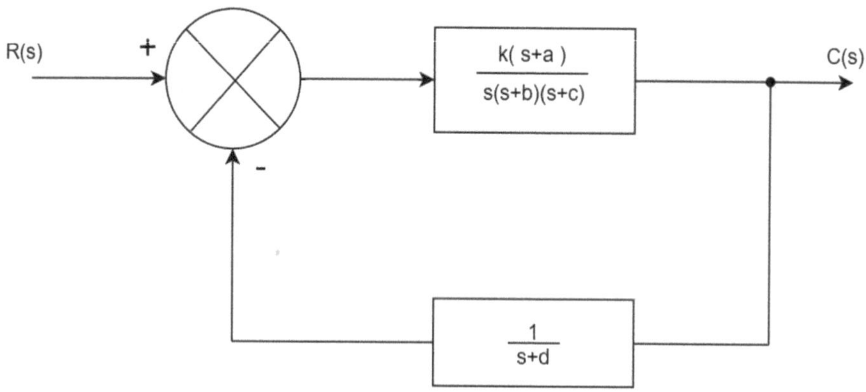


**FIGURE 9.17** Typical block diagram of feedback control system.

$$1+G(s)H(s)=1+\frac{s(s+a)}{s(s+b)(s+c)(s+d)}$$

$$=\frac{s(s+b)(s+c)(s+d)+k(s+a)}{s(s+b)(s+c)(s+d)} \tag{9.32}$$

### 9.6.1 Definitions

- **Encircled:** A point is said to be encircled by a closed path if it is found inside the path as shown in Figure 9.18.
- **Enclosed:** A point or region is said to be enclosed by a closed path if it is found to lie to the left of the path when the path is traversed in a prescribed direction as shown in Figure 9.19.

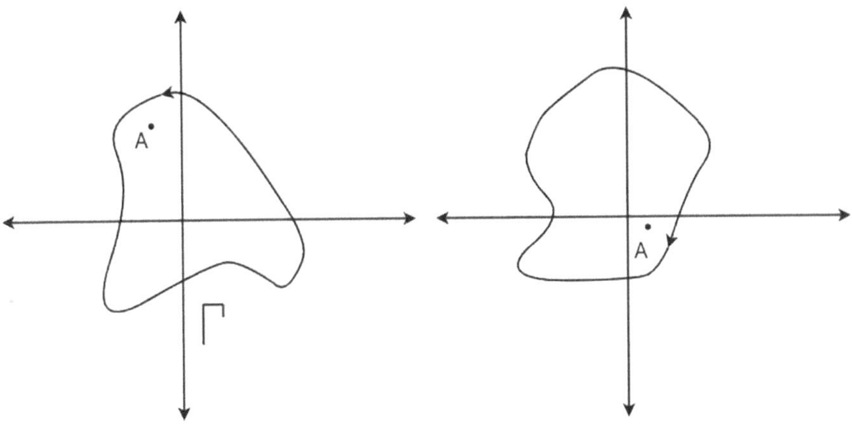


FIGURE 9.18 Diagrams defining the encirclement of a point.

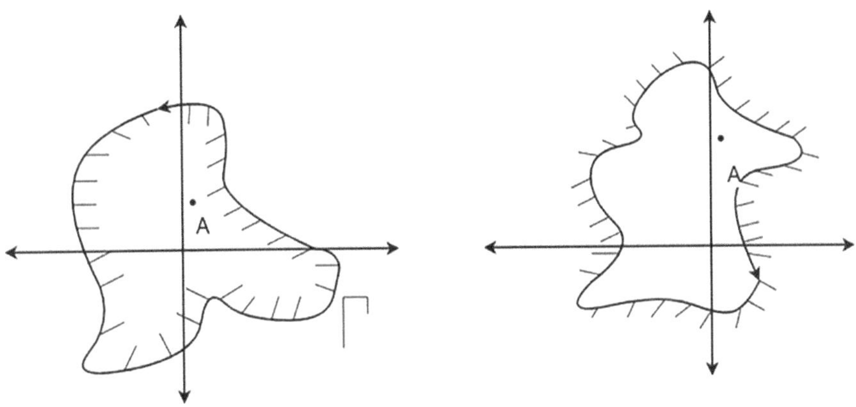


FIGURE 9.19 Diagrams showing enclosure of a point.

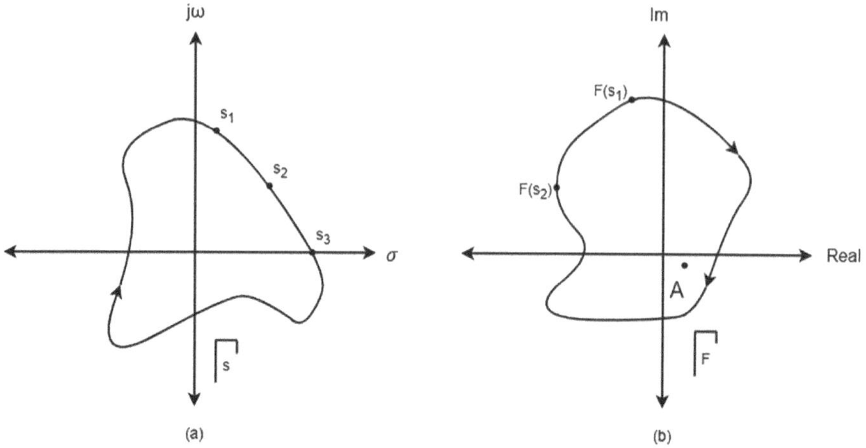


**FIGURE 9.20** The mapping of a function: (a) arbitrary chosen s-plane locus and (b) corresponding F(s) locus.

### 9.6.2 The Principle of Argument

Figure 9.20 shows the mapping of a function.

Let $F(s)$ be a single-valued rational function that is analytic everywhere in a specified region except at a finite number of points in the s-plane. For each point analytically in the selected area in the s-plane, suppose that a continuous closed path $s$ is chosen in the S-plane. If all the points on $s$ are in the specified region [see Figure 9.20(a)] in which $F(s)$ is analytic, the curve $F$ mapped by $F(s)$ into the F-plane is also a closed one [see Figure 9.20(b)]. Note that the direction may be clockwise or counterclockwise.

The corresponding $F(s)$ locus mapped in the F-plane will encircle the origin as many times as the difference between the number of zeros and the number of poles of $F(s)$ that are encircled by the s-plane locus $s$. In other words:

$$N = Z - P \tag{9.33}$$

Where $N$ =number of encirclements of the origin made by $F(s)$ locus $F$ in the F-plane.

$Z$ = number of zeros of $F(s)$ encircled by the s-plane locus $s$ in the s-plane.

$P$ = Number of poles of $F(s)$ encircled by the s-plane locus $s$ in the s-plane.

In general, $N$ can be positive ($Z > P$), zero ($Z = P$), or negative ($Z < P$).

1. $N > 0 (Z > P)$, if s-plane locus encircles more zeros than poles of $F(s)$ in a certain direction (*CW or CCW*). When $N$ is a positive integer, the $F(s)$ plane locus will encircle the origin $N$ times in the same direction as that of $s$.
2. $N = 0 (Z = P)$, if the s-plane locus encircles as many poles as zeros or encircles no poles or zeros of $F(s)$. The F-plane locus $F$ will not encircle the origin.

3. $N < 0 (Z < P)$, if s-plane locus encircles more poles than zeros of $F(s)$ in a certain direction (*CW or CCW*). When $N$ is a negative integer, the $F(s)$ plane locus will encircle the origin $N$ times in the opposite direction as that of $s$.

The principle of argument will become apparent from the following:

$$F(s) = \frac{(s+z_1)(s+z_2)}{(s+p_1)(s+p_2)(s+p_3)} \tag{9.34}$$

$$F(s) = \frac{\prod_{m=1}^{2} |(s+z_m)|}{\prod_{n=1}^{3} |(s+p_n)|} \left[ \sum_{m=1}^{2} \angle s + z_m - \sum_{n=1}^{3} \angle s + p_n \right] \tag{9.35}$$

$$F(s) = |F(s)| \angle F(s) \tag{9.36}$$

Let us now assume that the pole-zero configuration of $F(s)$ is as shown in Figure 9.21. $s$ is the s-plane locus, and $s_1$ is an arbitrary point on the closed path $s$. The factor $(s + z_1)$can be represented graphically by the vector drawn from $z_1$ to $z_2$.

g) Now point $s_1$ is moved along $s$ in the counterclockwise direction until it returns to the starting point. The angles generated by the vectors drawn from the poles and zeros are not encircled by $s$ when $s_1$ completes one round trip are zero, while the vector $s + z_2$ drawn from $-z_2$ which is encircled by $s$, generates a positive angle of $2\pi$ rad. Then, $\angle F(s) = 2\pi$. This means the corresponding $F(s)$ locus must go around the origin $2\pi$

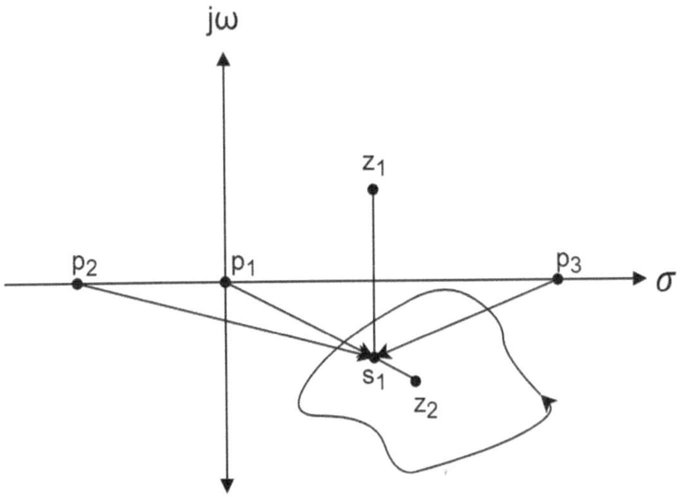


**FIGURE 9.21** The encirclement of pole or zero.

rad in the CCW direction. In general, if there are $N$ poles more zeros than poles of $F(s)$ which are encircled by the $s$ in the CCW direction, then

$$\angle F(s) = 2\pi(Z - P) = 2\pi N \tag{9.37}$$

This implies that $F$ will encircle the origin $N$ times in the CCW direction. Conversely, if $N$ more poles than zeros are encircled by the s-plane locus in the CCW direction, the $F$ locus must encircle the origin $N$ times in the CW direction. Hence,

$$\angle F(s) = 2\pi(Z - P) = 2\pi N \text{ (Negative number)} \tag{9.38}$$

Table 9.6 shows the summary of the argument principle.

### 9.6.3 The Nyquist Path

The main objective in the stability analysis of the feedback control system is to determine if any root of the characteristic equation lies in the right-hand plane (RHP). It is easy to see that the principle of arguments can be used for this purpose, provided that the s-plane locus is defined as shown in Figure 9.22.

**Section 1**: $s = +j\infty \text{ to } j\omega_1$ along the $j\omega$ axis.
**Section 2**: $s = +j\omega_1^+ \text{ to } j\omega_1^-$ along the small circle.
**Section 3**: $s = +j\omega_1^- \text{ to} + j0^+$ along the $j\omega$ axis.
**Section 4:** $s = +j0^+ \text{ to } j0^-$ along the semicircle.
**Section 5:** $s = -j0^- \text{ to} - j\omega_1^+$ along the $j\omega$ axis.
**Section 6:** $s = -j\omega_1^+ \text{ to} - j\omega_1^-$ along the semicircle.
**Section 7:** $s = -j\omega_1^- \text{ to} - j\infty$ along the $j\omega$ axis.
**Section 8:** $s = -j\infty \text{ to } j\infty$ along the infinite semicircle.

**TABLE 9.6**
**Summary of Argument Principle**

| | | F-plane Locus | |
|---|---|---|---|
| **N = Z–P** | **Sense of s-plane** | **Number of Encirclements of the Origin** | **Direction of Encirclement** |
| N > 0 | Clockwise | N | Clockwise |
| | Counterclockwise | | Counterclockwise |
| N < 0 | Clockwise | N | Counterclockwise |
| | Counterclockwise | | Clockwise |
| N = 0 | Clockwise | 0 | No encirclement |
| | Counterclockwise | | No encirclement |

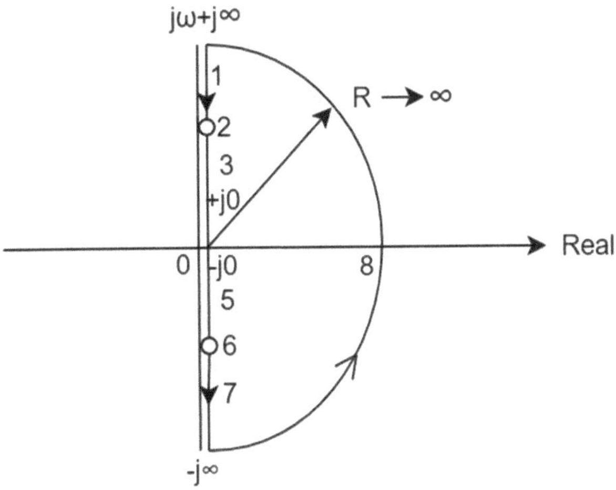


**FIGURE 9.22** Nyquist path.

Since the Nyquist Path should not pass through any singularity of $F(s)$, the small semicircle along the imaginary axis and at the origin are necessary if $F(s)$ has poles on the $j\omega$ axis and at the origin. It is apparent that the Nyquist path must enclose any poles and zeros in the RHP.

### 9.6.4 The Nyquist Criterion and the GH Plot

The Nyquist criterion directly applies to the principle of argument when the s-plane locus is the Nyquist path. Once the Nyquist path is specified, the stability of the closed-loop system can be determined by plotting $F(s) = 1 + G(s)H(s)$ locus when s takes on value along the Nyquist path and investigating the behavior of $F(s)$ plot with respect to the origin in the F-plane.

$$F(s) = 1 + G(s)H(s) = \frac{(s+z_1)(s+z_2)\ldots(s+z_m)}{sj(s+p_1)(s+p_2)\ldots(s+p_n)} \tag{9.39}$$

It is important to note that the poles of $F(s)$ are the poles of $G(s)H(s)$, and the zeros of $F(s)$ are the roots of thc characteristic equation.

$$G(s)H(s) = \frac{(s+3)}{s(s+1)(s+2)} \quad , F(s) = 1 + G(s)H(s)$$

$$= \frac{s(s+1)(s+2)(s+3)}{s(s+1)(s+2)} \tag{9.40}$$

The procedure for the determination of stability is the following:

a) Define Nyquist path
b) The $F(s) = 1 + G(s)H(s)$ locus is plotted in the F-plane corresponding to the Nyquist path.
c) If poles of $G(s)H(s)$ are all located in LHP or on $j\omega$ axis, the closed loop system is stable provided the Nyquist plot of $F(s)$ does not encircle the origin, since in this case, given that $P = 0$.

$$N = Z - P = Z$$

Therefore, for a stable system, there will be no zero of $F(s)$ for the Nyquist path to be enclosed $(N = Z = 0)$.

If $N \neq 0$, $N = Z$, it must be a positive integer, and the origin in the F-plane will be encircled in a positive direction (enclosed). However, if some of the poles of $G(s)H(s)$ are in the RHP, then

$$P \neq 0,\ N = Z - P$$

Since for a stable system, $Z$ must always be zero, the condition for a stable system is $N = -P$. This implies that the Nyquist plot of $F(s)$ must encircle the origin as many times as the number of $P$. The encirclement if any must be made in the clockwise direction.

It is easier to plot $G(s)H(s)$ than $F(s)$. The difference between $F(s)$ and $G(s)H(s)$ plots is simply a shift of the imaginary axis. The $(-1+j0)$ point in the GH-plane corresponds to the origin in the F-plane, as shown in Figure 9.23.

The Nyquist criterion is normally defined with respect to the GH plot. As for a stable closed loop system, the Nyquist plot of GH should encircle the $(-1+j0)$ point as many times as there are poles of GH in the RHP; the encirclement, if any, should

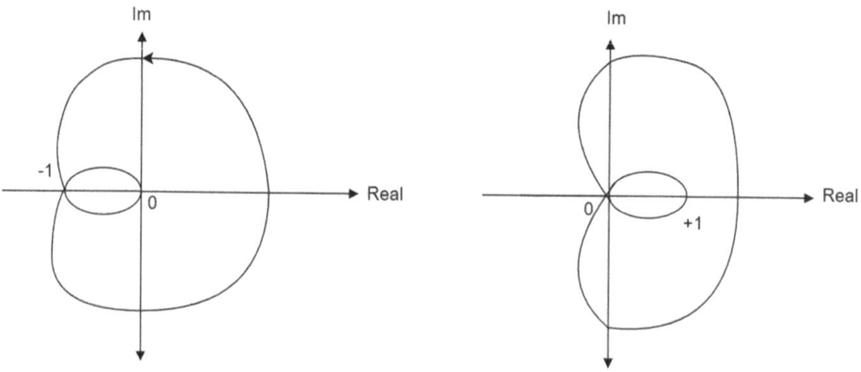


**FIGURE 9.23** Shifting the plot to $-1 + j0$.

be made in the clockwise direction. Furthermore, if $N$ = number of encirclements of the critical points $(-1+j0)$ in the GH-plane made by the Nyquist plot of GH, then $-N$ is for clockwise and $+N$ is for counterclockwise encirclement.

$$Z = Number\ of\ zeros\ of\ 1+G(s)H(s)\ in\,the\ RHP$$

$$P = Number\ of\ poles\ of\ 1+G(s)H(s)\ in\,the\ RHP$$

$$N = Z - P$$

For a stable system, $Z$ must be zero, and consequently $N = -P$.

In most cases, $G(s)$ and $H(s)$ are stable functions, so that $P = 0$ and $N = 0$. This implies that $G(s)H(s)$ must not encircle the $(-1+j0)$ point in the GH plane for a system to be stable. However, if in the case where the GH plot encircles the critical point $(-1+j0)$, $N = Z \neq 0$, the encirclement would be made in counterclockwise directions.

Let us now define that if the loop gain function, $G(s)H(s)$ is stable, $P = 0$, for a stable closed loop system, the Nyquist plot of $G(s)H(s)$ must not enclose the critical point $(-1+j0)$. Figure 9.24 shows a typical Nyquist plot which is for frequencies from $\omega = \infty$ to $\omega = 0$ and covers only the Nyquist path from $\omega = \infty$ to $\omega = 0$. Thus, the complete Nyquist plot is not required unless it is necessary to know how many zeros are located in the RHP. It is only necessary to investigate whether the $(-1+j0)$ point is to the left or the right of the GH curve going from infinite to zero frequency along the GH locus.

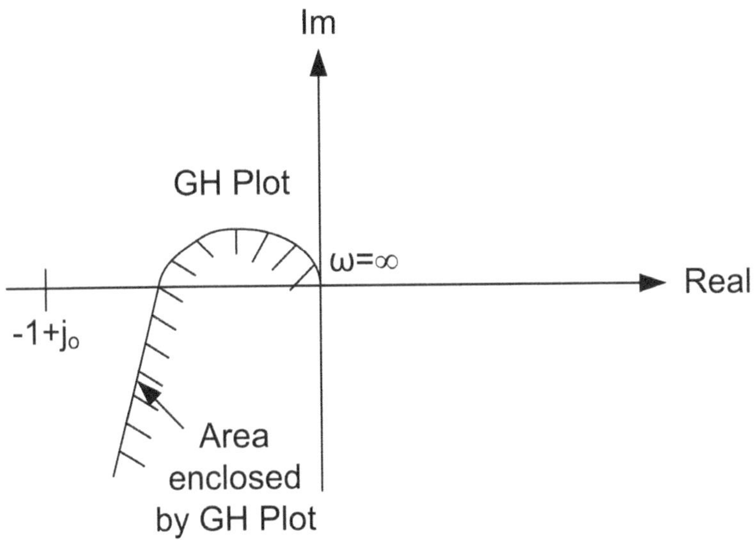


**FIGURE 9.24** A typical Nyquist plot.

## 9.7 APPLICATIONS OF NYQUIST CRITERION

Let us see some of the applications of the Nyquist Criterion as follows.

**Example 9.15:** Consider a single loop feedback control system with the loop transfer function

$$G(s)H(s) = \frac{k}{s(s+a)}$$

***Solution***

$G(s)H(s)$ does not have any poles in RHP. So $P = 0$. The Nyquist path is shown in Figure 9.25, and the phasor may represent the point on the semicircle at the origin.

$$s = \in e^{j\theta}$$

where $\in (\in \to 0)$ and $\theta$ denote the magnitude and phase of the phasor, respectively. As the Nyquist path is traversed from $+j0$ to $-j0$ along section 2, the phase rotates the clockwise direction through 180° (varies from +90° to −90°).

Thus

$$G(s)H(s)\big|_{s=\in e^{j\theta}} = \frac{k}{\in e^{j\theta}\left(\in e^{j\theta} + a\right)}$$

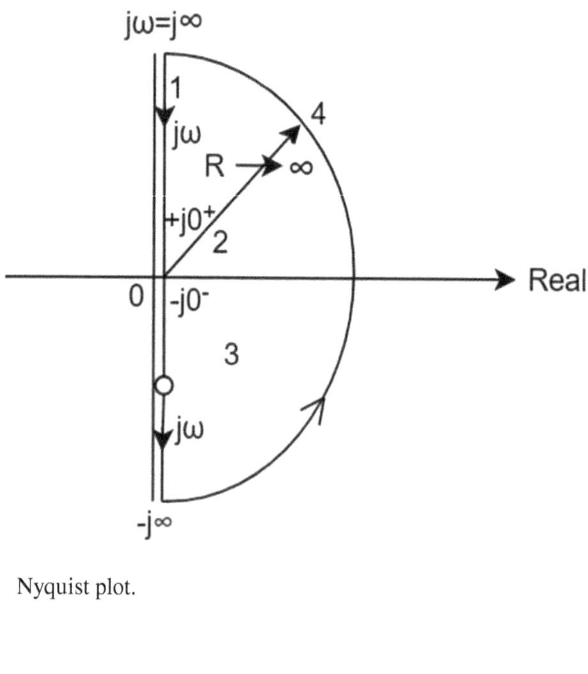


FIGURE 9.25 Nyquist plot.

Since $\in \rightarrow 0$

$$G(s)H(s)\Big|_{s=\in e^{j\theta}} = \frac{k}{a \in e^{j\theta}} = \infty e^{-j\theta}$$

This indicates that all the points on the Nyquist plot of $G(s)H(s)$ that correspond to section 2 of the Nyquist path have an infinite magnitude, and the corresponding phase is opposite to that of the s-plane locus.

So, at $+j0$, $\infty$ , $-90^0$

$-j0$ $\infty$, $+90^0$

Region 4 may also be computed as follows.

$$G(s)H(s)\Big|_{s=Re^{j\theta}_{R\to\infty}} = \frac{k}{Re^{j\theta}\left(Re^{j\theta}+a\right)}$$

$$= \frac{k}{R^2 e^{2j\theta}} = 0 * e^{-j2\theta}$$

This implies that the behavior of *the* $G(s)H(s)$ plot at infinite frequency is described by a phasor with an infinitesimally small magnitude that rotates around the origin $2*180^0 = 360^0$ in the clockwise direction $(+180^0\ to - 180^0)$.

The GH plots corresponding to section 2 and section 4 are shown in Figure 9.26.

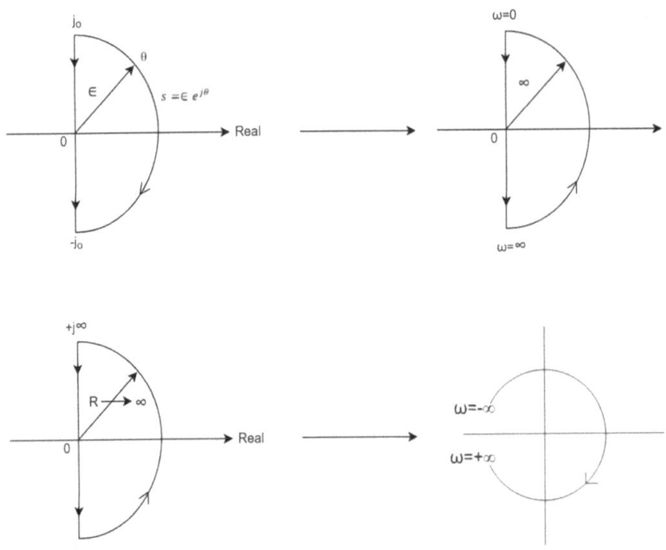


**FIGURE 9.26** The GH plot corresponding to sections 2 and 4.

**Example 9.16:** Let us consider the control system shown in Figure 9.27.

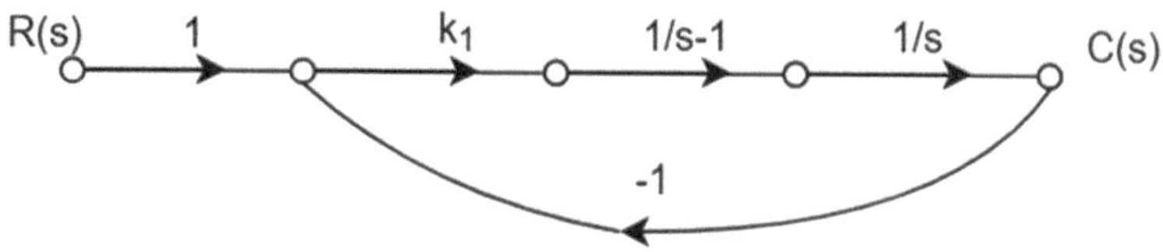


**FIGURE 9.27** A feedback control system.

***Solution***

$$GH(s) = \frac{k_1}{s(s-1)},\ P = 1$$

For this system to be stable, we require $N = -P = -1$, and one counterclockwise encirclement of the $-1$ point.

$$\lim_{\in \to 0} GH(s) = \lim_{\in \to 0} \frac{k_1}{- \in e^{j0}}$$

$$= \lim_{\in \to 0} \left| \frac{k_1}{\in} \right| \angle -180 - \varphi$$

Figure 9.28 shows the Nyquist plot of Figure 9.27.

$$GH(j\omega) = \frac{k_1}{j\omega(j\omega - 1)} = \frac{k_1}{\omega\left(\omega^2 + 1\right)^{\frac{1}{2}}} \angle -\frac{\pi}{2} - \tan^{-1}(-\omega)\left(\pi - \tan^{-1}(\omega)\right)$$

$$= \frac{k_1}{\omega\left(\omega^2 + 1\right)^{\frac{1}{2}}} \angle \frac{\pi}{2} + \tan^{-1}(\omega)$$

$$N = +1$$

$$P = +1$$

$$N = Z - 1$$

$$+1 = Z - 1$$

$$Z = 2$$

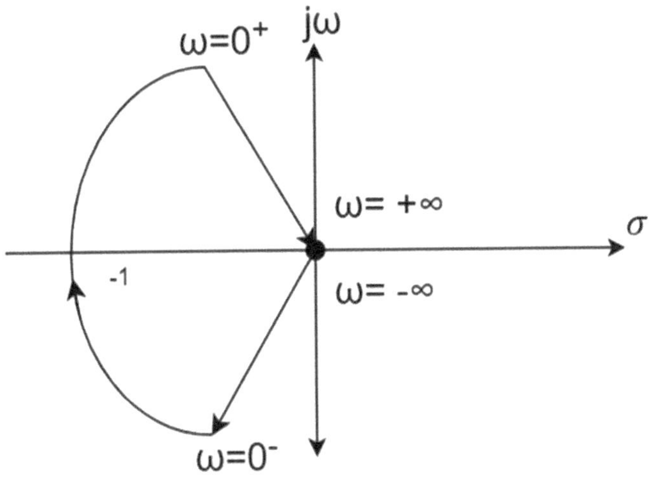


**FIGURE 9.28** Nyquist plot of Figure 9.27.

The system is unstable.

**Example 9.17:** Consider $G(s)H(s)=\dfrac{k(s-1)}{s(s+1)}$.

***Solution***

The desired Nyquist path is given in Figure 9.29. Consider section 2, $s=\in e^{j\theta}$

$$G(s)H(s)=\frac{k(s-1)}{s(s+1)}=\lim_{\in\to 0}\frac{k\left(\in e^{j\theta}-1\right)}{\in e^{j\theta}\left(\in e^{j\theta}+1\right)}$$

$$=-\frac{k}{\in e^{j\theta}}=\infty\in e^{-j(\theta+\pi)}$$

At $\omega=+0,\ \theta=+90^0$,

$$G\left(\in e^{j\theta}\right)H\left(\in e^{j\theta}\right)=\infty e^{-j(90+180)}$$

$$=\infty,\ e^{-j(90+180)}$$

$$=\infty,+90^0$$

At $\omega=-0,\ \theta=-90^0$,

$$G\left(\in e^{j\theta}\right)H\left(\in e^{j\theta}\right)=\infty e^{-j(-90+180)}$$

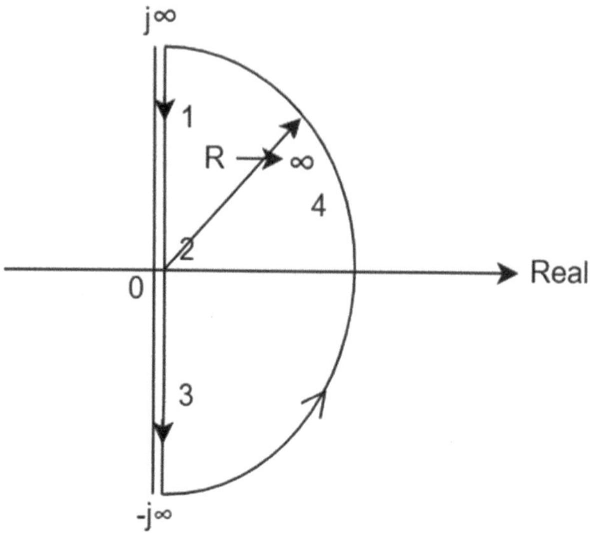


**FIGURE 9.29** Nyquist plot.

$$= \infty,\ e^{-j(90)}$$

$$= \infty, -90^0$$

For Section 4, $s = Re^{j\theta}$

$$G\left(Re^{j\theta}\right)H\left(Re^{j\theta}\right) = \frac{K\left(Re^{j\theta} - 1\right)}{Re^{j\theta}\left(Re^{j\theta} + 1\right)} = \frac{KRe^{j\theta}}{R^2 e^{j2\theta}} = 0, e^{-j\theta}$$

In section 4, $\theta$ varies from $-90^0$ to $+90$

$$G\left(Re^{j\theta}\right)H\left(Re^{j\theta}\right)\Big|_{\theta=-90} = 0,\left(90^0\right)$$

$$G\left(Re^{j\theta}\right)H\left(Re^{j\theta}\right)\Big|_{\theta=+90} = 0,\left(-90^0\right)$$

Open loop poles

$$z = 1,\ p = 0$$

$$N = Z - P$$

$$1 = Z - 0$$

$$Z = N$$

Figure 9.30 shows Nyquist plot for Example 9.17.

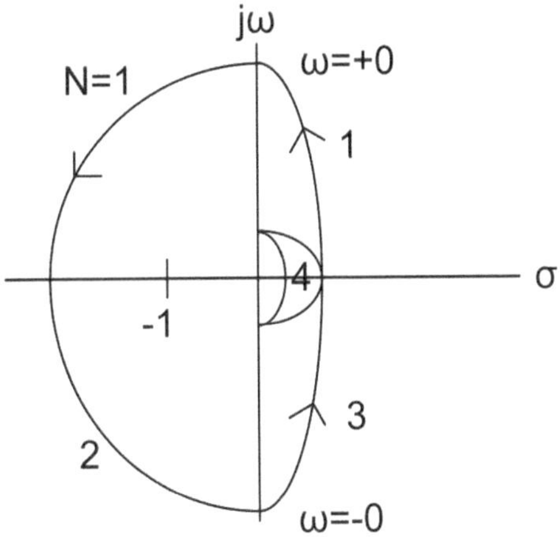


**FIGURE 9.30** Nyquist plot of Example 9.17.

**Example 9.18:** Consider a feedback control system

$$G(s)H(s)=\frac{k}{s(1+s)(1+2s)(1+3s)}$$

***Solution***

Table 9.7 shows the interpretation for G(s)H(s).
For section 2,

$$s=\in e^{j\theta}$$

$$=G\left(\in e^{j\theta}\right)H\left(\in e^{j\theta}\right)=\frac{k}{\in e^{j\theta}\left(1+\in e^{j\theta}\right)\left(1+2\in e^{j\theta}\right)\left(1+3\in e^{j\theta}\right)}$$

$$=\frac{k}{\in e^{j\theta}}=\infty * e^{-j\theta}$$

Figure 9.31 shows the Nyquist path.
When $\theta$ varies from $+90^0$ to $-90^0$, $\infty\angle-90^0$

$$\infty\angle+90^0$$

For Section 4, $s=\in e^{j\theta}$

$$GH\big|_{\left(s=Re^{j\theta}\right)_{R\to\infty}}=\frac{k}{Re^{j\theta}\left(1+Re^{j\theta}\right)\left(1+2Re^{j\theta}\right)\left(1+3Re^{j\theta}\right)}$$

**TABLE 9.7**
**Interpretation for G(s)H(s) for Example 9.18**

| Z | P | N | Interpretation |
|---|---|---|---|
| 0 | 0 | 0 | The GH plot does not encircle the origin |

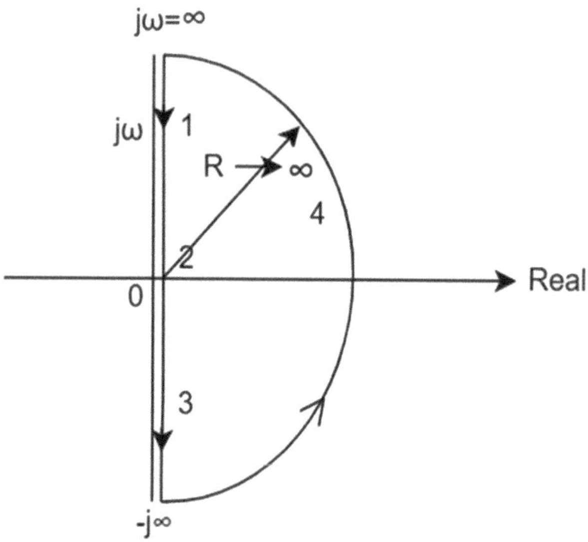


**FIGURE 9.31** Nyquist path.

$$= \frac{k}{R^4\, e^{j4\theta}} = 0 * e^{-4j\theta}$$

When $\theta = -90^0$ to $+90^0$ $\quad 0 * \angle 360^0$

$$0 * \angle -360^0$$

Table 9.8 shows four different areas of analysis after applying Nyquist criterion. Figure 9.32 shows the Nyquist plot. The Nyquist plot of $G(s)H(s)$ will enclose the $(-1 + j0)$ point twice if the value of $k$ is sufficiently large.

$$G(j\omega)H(j\omega) = \frac{k}{j\omega(1 + j\omega)(1 + 2j\omega)(1 + 3j\omega)}$$

$$= \frac{k}{\omega\sqrt{1+\omega^2} * \sqrt{1+4\omega^2} * \sqrt{1+9\omega^2}} \angle -90^0 - \tan^{-1}\omega - \tan^{-1}2\omega - \tan^{-1}3\omega$$

**TABLE 9.8**
**Four Different Areas of Analysis after Applying Nquist Criterion in Example 9.18**

| s | 0 | + j0 | − j0 | s-plane Nyquist path −180° |
|---|---|---|---|---|
| GH | $\frac{k}{s}$ | $\infty\angle-90$ | $\infty\angle+90$ | GH plane Nyquist plot +180 rotation |
| S | $\infty$ | $-j\infty$ | $j\infty$ | $+180^0$ |
| GH | $\frac{k}{s^4}$ | $0\angle 360^0$ | $0\angle-360^0$ | $-720^0$ rotation |

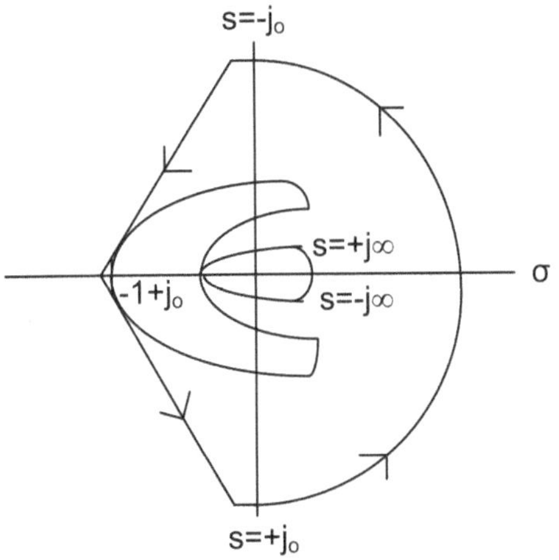


**FIGURE 9.32** Nyquist path.

**Example 9.19:** Consider a feedback control system with the loop transfer function, $G(s)H(s)$ given as:

$$G(s)H(s) = \frac{5}{s(1-s)}$$

$$1+G(s)H(s) = \frac{-s^2+s+5}{s(1-s)}$$

***Solution***

Table 9.9 shows the interpretation for G(s)H(s).
Table 9.10 shows the interpretation for 1+G(s)H(s).
Figure 9.33 shows the Nyquist path and Nyquist plot.

**TABLE 9.9**
**Interpretation for G(s)H(s) for Example 9.14**

| Z | P | N | Interpretation |
|---|---|---|---|
| 0 | 1 | −1 | The $G(s)H(s)$ plot encircles the origin once in the CCW direction. |

**TABLE 9.10**
**Interpretation for 1 + G(s)H(s) for Example 9.14**

| Z | P | N | Interpretation |
|---|---|---|---|
| 1 | 1 | 0 | The $G(s)H(s)$ plot does not encircle the $(-1+j0)$ point at all |

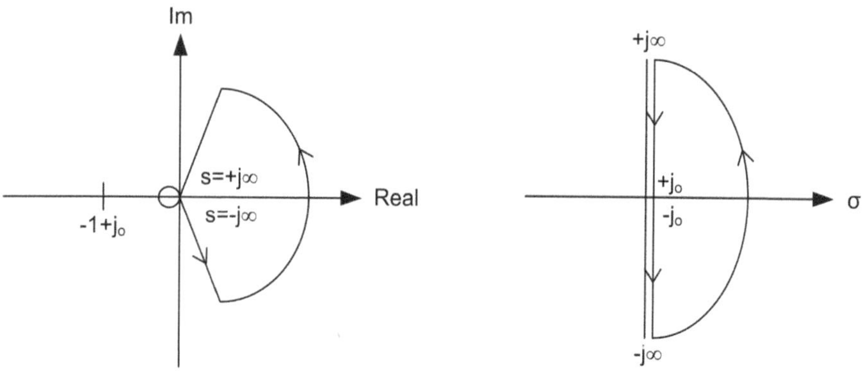


**FIGURE 9.33** Nyquist path and Nyquist plot.

$$N = -1$$

$$Z = 0$$

$$P = 1$$

At, $s = \in e^{j\theta}$

$$G\left(\in e^{j\theta}\right)H\left(\in e^{j\theta}\right) = \lim_{\in \to 0} \frac{5}{\in e^{j\theta}\left(1 - \in e^{j\theta}\right)} = \frac{5}{\in e^{j\theta}} = \infty * \in e^{-j\theta}$$

$\theta = 90^0$ $\qquad \infty, -90^0$

$\theta = -90^0$ $\qquad \infty, +90^0$

At, $s = Re^{j\theta}$

$$G\left(Re^{j\theta}\right)H\left(Re^{j\theta}\right) = \frac{5}{Re^{j\theta}\left(1 - Re^{j\theta}\right)} = \frac{5}{-R^2e^{j2\theta}} = 0 * e^{-2j\theta},\ 0\angle 180^0$$

**Example 9.20:** Consider the control system shown in Figure 9.34. It is desired to determine the range of $k$ for which the system is stable.

***Solution***

The open loop transfer function is:

$$\frac{C(s)}{E(s)} = G(s) = \frac{10k(s+2)}{s^3 + 3s^2 + 10}$$

Since the function does not have poles that are zeros on the $j\omega$ axis, the Nyquist path can consist of only three sections.

For Section 3, $s = Re^{j\theta}$

$$\lim_{R\to\infty} G(s) = \lim_{R\to\infty} \frac{10k\left(Re^{j\theta} + 2\right)}{R^3e^{j3\theta} + R^2e^{j2\theta} + 10}$$

$$= 0e^{-j2\theta}$$

Figure 9.35 shows the Nyquist path.

As the phasor for section 3 transverses from −90° to+90°, the corresponding Nyquist plot of $G(s)$ is traced by a phasor of +180° to −180° through a 360° clockwise sense.

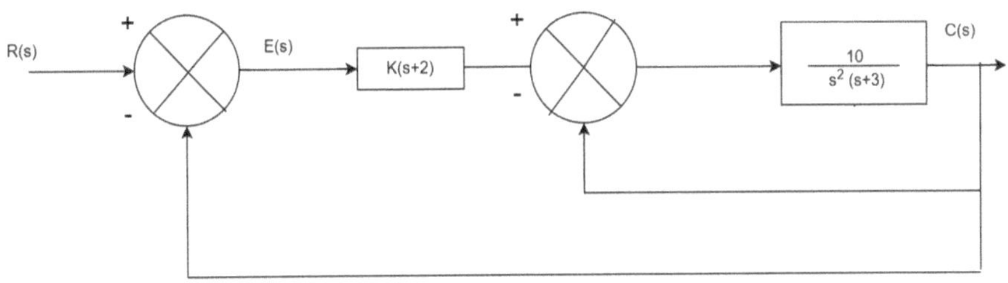


**FIGURE 9.34** A feedback control system.

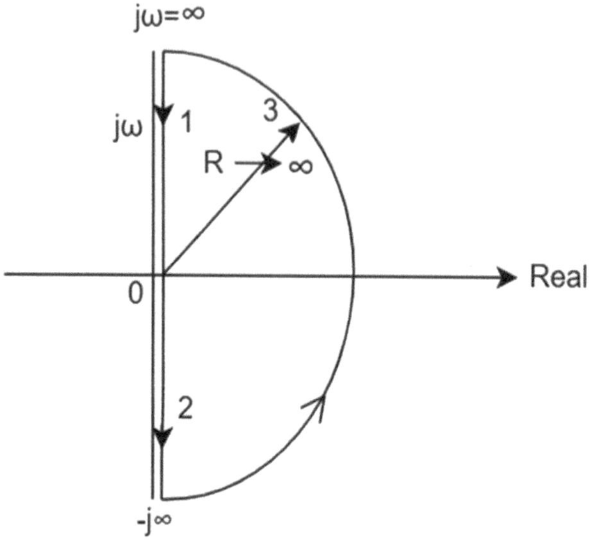


**FIGURE 9.35** Nyquist path.

For Section 1, $s = j\omega$

$$G(j\omega) = \frac{10(j\omega+2)}{-j\omega^3 - 3\omega^2 + 10} = \frac{10(j\omega+2)}{\left(10-3\omega^2\right) - j\omega^3}$$

$$= 10\left\{\frac{(j\omega+2)\left(10-3\omega^2+j\omega^3\right)}{\left(10-2\omega^2\right)+\omega^6}\right\}$$

$$= \frac{10k\left\{2\left(10-3\omega^2\right)-\omega^4+j\omega\left(10-3\omega^2\right)+2j\omega^3\right\}}{\left(10-3\omega^2\right)+\omega^6}$$

Setting the imaginary part to zero, $\omega = 0$

$$j\omega\left(10-3\omega^2\right)+j2\omega^3 = 0$$

$$10-3\omega^2+2\omega^3 = 0$$

$$\omega^2 = 10$$

$$\omega = \pm\sqrt{10}\, rad / sec$$

Which corresponds to frequencies at the intersects of the real axis of the $G(s)$ plane. For determining the intersect at the imaginary axis, let the real part be zero.

$$\omega^4 + 6\omega^2 - 20 = 0$$

$$\omega = \pm 1.54\, rad / sec$$

The intersect on the imaginary axis is

$$G(j1.54) = j5.43k$$

And the intersect on the real axis

- $G(j0) = 2k$
- $G(j\sqrt{10}) = -k$

The plot will then be as shown in Figure 9.36.

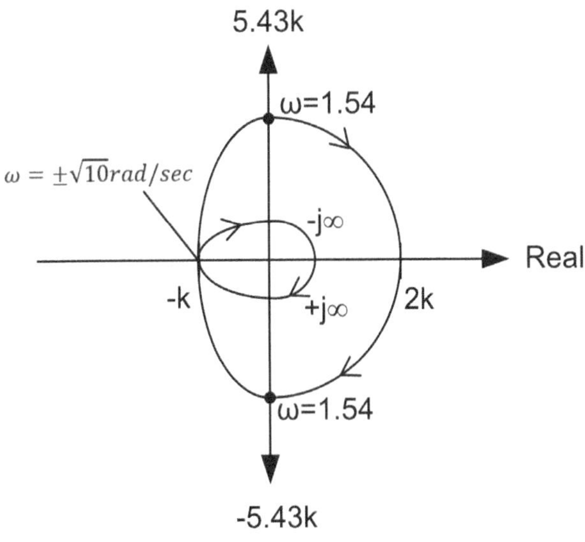


**FIGURE 9.36** Nyquist path.

An inspection of the Nyquist plot reveals that $N = -2$.

$$N = Z - P$$

$$-2 = Z - 0$$

$$Z = -2$$

So the system is stable.

$$1 + \frac{10k(s+2)}{s^3 + 3s^2 + 10}$$

$$s^3 + 2s^2 + 10 + 10ks + 20k = 0$$

$$s^3 + 3s^2 + 10ks + 10(2k+1) = 0$$

$$\begin{array}{c|cc} s^3 & 1 & 10k \\ s^2 & 3 & 10(2k+1) \\ s^1 & \multicolumn{2}{c}{\dfrac{30k - 10(2k+1)}{3}} \\ s^0 & 10(2k+1) & \end{array}$$

## 9.8 RELATIVE STABILITY

### 9.8.1 Phase Crossover Frequency

The phasor crossover frequency, $\omega_\pi$ is the frequency at which the phase angle, $GH\ (j\omega)$ is $-180^0$, that is, it is the frequency at which the polar plot crossovers the negative real axis.

### 9.8.2 Gain Margin

It is defined as

$$GM = \frac{1}{|GH(j\omega)|} \tag{9.41}$$

$$= \frac{1}{a}$$

Gain crossover frequency $\omega$ is defined as the frequency at which $|GH(j\omega)| = 1$.

### 9.8.3 Phase Margin

It is the angle by which the phasor plot must be rotated to cause it to pass through $-1 + j0$ points.

$$\varphi_{PM} = \left[180 + \arg GH\left(j\omega_1\right)\right] \tag{9.42}$$

Figure 9.37 shows the gain margin and phase margin by Nyquist method.

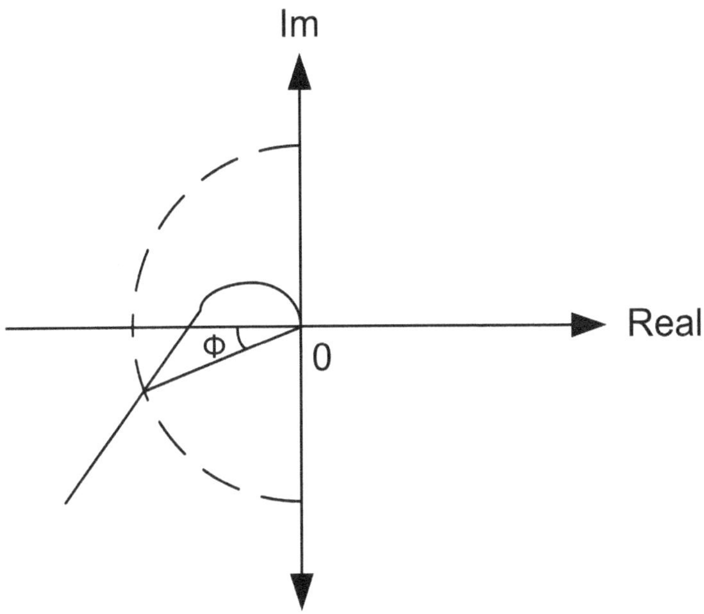


**FIGURE 9.37** Gain margin and phase margin by Nyquist method.

**Problems**

P9.1 Determine the frequency domain specifications for a second-order system with unity feedback and $G(s) = \frac{100}{s(s+6)}$.

P9.2 A unity feedback system having $G(s) = \frac{k}{s(1+Ts)}$ is subjected to a unit step input. The overshoot is 16.2%, and the resonant frequency is 10 $rad/sec$ in the frequency domain. Calculate $k$ and $T$. Also, determine (i) gain crossover frequency, (ii) phase margin, and (iii) resonant peak.

P9.3 Determine the frequency domain specifications for a unity feedback control system having $G(s) = \frac{25}{s(s+6)}$.

P9.4 Using the Nyquist stability criterion, comment on the closed-loop system stability for a system whose open loop transfer function is given by $G(s)H(s)=\frac{k(s+3)}{s(s-1)}$.

P9.5 Using Nyquist stability criterion comment on the stability for a system whose open loop transfer function is $G(s)H(s)=\frac{100}{(1+0.1s)(1+0.25)(1+0.3s)}$.

P9.6 The open loop transfer function is $G(s)=\frac{k}{s(s+1)(s+2)}$.

i) Determine if the system is stable for $k = 2$.
ii) Using the Nyquist stability criterion, determine the critical value of the gain $k$ for stability.

# 10 State Variable Method for Analysis of Control Systems

## 10.1 INTRODUCTION

The modelling, analysis, and design concept is based on the transfer function, which provides the output information for a given input. However, these methods do not provide precise information about all variables or the effects of incorporating initial conditions. In addition to this, classical methods are not adequate for nonlinear, multiple input multiple output (MIMO) systems, and time-varying systems. They are only applicable to a few specific inputs and are not feasible for obtaining an optimal solution because they require a trial-and-error technique.

To overcome the previous drawbacks of the transfer function approach, R. E. Kalman presented a more generalized and powerful technique of state variables in the time domain in 1960. The state variable method of modelling analysis and design is applicable to linear and nonlinear, time-invariant and time-varying MIMO systems. State feedback can easily be used to place closed-loop poles to improve system performance.

## 10.2 THE FORMULATION OF STATE VARIABLE TECHNIQUE

Consider the mechanical system shown in Figure 10.1, where K is the spring constant, B is the damping ratio, M is the mass, and f(t) is the force applied, which results in y displacement.

Modelling this system results in the following second-order differential equation.

$$M\ddot{y} + B\dot{y} + Ky = f(t)$$

In the state space method, this second-order differential equation can be represented by two first-order differential equations, as follows.

Let

$$y = x_1$$

$$\dot{x}_1 = x_2$$

$$\dot{x}_2 = -\frac{B}{M} - \frac{K}{M}x_1 + f(t) \tag{10.1}$$

DOI: 10.1201/9781003484202-10

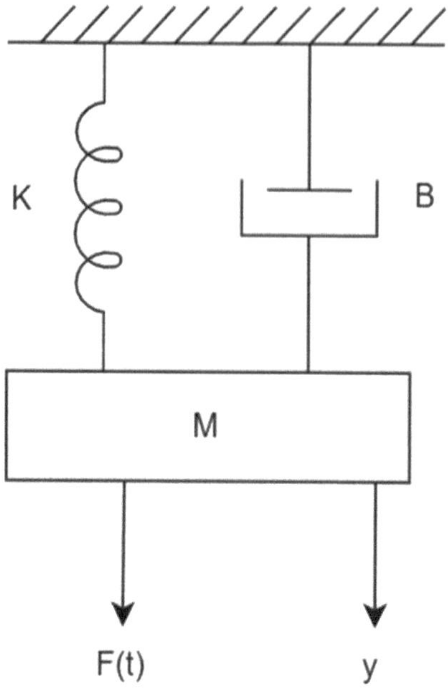


**FIGURE 10.1** A second-order mechanical system.

The following matrix form (Eq. 10.2) can represent the set of differential equations in Eq. 10.1. Such formulation of the problem is called the state variable technique.

$$\begin{bmatrix} \dot{x}_1 \\ \dot{x}_2 \end{bmatrix} = \begin{bmatrix} 0 & 1 \\ -\frac{K}{M} & -\frac{B}{M} \end{bmatrix} + \begin{bmatrix} 0 \\ \frac{1}{M} \end{bmatrix} f(t) \tag{10.2}$$

The matrix representation can be extended to a higher order differential equation as follows. Consider an nth-order differential equation

$$y^{(n)} + a_1 y^{(n-1)} + a_2 y^{(n-2)} + \ldots + a_{n-1} y^{(1)} + a_n y = u \tag{10.3}$$

Let

$$\begin{aligned} x_1 &= y \\ \dot{x}_1 &= x_2 \\ \dot{x}_2 &= x_3 \\ &\vdots \end{aligned}$$

$$\dot{x}_n = a_n x_1 - a_{n-1} x_2 - \cdots - a_1 x_n + u \tag{10.4}$$

The matrix form of the previous set of differential equations (Eq. 10.4) is as follows:

$$\begin{bmatrix} \dot{x}_1 \\ \dot{x}_2 \\ \vdots \\ \dot{x}_{n-1} \\ \dot{x}_n \end{bmatrix} = \begin{bmatrix} 0 & 1 & 0 & . & . & 0 \\ 0 & 0 & 1 & . & . & 0 \\ . & . & . & . & . & . \\ . & . & . & . & . & . \\ . & . & . & . & . & . \\ 0 & 0 & 0 & . & . & 1 \\ -a_n & -a_{n-1} & . & . & . & a_1 \end{bmatrix} \begin{bmatrix} x_1 \\ x_2 \\ . \\ . \\ . \\ x_{n-1} \\ x_n \end{bmatrix} + \begin{bmatrix} 0 \\ 0 \\ . \\ . \\ . \\ . \\ 1 \end{bmatrix} u$$

$$\dot{X} = AX + Bu \tag{10.5}$$

$$Y = CX$$

and represented as,

$$Y = \begin{bmatrix} 1 & 0 & 0 & 0 & 0 & \cdots & 0 \end{bmatrix} \begin{bmatrix} x_1 \\ x_2 \\ \vdots \\ x^n \end{bmatrix}$$

The general block diagram of the system is as shown in Figure 10.2.

where,

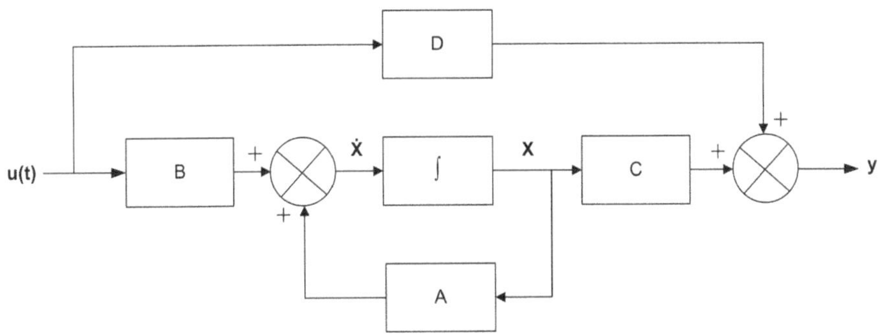


**FIGURE 10.2** Block diagram of general state space representation of a control system.

x = state vector, $n \times 1$
u = input vector, $r \times 1$
y = output vector, $m \times 1$
A = system matrix, $n \times n$
B = input matrix, $n \times r$

C = output matrix, $n \times m$
D = disturbance matrix, $m \times r$

The state of the system at any time $t_0$ is the amount of information at $t_0$ that, together with all inputs for $t \geq t_0$, uniquely determines the behavior of the system for all $t \geq t_0$.

## 10.3 DEFINITIONS

a) **State**
The state of a dynamical system is the smallest number of variables (called state variables) such that the knowledge of these variables at $t = t_0$, together with the knowledge of the input for $t \geq t_0$, completely determines the behavior of the system for anytime $t \geq t_0$.

b) **State variables**
The state variables of dynamic systems are the variables that make up the smallest set of variables that determine the state of a dynamic system. If at least n variables, $x_1, x_2, \cdots x_n$, are needed to describe the behavior of a dynamic system completely (so that once the input is given for $t \geq t_0$), the specified future state of the system is completely determined. Then, such variables are referred to as a set of state variables.

c) **State vector**
If 'n' state variables are needed to completely describe the behavior of a given system, then these 'n' state variables can be considered the 'n' components of a vector x. Such a vector is called a state vector. A state vector determines uniquely the system state $X(t)$ at any time $t \geq t_0$, once the state at $t = t_0$ is given an input $u(t)$ for a specified $t \geq t_0$.

d) **State Space**
The n-dimensional space whose coordinate axes consist of the $x_1\ axis, x_2\ axis, \cdots, x_n\ axis$ is called state space. A point in state space can represent any state.

## 10.4 DIFFERENT REPRESENTATIONS OF STATE MODEL

### 10.4.1 State Space Representation Using Phase Variables in Controller Canonical Form (CCF)

Consider a transfer function as shown in Eq. 10.6.

$$\frac{Y(s)}{U(s)} = \frac{b}{s^n + a_1 s^{n-1} + as^{n-2} + \cdots + a_{n-1}s + a_n} \tag{10.6}$$

The transfer functions given in Eq. 10.6 can be represented by the following differential equations:

$$\frac{d^n}{dt^n} + a\frac{d^{n-1}y}{dt^{n-1}} + \cdots + a_{n-1}\frac{dy}{dt} + a_n y = bu \tag{10.7}$$

$$x_1 = y$$

$$\dot{x}_1 = x_2$$

$$\dot{x}_n = x_3$$

$$\vdots$$

$$x_{n-1} = x_n$$

With the help of the previous representation, Eq 10.7 can be written as:

$$x_n = -a_n x_1 - a_{n-1}x_2 - a_{n-2}x_3 - \cdots - a_1 x_n + bu$$

The state equation is represented by Figure 10.3 which shows a signal flow graph of CCF.

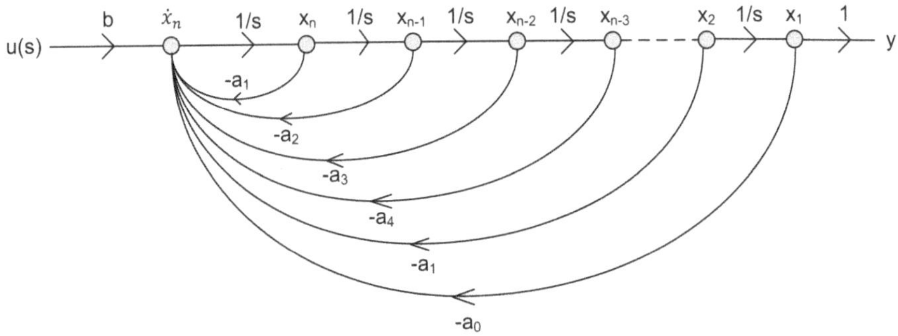


**FIGURE 10.3** Signal flow graph of controller canonical form (CCF).

Using Laplace transform,

$$\begin{bmatrix} \dot{x}_1 \\ \dot{x}_2 \\ \vdots \\ \dot{x}_{n-1} \\ \dot{x}_n \end{bmatrix} = \begin{bmatrix} 0 & 1 & 0 & . & . & 0 \\ 0 & 0 & 1 & . & . & 0 \\ . & . & . & . & . & . \\ . & . & . & . & . & . \\ . & . & . & . & . & . \\ 0 & 0 & 0 & . & . & 1 \\ -a_n & -a_{n-1} & . & . & . & a_1 \end{bmatrix} \begin{bmatrix} x_1 \\ x_2 \\ . \\ . \\ . \\ x_{n-1} \\ x_n \end{bmatrix} + \begin{bmatrix} 0 \\ 0 \\ . \\ . \\ . \\ . \\ 1 \end{bmatrix} u$$

$$\dot{X} = AX + Bu \tag{10.8}$$

The matrix for the input equation of the system is given here:

$$y = \begin{bmatrix} 1 & 0 & 0 & 0 & 0 & \cdots & 0 \end{bmatrix} \begin{bmatrix} x_1 \\ x_2 \\ . \\ . \\ x_n \end{bmatrix} + [0][\mu] \tag{10.9}$$

$$y = Cx + Du \tag{10.10}$$

The matrix $A$ has all the ones in the upper diagonal and other elements. The coefficients of the last row are negative. This form is also known as 'bush' or comparison form.

**Example 10.1:** Express the following transfer function in a CCF. Draw a signal flow graph.

$$\frac{Y(s)}{U(s)} = \frac{3}{s^4 + 2s^3 + 3s + 2}$$

***Solution***

From the given transfer function, we have:

$$(s^4 + 2s^3 + 3s + 2)Y(s) = 3U(s)$$

$$\frac{d^{4y}}{dt^4} + 2\frac{d^{3y}}{dt^3} + 3\frac{d^y}{dt} + 2y = 3\mu$$

$$x_1 = y$$

$$\dot{x}_1 = \frac{dy}{dt} = x_2$$

$$\dot{x}_2 = \frac{d^2y}{dt^2} = x_3$$

$$\dot{x}_3 = \frac{d^3y}{dt^3} = x_4$$

$$\dot{x}_4 = \frac{d^4y}{dt^4} = 3u - 2\frac{d^3y}{dt^3} - 3\frac{dy}{dt} - 2y = 3u - 2x_4 - 3x_2 - 2x_1$$

The signal flow graph is given in Figure 10.4.

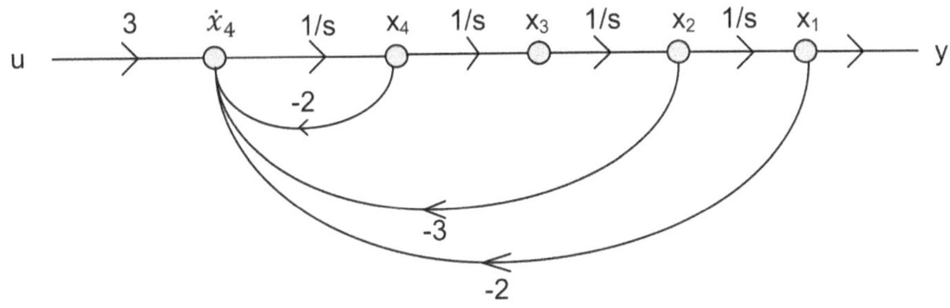


**FIGURE 10.4** Signal flow graph of Example 10.1.

$$\begin{bmatrix} \dot{x}_1 \\ \dot{x}_2 \\ \dot{x}_3 \\ \dot{x}_4 \end{bmatrix} = \begin{bmatrix} 0 & 1 & 0 & 0 \\ 0 & 0 & 1 & 0 \\ 0 & 0 & 0 & 1 \\ -2 & -3 & 0 & -2 \end{bmatrix} \begin{bmatrix} x_1 \\ x_2 \\ x_3 \\ x_4 \end{bmatrix} + \begin{bmatrix} 0 \\ 0 \\ 0 \\ 3 \end{bmatrix} u$$

and

$$y = \begin{bmatrix} 1 & 0 & 0 & 0 \end{bmatrix} \begin{bmatrix} x_1 \\ x_2 \\ x_3 \\ x_4 \end{bmatrix} + [0][\mu] \tag{10.11}$$

The previous equations represent the required CCF.

## 10.5 SOLUTION OF STATE EQUATION BY LAPLACE TRANSFORM

The general equation of the state space method is given here:

$$x(t) = Ax(t) + Bu(t)$$

$$y(t) = Cx + Du \tag{10.12}$$

The Laplace transform of Eq. 10.12 is:

$$SX(s) - x(0) = AX(s) + BU(s)$$

$$SX(s) - AX(s) = x(0) + BU(s)$$

$$(SI - A)X(s) = x(0) + BU(s)$$

$$X(s) = (SI - A)^{-1} x(0) + (SI - A)^{-1} BU(s) \tag{10.13}$$

The state vector $\overline{x}(t)$ is the inverse Laplace transform of this equation. The state transition matrix or fundamental matrix is:

$$\varphi(t) = \mathcal{L}^{-1}[SI - A]^{-1} \tag{10.14}$$

and

$\mathcal{L}^{-1}[SI - A]^{-1}$ is called a normal matrix.

$$\overline{x}(t) = \varphi(t)x(0) + \mathcal{L}^{-1}[SI - A]^{-1} BU(s) \tag{10.15}$$

**Example 10.2:** Find the solution of the feedback control system whose open loop transfer function is as given here:

$$G(s) = \frac{1}{(s+1)(s+2)} = \frac{1}{s^2 + 3s + 2}$$

***Solution***

Let

$$G(s) = \frac{1}{s^2 + 3s + 2} = \frac{Y(s)}{U(s)} \tag{10.16}$$

Assuming all initial conditions to be zero,

$$y = x_1, \quad \dot{x}_1 = x_2 \tag{10.17}$$

$$\dot{x}_2 = -2x_1 - 3x_2 + u$$

$$\begin{bmatrix} \dot{x}_1 \\ \dot{x}_2 \end{bmatrix} = \begin{bmatrix} 0 & 1 \\ -2 & -3 \end{bmatrix} \begin{bmatrix} x_1 \\ x_2 \end{bmatrix} + \begin{bmatrix} 0 \\ 1 \end{bmatrix} u$$

$$\dot{X} = AX + BU \tag{10.18}$$

$$[SI - A] = S\begin{bmatrix} 1 & 0 \\ 0 & 1 \end{bmatrix} - \begin{bmatrix} 0 & 1 \\ -2 & -3 \end{bmatrix} = \begin{bmatrix} s & -1 \\ 2 & S+3 \end{bmatrix}$$

$$Y = \begin{bmatrix} 1 & 0 \end{bmatrix} \begin{bmatrix} x_1 \\ x_2 \end{bmatrix} \tag{10.19}$$

$$Adj[SI - A] = \begin{bmatrix} s+3 & 1 \\ -2 & s \end{bmatrix}$$

The determinant of $\begin{bmatrix} s & -1 \\ 2 & s+3 \end{bmatrix} = (s+1)(s+2)$

$$[SI - A]^{-1} = \begin{bmatrix} \dfrac{s+3}{(s+1)(s+2)} & \dfrac{1}{(s+1)(s+2)} \\ -\dfrac{2}{(s+1)(s+2)} & \dfrac{s}{(s+1)(s+2)} \end{bmatrix}$$

$$[SI - A]^{-1} = \begin{bmatrix} \dfrac{2}{(s+1)} - \dfrac{1}{(s+2)} & \dfrac{1}{(s+1)} - \dfrac{1}{(s+2)} \\ -\dfrac{2}{(s+1)} + \dfrac{2}{(s+2)} & \dfrac{-1}{(s+1)} + \dfrac{2}{(s+2)} \end{bmatrix}$$

$$\mathcal{L}^{-1}[SI - A]^{-1} = \varphi(t) = \begin{bmatrix} 2e^{-t} - e^{-2t} & e^{-t} - e^{-2t} \\ -2e^{-t} + 2e^{-2t} & -e^{-t} + 2e^{-2t} \end{bmatrix} \tag{10.20}$$

If $u(t)$ is a unit step input, then

$$[sI - A]^{-1} BU(s) = \begin{bmatrix} \dfrac{s+3}{(s+1)(s+2)} & \dfrac{1}{(s+1)(s+2)} \\ -\dfrac{2}{(s+1)(s+2)} & \dfrac{s}{(s+1)(s+2)} \end{bmatrix} \begin{bmatrix} 0 \\ \dfrac{1}{s} \\ 1 \end{bmatrix}$$

$$[sI - A]^{-1} BU(s) = \begin{bmatrix} \dfrac{1}{s(s+1)(s+2)} \\ \dfrac{1}{(s+1)(s+2)} \end{bmatrix} = \begin{bmatrix} \dfrac{1/2}{s} - \dfrac{1}{(s+1)} + \dfrac{1/2}{(s+2)} \\ \dfrac{1}{(s+1)} - \dfrac{1}{(s+2)} \end{bmatrix} \tag{10.21}$$

$$x_1(t) = \frac{1}{2}u(t) - e^{-t} + \frac{1}{2}e^{-2t}, x_2(t) = e^{-t} - e^{-2t} \tag{10.22}$$

If the initial conditions are not zero, and $x_1(0)=1$ and $x_2(0)=0$, then part one of Eq. 10.16 must be included in Eq. 10.23 as follows:

$$\varphi(t)x(0)=\begin{bmatrix} 2e^{-t}-e^{-2t} & e^{-t}-e^{-2t} \\ -2e^{-t}+2e^{-2t} & -e^{-t}+2e^{-2t} \end{bmatrix}\begin{bmatrix} 1 \\ 0 \end{bmatrix}$$

$$\varphi(t)x(0)=\begin{bmatrix} 2e^{-t}-e^{-2t} \\ -2e^{-t}+2e^{-2t} \end{bmatrix}$$

The total solution of Eq. 10.16 is as shown here:

$$\varphi(t)x(0)+\mathcal{L}^{-1}[sI-A]^{-1}BU(s)=\begin{bmatrix} 2e^{-t}-e^{-2t} \\ -2e^{-t}+2e^{-2t} \end{bmatrix}+\begin{bmatrix} \frac{1}{2}u(t)-e^{-t}+\frac{1}{2}e^{-2t} \\ e^{-t}-e^{-2t} \end{bmatrix} \tag{10.23}$$

$$=\begin{bmatrix} \frac{1}{2}u(t)+e^{-t}-\frac{1}{2}e^{-2t} \\ -e^{-t}+e^{-2t} \end{bmatrix} \tag{10.24}$$

and

$$x_1(t)=\frac{1}{2}u(t)+e^{-t}-\frac{1}{2}e^{-2t}$$

$$x_2(t)=-e^{-t}+e^{-2t} \tag{10.25}$$

**Example 10.3:** Consider the control system in Figure 10.5 and solve it using the state space technique.

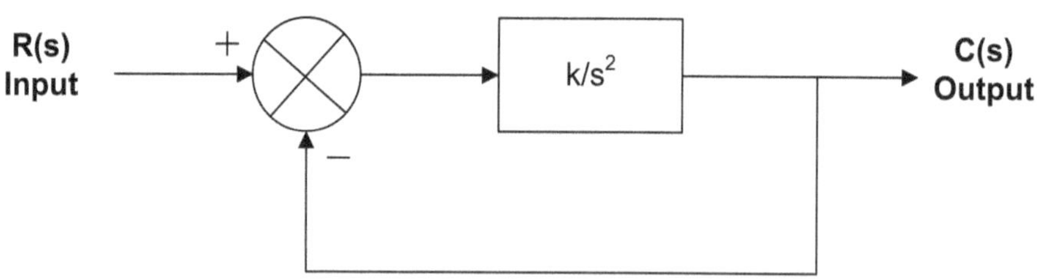


**FIGURE 10.5** Block diagram of feedback control system.

***Solution***

$$\frac{d^2c}{dt^2} = k(r-c) \tag{10.26}$$

$$\frac{d^2c}{dt^2} + kc = kr$$

Let $c = x_1$

$$\dot{x}_1 = x_2$$

$$\dot{x}_2 = -kx_1 + kr$$

$$\begin{bmatrix} \dot{x}_1 \\ \dot{x}_2 \end{bmatrix} = \begin{bmatrix} 0 & 1 \\ -k & 0 \end{bmatrix} \begin{bmatrix} x_1 \\ x_2 \end{bmatrix} + \begin{bmatrix} 0 \\ k \end{bmatrix} \tag{10.27}$$

The signal flow graph is given in Figure 10.6.

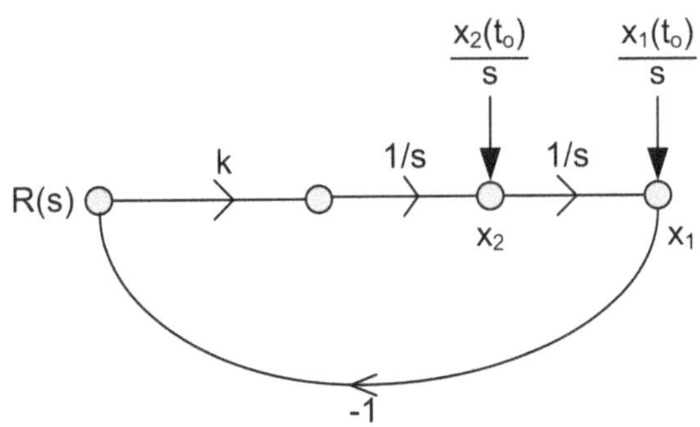


**FIGURE 10.6** Signal flow graph for Example 10.3.

$$[SI - A] = \begin{bmatrix} s & -1 \\ k & s \end{bmatrix}$$

The determinant of $\begin{bmatrix} s & -1 \\ k & s \end{bmatrix} = s^2 + k$

$$\text{Adj} \begin{bmatrix} s & -1 \\ k & s \end{bmatrix} = \begin{bmatrix} s & k \\ -1 & s \end{bmatrix}$$

$$[SI-A]^{-1}=\begin{bmatrix}\frac{s}{s^2+k} & \frac{1}{s^2+1}\\ -\frac{k}{s^2+k} & \frac{s}{s^2+k}\end{bmatrix}$$

$$\begin{bmatrix}x_1(t)\\ \\ x_2(t)\end{bmatrix}=\mathcal{L}^{-1}\begin{bmatrix}\frac{s}{s^2+k} & \frac{1}{s^2+k}\\ -\frac{k}{s^2+k} & \frac{s}{s^2+k}\end{bmatrix}\begin{bmatrix}x_1(t_0)\\ \\ x_2(t_0)\end{bmatrix}+\mathcal{L}^{-1}\begin{bmatrix}\frac{k}{s^2+k}\\ \frac{ks}{s^2+k}\end{bmatrix}R(s) \qquad (10.28)$$

**Example 10.4:** Determine the transition matrix of the following system.

$$A=\begin{bmatrix}1 & 2\\ 3 & 2\end{bmatrix}$$

***Solution***

$$[sI-A]=\begin{bmatrix}s-1 & -2\\ -3 & s-2\end{bmatrix}$$

Determine, $[\mathrm{sI}-\mathrm{A}]=(\mathrm{s}-1)(\mathrm{s}-2)-6=(\mathrm{s}-4)(\mathrm{s}+1)$

Adj [SI-A] $=\begin{bmatrix}s-2 & 3\\ 2 & s-1\end{bmatrix}$

$$[sI-A]^{-1}=\begin{bmatrix}\frac{s-2}{(s-4)(s+1)} & \frac{3}{(s-4)(s+1)}\\ \frac{2}{(s-4)(s+1)} & \frac{s-1}{(s-4)(s+1)}\end{bmatrix} \qquad (10.29)$$

$$[sI-A]^{-1}=\begin{bmatrix}\frac{2}{5(s-4)}+\frac{3}{5(s+1)} & \frac{2}{5(s-4)}-\frac{2}{5(s+1)}\\ \frac{3}{5(s-4)}-\frac{3}{5(s+1)} & \frac{3}{5(s-4)}+\frac{2}{5(s+1)}\end{bmatrix}$$

$$\mathcal{L}^{-1}[sI-A]^{-1}=\varphi(t)=\begin{bmatrix}\frac{1}{5}\left(2e^{4t}+3e^{-t}\right) & \frac{2}{5}\left(e^{4t}-2e^{-t}\right)\\ \frac{3}{5}\left(e^{4t}-e^{-t}\right) & \frac{1}{5}\left(3e^{4t}+2e^{-t}\right)\end{bmatrix} \tag{10.30}$$

**Example 10.5:** Compute the transition matrix for the following system.

$$A=\begin{bmatrix}1 & 0 & 1\\ 0 & 1 & 0\\ 0 & -2 & -1\end{bmatrix}$$

***Solution***

$$[sI-A]=\begin{bmatrix}s-1 & 0 & -1\\ 0 & s-1 & 0\\ 0 & 2 & s+1\end{bmatrix}$$

$$\det[sI-A]=(s-1)(s-1)(s+1)$$

$$Adj[sI-A]=\begin{bmatrix}(s-1)(s+1) & 2 & (s-1)\\ 0 & (s-1)(s+1) & 0\\ 0 & 2 & (s+1)^2\end{bmatrix}$$

$$[sI-A]^{-1}=\begin{bmatrix}\frac{1}{(s-1)} & \frac{2}{(s-1)^2(s+1)} & (s-1)\\ 0 & \frac{1}{(s-1)} & 0\\ 0 & \frac{-2}{(s+1)(s-1)} & \frac{1}{(s+1)}\end{bmatrix} \tag{10.31}$$

$$\varnothing(t)=\begin{bmatrix}e^{t} & \frac{1}{2}e^{-t}-\frac{1}{2}e^{t}+e^{t} & \frac{1}{2}e^{t}-\frac{1}{2}e^{-t}\\ 0 & e^{t} & 0\\ 0 & -e^{t}+e^{-t} & e^{-t}\end{bmatrix} \tag{10.32}$$

**Example 10.6:** A unity feedback control system is governed by open loop transfer function as given here:

$$G(s) = \frac{1}{s(s+2)}$$

a) Show the block diagram.
b) Derive the signal flow diagram from the block diagram.
c) Write down the state variable equations.
d) Write down the state equation in matrix form.
e) Compute the state variable $x_1(t)$ and $x_2(t)$ for initial condition $x_1(0) = 1$, $x_2(0) = 0$

***Solution***

a) The block diagram is as shown in Figure 10.7.
b) The signal flow graph of control system is shown in Figure 10.8.

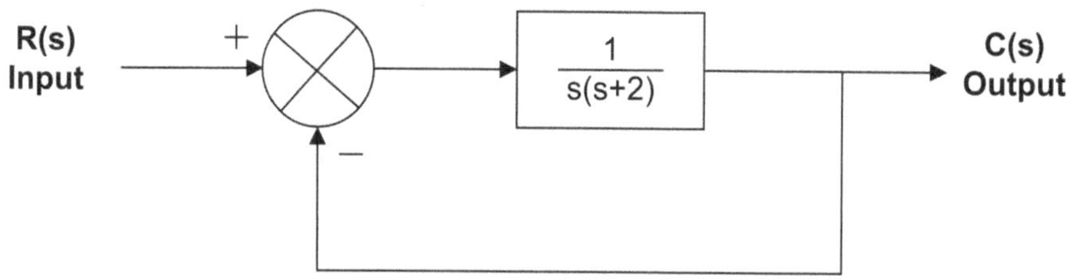


FIGURE 10.7 Block diagram of feedback control system of Example 10.6.

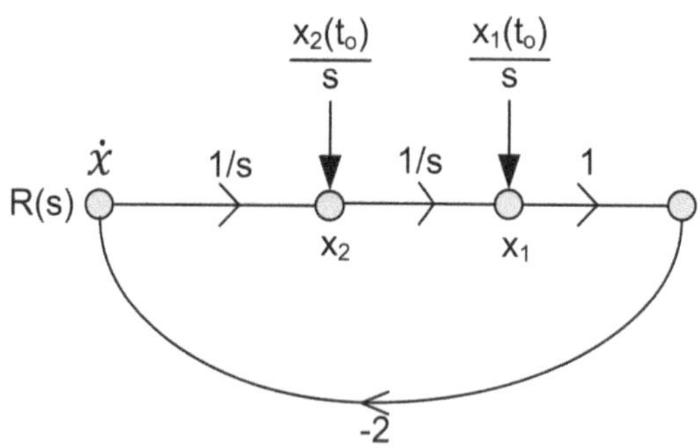


FIGURE 10.8 Signal flow graph of control system of Example 10.6.

c) $\dot{x}_2 = x_2$

$$\dot{x}_2 = -x_1 - 2x_2 + r(t) \tag{10.33}$$

d) $\begin{bmatrix} \dot{x}_1 \\ \dot{x}_2 \end{bmatrix} = \begin{bmatrix} 0 & 1 \\ -1 & -2 \end{bmatrix} \begin{bmatrix} x_1 \\ x_2 \end{bmatrix} + \begin{bmatrix} 0 \\ 1 \end{bmatrix} r(t)$

$$A = \begin{bmatrix} 0 & 1 \\ -1 & -2 \end{bmatrix}, B = \begin{bmatrix} 0 \\ 1 \end{bmatrix}$$

$$A = \begin{bmatrix} 0 & 1 \\ -1 & -2 \end{bmatrix}, B = \begin{bmatrix} 0 \\ 1 \end{bmatrix}$$

e) $[sI - A] = \begin{bmatrix} s & -1 \\ 1 & (s+2) \end{bmatrix}$

$$det = s(s+1) + 1 = (s+1)^2$$

$$[sI - A]^{-1} = \begin{bmatrix} \dfrac{1}{(s+1)^2} + \dfrac{1}{(s+1)} & \dfrac{1}{(s+1)^2} \\ -\dfrac{1}{(s+1)^2} & -\dfrac{1}{(s+1)^2} + \dfrac{1}{(s+1)} \end{bmatrix} \tag{10.34}$$

$$\overline{x}(t) = j(t)\overline{x}(0) + \mathcal{L}^{-1}\left[(sI - A)^{-1} \begin{bmatrix} 0 \\ 1 \end{bmatrix} \frac{1}{s}\right]$$

$$= \begin{bmatrix} te^{-t} + e^{-t} & te^{-t} \\ -te^{-t} & -te^{-t} + e^{-t} \end{bmatrix} \begin{bmatrix} 1 \\ 0 \end{bmatrix} + \mathcal{L}^{-1} \begin{bmatrix} \dfrac{1}{s(s+1)^2} \\ \dfrac{1}{(s+1)} \end{bmatrix} \tag{10.35}$$

$$= \begin{bmatrix} te^{-t} - e^{-t} \\ -te^{-t} \end{bmatrix} + \begin{bmatrix} u(t) - te^{-t} - e^{-t} \\ te^{-t} \end{bmatrix} = u(t) \tag{10.36}$$

$$x_1 = u(t)$$

$$x_2 = 0 \tag{10.37}$$

**Example 10.7:** The system matrix of the feedback control system is as shown in Eq. 10.38.

$$A = \begin{bmatrix} 0 & 1 \\ -6 & -5 \end{bmatrix} \tag{10.38}$$

Determine $\overline{x}(t)$ if $x(0) = \begin{bmatrix} 1 \\ 0 \end{bmatrix}$.

***Solution***

$$[sI - A] = s\begin{bmatrix} 1 & 0 \\ 0 & 1 \end{bmatrix} - \begin{bmatrix} 0 & 1 \\ -6 & -5 \end{bmatrix} = \begin{bmatrix} s & -1 \\ 6 & s+5 \end{bmatrix}$$

$$\text{Det}\,[sI - A] = s(s + 5) + 6 = (s + 2)\,(s + 3)$$

$$\varphi(s) = [sI - A]^{-1} = \begin{bmatrix} \dfrac{S+5}{(S+2)(S+3)} & \dfrac{1}{(S+2)(S+3)} \\ \dfrac{-6}{(S+2)(S+3)} & \dfrac{5}{(S+2)(S+3)} \end{bmatrix}$$

$$\varphi(t) = \begin{bmatrix} \dfrac{3}{(S+2)} - \dfrac{2}{(S+3)} & \dfrac{1}{(S+2)} - \dfrac{1}{(S+3)} \\ \dfrac{-6}{(S+2)} + \dfrac{6}{(S+3)} & \dfrac{-2}{(S+2)} - \dfrac{3}{(S+3)} \end{bmatrix} \tag{10.39}$$

$$\varphi(t) = \begin{bmatrix} 3e^{-2t} - 2e^{-3t} & e^{-2t} - e^{-3t} \\ -6e^{-2t} + 6e^{-3t} & -2e^{-2t} - e^{-3t} \end{bmatrix} \tag{10.40}$$

The first point of $x(t)$ is

$$\varphi(t)x(0)$$

Therefore,

$$\begin{bmatrix} 3e^{-2t} - 2e^{-3t} & e^{-2t} - e^{-3t} \\ -6e^{-2t} + 6e^{-3t} & -2e^{-2t} - e^{-3t} \end{bmatrix} \begin{bmatrix} 1 \\ 0 \end{bmatrix} = \begin{bmatrix} 3e^{-2t} - 2e^{-3t} \\ -6e^{-2t} + 6e^{-3t} \end{bmatrix} \tag{10.41}$$

$$x_1(0) = 3e^{-2t} - 2e^{-3t}$$

$$x_2(0) = -6e^{-2t} + 6e^{-3t} \tag{10.42}$$

## 10.6 TRANSFER FUNCTION DERIVATION FROM STATE MODEL

A transfer can be given by the following relationship:

$$\text{Transfer function} = \frac{\textit{Laplace transform of output}}{-\textit{Laplace transform of input}}$$

For time-invariant systems, the dynamics can be written in a vector differential form as follows:

Note that the Initial condition is zero.

$$\mathrm{X} = \mathrm{Ax} + \mathrm{Bu}$$

$$\mathrm{Y} = \mathrm{Cx} + \mathrm{Du} \tag{10.43}$$

Taking Laplace transform of Eq. 10.43,

$$sX(s) - x(0) = Ax(s) + Bu(s)$$

$$[sI - A]x(s) = Bu(s), x(0) = 0$$

$$\mathrm{X}(\mathrm{s}) = [sI - A]^{-1} B\,u(s)$$

$$\mathrm{Y}(\mathrm{s}) = \mathrm{Cx}(\mathrm{s}) + \mathrm{D}\,\mathrm{u}(\mathrm{s}) = \mathrm{C}\,[sI - A]^{-1}\,\mathrm{BU}(\mathrm{s}) + \mathrm{DU}(\mathrm{s}) \tag{10.44}$$

Transfer function $= \frac{Y(s)}{U(s)} = C[sI - A]^{-1}\mathrm{B} + \mathrm{D}$

If there is direct compiling of input and output, then $D = 0$.

$$\text{Transfer function} = \frac{Y(s)}{U(s)} = C[SI - A]^{-1}B \tag{10.45}$$

**Example 10.8:** Find the transfer function when

$$\mathrm{A} = \begin{bmatrix} -2 & 1 \\ 0 & -3 \end{bmatrix}, \mathrm{B} = \begin{bmatrix} 0 \\ 1 \end{bmatrix}, \textbf{ and } \mathrm{C} = \begin{bmatrix} 1 & 1 \end{bmatrix}$$

***Solution***

$$[sI \; - \; A] = s\begin{bmatrix} 1 & 0 \\ 0 & 1 \end{bmatrix} - \begin{bmatrix} -2 & 1 \\ 0 & 3 \end{bmatrix} = \begin{bmatrix} s+1 & -1 \\ 0 & s+3 \end{bmatrix} \tag{10.46}$$

$$Det\left[sI \quad - \quad A\right] = Det\begin{bmatrix} s+1 & -1 \\ 0 & s+3 \end{bmatrix} = (s+2)\ (s+3)$$

$$\left[sI - A\right]^{-1} = \frac{Adj(SI - A)}{\text{Det}\left[SI - A\right]} = \frac{\begin{bmatrix} s+3 & 1 \\ 0 & s+2 \end{bmatrix}}{(s+2)(s+3)} = \begin{bmatrix} \dfrac{1}{s+2} & \dfrac{1}{(s+2)(s+3)} \\ 0 & \dfrac{1}{s+3} \end{bmatrix}$$

$$\text{C}\ \left[sI - A\right]^{-1} = \begin{bmatrix} 1 & 1 \end{bmatrix} \begin{bmatrix} \dfrac{1}{s=2} & \dfrac{1}{(s+2)(s+3)} \\ 0 & \dfrac{1}{s+3} \end{bmatrix} \begin{bmatrix} 0 \\ 1 \end{bmatrix} \tag{10.47}$$

$$= \begin{bmatrix} \dfrac{1}{s+2} & \dfrac{1}{(s+2)(s+3)} \\ 0 & \dfrac{1}{s+3} \end{bmatrix} \begin{bmatrix} 0 \\ 1 \end{bmatrix} = \frac{s+3}{(s+2)(s+3)} = \frac{1}{s+2} \tag{10.48}$$

## 10.7 EQUIVALENCE OF $\varphi(t)$ AND $e^{At}$

$$\phi(t) = e^{At} = 1 + \text{At} + \frac{A^2t^2}{2!} + \frac{A^3t^2}{3!} + \ldots$$

Let $\dot{x}(t) = \text{Ax}(\text{t}) + \text{Bu}(\text{t})$

$$\dot{x}(t) = \begin{bmatrix} 0 & 1 \\ 0 & 0 \end{bmatrix} x(t) + \begin{bmatrix} 0 \\ 1 \end{bmatrix} u(t) \tag{10.49}$$

$$A = \begin{bmatrix} 0 & 1 \\ 0 & 0 \end{bmatrix}$$

$$A^2 = \begin{bmatrix} 0 & 1 \\ 0 & 0 \end{bmatrix} \begin{bmatrix} 0 & 1 \\ 0 & 0 \end{bmatrix} = \begin{bmatrix} 0 & 0 \\ 0 & 0 \end{bmatrix}$$

$$A^3 = AA^2 = 0$$

$$A^n = 0$$

$$\varphi(t) = I + At$$

$$\varphi(t) = \begin{bmatrix} 1 & 0 \\ 0 & 1 \end{bmatrix} + \begin{bmatrix} 0 & 1 \\ 0 & 0 \end{bmatrix} t = \begin{bmatrix} 1 & t \\ 0 & 1 \end{bmatrix} \tag{10.50}$$

To confirm the previous solution

$$[sI - A] = \begin{bmatrix} s & 0 \\ 0 & s \end{bmatrix} - \begin{bmatrix} 0 & 1 \\ 0 & 0 \end{bmatrix} = \begin{bmatrix} s & -1 \\ 0 & s \end{bmatrix}$$

$$Det \begin{bmatrix} s & -1 \\ 0 & s \end{bmatrix} = s^2$$

$$[sI - A]^{-1} = \frac{Adj(SI - A)}{\text{Det}[SI - A]} = \frac{\begin{bmatrix} s & 1 \\ 0 & s \end{bmatrix}}{s^2} = \begin{bmatrix} \frac{1}{s} & \frac{1}{s^2} \\ 0 & \frac{1}{s} \end{bmatrix}$$

$$\mathcal{L}^{-1}[sI - A]^{-1} = \phi(t) = \begin{bmatrix} 1 & t \\ 0 & 1 \end{bmatrix} \tag{10.51}$$

Therefore, $\phi(t) = e^{At}$

### 10.7.1 Equivalence of $n \times n$ Matrix A

The eigenvalues of $n \times n$ matrix A are the roots of the characteristic equation

$$|\lambda I - A| = 0$$

$\lambda's$ are called eigenvalues.

**Example 10.9:** Similarity transformation to a diagonal matrix.

***Solution***

Consider

$$\dot{X} = \text{Ax} + \text{Bu}$$
$$\text{Y} = \text{Cx}$$

where,

$$\text{A} = \begin{bmatrix} 0 & 1 \\ -2 & -3 \end{bmatrix} \tag{10.52}$$

$$[\lambda I - A] = \begin{bmatrix} \lambda & -1 \\ 2 & \lambda + 3 \end{bmatrix} = \lambda(\lambda + 3) + 2 = 0$$

$$= (\lambda + 1)(\lambda + 2) = 0, \ \lambda = -1, \qquad \lambda = -2$$

Consider another transformation,

$$
\begin{aligned}
x &= PZ \\
\mathrm{P}\dot{\mathrm{Z}} &= \mathrm{APZ} + \mathrm{Bu} \\
Y &= CPZ \\
\mathrm{Z}' &= \mathrm{P}^{-1}\mathrm{APZ} + \mathrm{P}^{-1}\ \mathrm{Zu} \\
Y &= CPZ
\end{aligned}
\tag{10.53}
$$

To determine P, knowing that the characteristic equation should be the same,

$$
\mathrm{P}^{-1}\mathrm{AP} = \lambda
$$

$$
AP = P\lambda
$$

Now,

$$
P = \begin{bmatrix} p_{11} & p_{12} \\ p_{21} & p_{22} \end{bmatrix}
$$

$$
\begin{bmatrix} 0 & 1 \\ -2 & -3 \end{bmatrix}\begin{bmatrix} p_{11} & p_{12} \\ p_{21} & p_{22} \end{bmatrix} = \begin{bmatrix} p_{11} & p_{12} \\ p_{21} & p_{22} \end{bmatrix}\begin{bmatrix} -1 & 0 \\ 0 & -2 \end{bmatrix}
$$

$$
\begin{bmatrix} p_{21} & p_{22} \\ -2p_{11} - 3p_{21} - 2p_{12} & -3p_{22} \end{bmatrix} = \begin{bmatrix} -p_{11} & -2p_{12} \\ -p_{21} & -2p_{22} \end{bmatrix} \tag{10.54}
$$

Or

$$
p_{12} = -p_{11}, p_{22} = -2p_{12}, -2p_{11} - 3p_{21} = -p_{21}, -2p_{12} - 3p_{22} = -2p_{22}
$$

Let

$$
p_{11} = -1,\quad p_{12} = -1, p_{21} = 1, p_{22} = -2
$$

Thus,

$$
P = \begin{bmatrix} -1 & 1 \\ 1 & -2 \end{bmatrix}, \quad Det\ \ P = 1
$$

$$
P^{-1}\begin{bmatrix} -2 & -1 \\ -1 & -1 \end{bmatrix}
$$

$$p^{-1}Ap = \begin{bmatrix} -2 & -1 \\ -1 & -1 \end{bmatrix} \begin{bmatrix} 0 & 1 \\ -2 & -3 \end{bmatrix} \begin{bmatrix} -1 & 1 \\ 1 & -2 \end{bmatrix} = \begin{bmatrix} -1 & 0 \\ 0 & -2 \end{bmatrix} \tag{10.55}$$

$$\begin{bmatrix} \dot{z}_1 \\ \dot{z}_2 \end{bmatrix} = \begin{bmatrix} -1 & 0 \\ 0 & -2 \end{bmatrix} \begin{bmatrix} z_1 \\ z_2 \end{bmatrix} + \begin{bmatrix} -2 & -1 \\ -1 & -1 \end{bmatrix} \begin{bmatrix} 0 \\ u \end{bmatrix}$$

$$Y = \begin{bmatrix} 0 & 1 \end{bmatrix} \begin{bmatrix} -1 & 1 \\ 1 & -2 \end{bmatrix} \begin{bmatrix} z_1 \\ z_2 \end{bmatrix} \tag{10.56}$$

**Example 10.10:** Find the eigenvalues of following:

$$A = \begin{bmatrix} 0 & 1 & 0 \\ 0 & 0 & 1 \\ -6 & 11 & -6 \end{bmatrix}$$

***Solution***

The characteristic equation is

$$[\lambda I - A] = \begin{bmatrix} \lambda & -1 & 0 \\ 0 & \lambda & -1 \\ 6 & 11 & \lambda + 6 \end{bmatrix} \tag{10.57}$$

$$= \lambda^3 + 6\lambda^2 + 11\lambda + 6 = 0$$

$$= (\lambda + 1)(\lambda + 2)(\lambda + 3) \tag{10.58}$$

Eigen values are

$$\lambda_1 = -1 \ , \lambda_2 = -2 \ , \lambda_3 = -3$$

**Example 10.11:** In reference to Matrix A in Example 10.10, let's define a set of new state variables, $z_1, z_2$, and $z_3$ by transformation.

***Solution***

$$\begin{bmatrix} x_1 \\ x_2 \\ x_3 \end{bmatrix} = \begin{bmatrix} 1 & 1 & 1 \\ 1 & -2 & -3 \\ 1 & 4 & 9 \end{bmatrix} \begin{bmatrix} z_1 \\ z_2 \\ z_3 \end{bmatrix} \tag{10.59}$$

$$x = PZ$$

Also,

$$\dot{X} = A_X + B_U$$

$$P\dot{z} = AP_Z + B_U$$

$$\dot{z} = P^{-1}APZ + P^{-1}BU$$

$$\begin{bmatrix} \dot{z}_1 \\ \dot{z}_2 \\ \dot{z}_3 \end{bmatrix} = \begin{bmatrix} 3 & 2.5 & 0.5 \\ -3 & -4 & -1 \\ 1 & 1.5 & 0.5 \end{bmatrix} \begin{bmatrix} 0 & 1 & 0 \\ 0 & 0 & 1 \\ -6 & -11 & -6 \end{bmatrix} \begin{bmatrix} 1 & 1 & 1 \\ -1 & -2 & -3 \\ 1 & 4 & 9 \end{bmatrix} \begin{bmatrix} z_1 \\ z_2 \\ z_3 \end{bmatrix} + \begin{bmatrix} 3 & 2.5 & 0.5 \\ -3 & -4 & -1 \\ 1 & 1.5 & 0.5 \end{bmatrix} \begin{bmatrix} 0 \\ 0 \\ 6 \end{bmatrix} U$$

(10.60)

$$= \begin{bmatrix} \dot{z}_1 \\ \dot{z}_2 \\ \dot{z}_3 \end{bmatrix} = \begin{bmatrix} -1 & 0 & 0 \\ 0 & -2 & 0 \\ 0 & 0 & -3 \end{bmatrix} \begin{bmatrix} z_1 \\ z_2 \\ z_3 \end{bmatrix} + \begin{bmatrix} 3 \\ -6 \\ 3 \end{bmatrix} u \qquad (10.61)$$

Eq. 10.61 is also a state equation that describes the same system. The output equations are then as follows:

$$Y = \begin{bmatrix} 1 & 0 & 0 \end{bmatrix} \begin{bmatrix} 1 & 1 & 1 \\ -1 & -2 & -3 \\ 1 & 4 & 9 \end{bmatrix} \begin{bmatrix} z_1 \\ z_2 \\ z_3 \end{bmatrix} \qquad (10.62)$$

$$= \begin{bmatrix} 1 & 0 & 0 \end{bmatrix} \begin{bmatrix} z_1 \\ z_2 \\ z_3 \end{bmatrix}$$

It may be noticed that transformation matrix $P$ modifies the coefficient matrix of $z$ into a diagonal matrix. Also, notice that diagonal elements of $P^{-1}AP$ are identical to the eigenvalues of $A$.

### 10.7.2 Invariance of Eigenvalues

To prove the eigenvalues under the linear transformation, it must be shown that characteristic polynomials $|\lambda I - A|$ and $|\lambda^{-1} - P^{-1}AP|$ are identical since the determinant of a product is the product of the determinant. Hence, following is obtained.

$$|\lambda^{-1} - P^{-1}AP| = |\lambda P^{-1}P - P^{-1}AP| \qquad (10.63)$$

$$= |P^{-1} - (\lambda I - A)P|$$

$$= |P^{-1} - \lambda I - AP|$$

$$= \left|P^{-1} - P\lambda I - A\right|$$

$$\left|\lambda^{-1} - P^{-1}AP\right| = \left|P^{-1} - P\lambda I - A\right|$$

$$= \left|\lambda I - A\right| \tag{10.64}$$

The eigenvalues of $A$ are invariant under linear transformation.

### 10.7.3 Diagonalization of $n \times n$ Matrix

Let,

$$A = \begin{bmatrix} 0 & 1 & 0 & 0 & \ldots & 0 \\ 0 & 0 & 1 & 0 & \ldots & 0 \\ . & . & . & . & \ldots & 1 \\ 0 & 0 & 0 & 0 & \ldots & 1 \\ -a_n & -a_{n-1} & . & . & \ldots & a_1 \end{bmatrix} \tag{10.65}$$

with $x = Pz$

$$P = \begin{bmatrix} 1 & 1 & . & . & .. & 1 \\ \lambda_1 & \lambda_2 & 1 & 0 & .. & \lambda_n \\ \lambda_1^2 & \lambda_2^2 & . & . & .. & \lambda_n^2 \\ \lambda_1^{n-1} & \lambda_2^{n-1} & 0 & 0 & .. & \lambda_n^{n-1} \end{bmatrix} \tag{10.66}$$

$\lambda_1, \lambda_2, \ldots, \lambda_n$ are n distinct values of $A$

$$P^{-1}AP = \begin{bmatrix} \lambda & 0 & \ldots & 1 \\ 0 & \lambda_2 & \ldots & 0 \\ . & . & \ldots & . \\ . & . & \ldots & . \\ 0 & 0 & \ldots & \lambda_n \end{bmatrix} \tag{10.67}$$

If there are multiple eigenvalues, then

$$\left|\lambda I - q^{-1}Aq\right| \text{ and } \left|\lambda I - A\right|$$

where, $A = \begin{bmatrix} 0 & 1 & 0 \\ 0 & 0 & 1 \\ -a_3 & -a_2 & -a_1 \end{bmatrix}$

Let $x = sz$

$$sA = \begin{bmatrix} 0 & 1 & 0 \\ \lambda_1 & 1 & \lambda_3 \\ \lambda_1^2 & 2\lambda_1 & \lambda_3^2 \end{bmatrix}$$

will yield,

$$s^{-1}As = \begin{bmatrix} \lambda_1 & 1 & 0 \\ 0 & \lambda_1 & 0 \\ 0 & 0 & \lambda_3 \end{bmatrix} \tag{10.68}$$

is called Jordan canonical form.

## 10.8 METHOD OF DECOUPLING A TRANSFER FUNCTION

A feedback control system has a transfer function

$$\frac{Y(s)}{U(s)} = \frac{6}{s^3 + 6s^2 + 11s + 6} = \frac{6}{(s+1)(s+2)(s+3)} \tag{10.69}$$

$$\frac{Y(s)}{U(s)} = \frac{3}{s+3} - \frac{6}{s+2} + \frac{3}{s+3}$$

$$Y(s) = \frac{3}{s+1}U(s) + \frac{-6}{s+2}U(s) + \frac{3}{s+1}U(s) \tag{10.70}$$

Define

$$x_1(s) = \frac{3}{s+1}U(s),\ x_2(s) = \frac{-6}{s+2}U(s),\ x_3(s) = \frac{3}{s+1}U(s) \tag{10.71}$$

$$\dot{x}_1 = -x_1 + 3u$$

$$\dot{x}_2 = -2x_2 - 6u$$

$$\dot{x}_3 = -3x_3 + 3u \tag{10.72}$$

Previous equation can also be written as

$$= \begin{bmatrix} \dot{x}_1 \\ \dot{x}_2 \\ \dot{x}_3 \end{bmatrix} = \begin{bmatrix} -1 & 0 & 0 \\ 0 & -2 & 0 \\ 0 & 0 & -3 \end{bmatrix}\begin{bmatrix} x_1 \\ x_2 \\ x_3 \end{bmatrix} + \begin{bmatrix} 3 \\ -6 \\ 3 \end{bmatrix}u \tag{10.73}$$

$$Y(s) = x_1(s) + x_2(s) + x_3(s) \tag{10.74}$$

$$Y = X_1 + X_2 + X_3$$

The block diagram of Eq. 10.71 is shown in Figure 10.9.

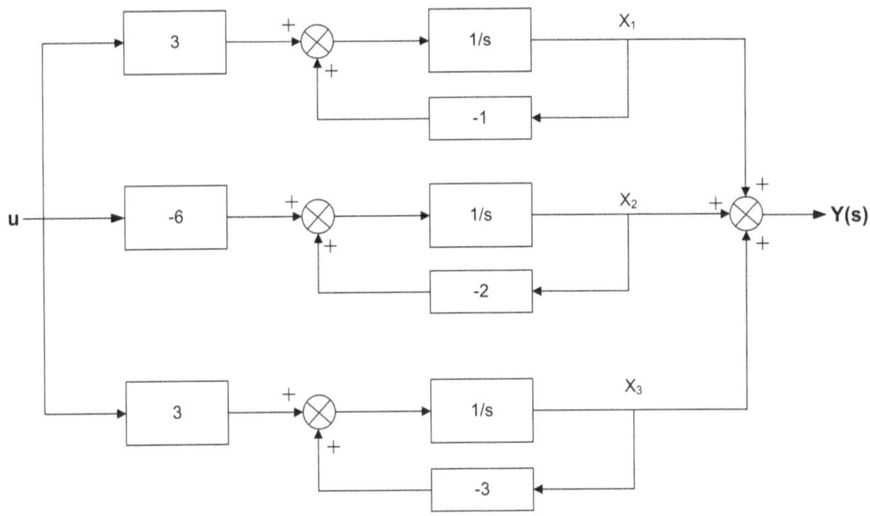


**FIGURE 10.9** Block diagram of Eq. 10.71.

$$Y = \begin{bmatrix} 1 & 1 & 1 \end{bmatrix} \begin{bmatrix} X_1 \\ X_2 \\ X_3 \end{bmatrix} \tag{10.75}$$

## 10.9 CAYLEY HAMILTON METHOD

Caley Hamilton's theorem states that square matrix A satisfies its own characteristic equation. Let $f(\lambda)$ be the characteristic equation which is represented by

$$f(\lambda) = |\lambda I - A| = 0$$

$$f(\lambda) = \lambda^n + a_1\lambda^{n-1} + a_2\lambda^{n-2} + \ldots a_{n-1}\lambda + a_n \mathrm{I} = 0 \tag{10.76}$$

The matrix A has to satisfy the characteristic equation as per Caley Hamilton's theorem. Therefore, we have

$$f(A) = A^n + a_1 A^{n-1} + a_2 A^{n-2} + \ldots + a_{n-1}A + a_n \mathrm{I} = 0 \tag{10.77}$$

where $I$ is the identity.

Let's consider the following matrix polynomial.

$$p(A) = \beta_0 I + \beta_1 A + \beta_2 A^2 + \ldots + \beta_n A^n + \beta_{n+1} A^{n+1} + \ldots \tag{10.78}$$

The polynomial represented by Eq 10.78 has a higher degree than A. The scalar polynomial is considered to calculate the previous polynomial.

$$p(\lambda) = \beta_0 + \beta_1\lambda + \beta^2\lambda^2 + \ldots + \beta^n\lambda^n + \beta^{n+1}\lambda^{n+1} + \ldots \tag{10.79}$$

From Eq 10.77 and Eq 10.79, we have

$$\frac{p(\lambda)}{f(\lambda)} = Q(\lambda) + \frac{R(\lambda)}{f(\lambda)} \tag{10.80}$$

where R(λ) is the remainder of polynomial.

$$p(\lambda) = f(\lambda)Q(\lambda) + R(\lambda) \tag{10.81}$$

The order of the remainder polynomial is less than that of the characteristic polynomial, and it can be represented as shown in Eq. 10.82.

$$R(\lambda) = a_0 + a_1\lambda + a_2\lambda^2 + \ldots + a_n\lambda^{n-1} \tag{10.82}$$

Let $\lambda_1, \lambda, \ldots, \lambda_n$ be the distinct Eigenvalues of matrix A for $fi\lambda_i = 0$ where $i = 1, 2, \ldots, n$ are roots of $f(\lambda) = 0$. These need to be evaluated for all Eigenvalues i.e $\lambda_{1,}\lambda_2, \ldots, \lambda_n$.

Therefore, we have

$$p(\lambda_i) = f(\lambda_i) + \varphi(\lambda_i) + R(\lambda_i)\, for\;\; i = 1, 2, \ldots, n \tag{10.83}$$

Since $f(\lambda_i) = 0$, we have from Eq. 10.82

$$p(\lambda_i) = R(\lambda_i) = \alpha_0 + \alpha_1\lambda_i + \alpha_2\lambda_i^2 + \ldots + \alpha_{n-1}\lambda^{n-1} \tag{10.84}$$

which is to be solved to obtain $\alpha_0, \alpha_1, \alpha_2 + \ldots + \alpha_n$.

Replacing λ in scalar polynomial by matrix A in Eq 10.84, we have

$$p(A) = f(A)Q(A) + R(A).$$

Since $f(A) = 0$ by Cayley Hamilton theorem, we have

$$p(A) = R(A) = \alpha_0 + \alpha_1 A + \alpha_2 A^2 + \ldots + \alpha_{n-1}A^{n-1} \tag{10.85}$$

The matrix polynomial can be evaluated.

**Example 10.12:** Find $\mathrm{f(A) = A^{10}}$ for $\mathrm{A} = \begin{bmatrix} 0 & 1 \\ -3 & -4 \end{bmatrix}$ using the Cayley Hamilton theorem.

***Solution***

The function is $f(A) = A^{10}$. Therefore,

$$f(\lambda) = \lambda^{10}$$

Step 1. Find eigenvalues.

$$|\lambda I - A| = \begin{vmatrix} \lambda & -1 \\ 3 & \lambda+4 \end{vmatrix} = 0 \tag{10.86}$$

$$\lambda(\lambda+4)+3=0$$

$$(\lambda+1)(\lambda+3)=0,\ \lambda_1 = -1 \text{ and } \lambda_2 = -3$$

Step 2. $R(\lambda) = \alpha_0 + \alpha_1\lambda = \mathrm{p}(\lambda)$, since $n = 2$

Step 3. Substitute $\lambda_1$ *and* $\lambda_2$ in $R(\lambda)$

$$\alpha_0 - \alpha_1 = (-1)^{10} = 1 \tag{10.87}$$

and

$$\alpha_0 - 3\alpha_1 = (-3)^{10} = 59049 \tag{10.88}$$

Step 4: Solving Eq 10.87 and Eq 10.88, we get

$$\alpha_0 = -29523 \quad \textit{and} \quad \alpha_1 = -2952$$

Step 5: $f(A) = A^{10} = R(A) = \alpha_0 I + \alpha_1 A$

$$-29523\begin{bmatrix} 1 & 0 \\ 0 & 1 \end{bmatrix} - -29524\begin{bmatrix} 0 & 1 \\ -3 & -4 \end{bmatrix} = \begin{bmatrix} -2523 & -29524 \\ 88572 & 88573 \end{bmatrix}$$

$$A^{10} = \begin{bmatrix} -29523 & -29524 \\ 88572 & 88573 \end{bmatrix}$$

**Example 10.13:** Find the transition matrix for $A = \begin{bmatrix} 0 & 1 \\ -2 & -3 \end{bmatrix}$ using the Cayley Hamilton theorem.

***Solution***

The function to be evaluated is

$$f(A) = e^{AT}$$

$$f(\lambda) = e^{AT}$$

Step 1. Find eigenvalues

$$|\lambda I - A| = \begin{vmatrix} \lambda & -1 \\ 2 & \lambda+3 \end{vmatrix} = 0$$

$$\lambda(\lambda+3)+2=0$$

$$(\lambda+1)(\lambda+2)=0$$

$$\lambda=-1,\ \lambda=-2.$$

Step 2. $R(\lambda)=\alpha_0+\alpha_1\lambda=\mathrm{p}(\lambda)\quad ,as\ n=2$

Step 3. Substitute $\lambda_1\ and\ \lambda_2$ in $R(\lambda)$

$$\alpha_0-\alpha_1=e^{-t} \tag{10.89}$$

$$\alpha_0-3\alpha_1=e^{-3t} \tag{10.90}$$

Step 4. Solving Eq 10.89 and Eq 10.90 we have

$$\alpha_0=2e^{-t}-e^{-2t}\quad and\quad \alpha_1=e^{-t}-e^{-2t}$$

Step 5. $f(A)=A^{10}=R(A)=\alpha_0 I+\alpha_1 A$

$$f(A)=\left(2e^{-t}-e^{-2t}\right)\begin{bmatrix}1&0\\0&1\end{bmatrix}+\left(e^{-t}-e^{-2t}\right)\begin{bmatrix}0&1\\-3&-4\end{bmatrix}$$

$$f(A)=\begin{bmatrix}2e^{-t}-e^{-2t} & e^{-t}-e^{-2t}\\ -2e^{-t}+e^{-2t} & -e^{-t}+2e^{-2t}\end{bmatrix} \tag{10.91}$$

## 10.10 SIMILARITY TRANSFORMATION METHOD

A similarity transformation which diagonalizes 'A' can be created using a set of eigenvectors, one for each eigenvalue (eigen means characteristic). Each eigenvector can be multiplied by a constant that works just as well. The following procedure can be used to get P.

**Example 10.14:** Evaluate the following.

a) Find the eigenvalues $\lambda_i$ where

$$|SI - A| = 0$$

b) Find the eigenvectors $x_i$ for each $\lambda_i$
c) Let $P^{-1}$ be a matrix consisting of eigenvectors

$$P^{-1} = [x_1, x_2, \ldots, x_n]$$

$$PAP^{-1} = \begin{bmatrix} \lambda_1 & 0 & \ldots\ldots \\ 0 & \lambda_2 & \ldots\ldots \end{bmatrix}$$

***Solution***

For example,

$$\begin{bmatrix} \dot{x}_1 \\ \dot{x}_2 \\ \dot{x}_3 \end{bmatrix} = \begin{bmatrix} -1 & -2 & 0 \\ 1 & 2 & 0 \\ -2 & -1 & -3 \end{bmatrix} \begin{bmatrix} x_1 \\ x_2 \\ x_3 \end{bmatrix} + \begin{bmatrix} 1 \\ 0 \\ 0 \end{bmatrix} r = AX + BU \tag{10.92}$$

$$\begin{bmatrix} y_1 \\ y_2 \end{bmatrix} = \begin{bmatrix} 1 & 0 & 1 \\ 1 & -1 & 0 \end{bmatrix} \begin{bmatrix} x_1 \\ x_2 \\ x_3 \end{bmatrix} = cx \tag{10.93}$$

The characteristic polynomial is,

$$\begin{bmatrix} \lambda+1 & 2 & 0 \\ -1 & \lambda-2 & 0 \\ 2 & 1 & \lambda+3 \end{bmatrix} = 0, \lambda(\lambda-1)(\lambda+3) \tag{10.94}$$

The eigenvalues can be selected in any order.

$$\lambda_1 = 0, \lambda_2 = 1, \lambda_3 = -3$$

Then, the first eigenvector is found by solving,

$$\begin{bmatrix} \lambda+1 & 2 & 0 \\ -1 & \lambda-2 & 0 \\ 2 & 1 & \lambda+3 \end{bmatrix} \begin{bmatrix} x_{11} \\ x_{21} \\ x_{31} \end{bmatrix} = \begin{bmatrix} 0 \\ 0 \\ 0 \end{bmatrix} \tag{10.95}$$

$$X_1 = \begin{bmatrix} x_{11} \\ x_{21} \\ x_{31} \end{bmatrix} \tag{10.96}$$

With the first eigenvalue being zero,

$$\begin{aligned} x_{11} + 2x_{21} &= 0 \\ -x_{11} - 2x_{21} &= 0 \\ 2x_{11} + x_{21} + 3x_{31} &= 0 \end{aligned} \tag{10.97}$$

The first two equations are equivalent, so infinite solutions exist. It is possible to select $x_{21}$ arbitrarily (say −1), then $x_{11} = 2$ and $x_{31} = -1$. By proceeding similarly to these eigenvectors results in:

$$x_1 = \begin{bmatrix} 2 \\ -1 \\ -1 \end{bmatrix}, x_2 = \begin{bmatrix} 4 \\ -4 \\ -1 \end{bmatrix}, x_3 = \begin{bmatrix} 0 \\ 0 \\ 1 \end{bmatrix}$$

Any non-zero multiply of an eigenvector also works by collecting these eigenvectors into $P^{-1}$.

$$P = \begin{bmatrix} 1 & 1 & 0 \\ -\frac{1}{4} & -\frac{1}{2} & 0 \\ \frac{3}{4} & \frac{3}{2} & 1 \end{bmatrix}, P^{-1} = \begin{bmatrix} 2 & 4 & 0 \\ -1 & -4 & 0 \\ -1 & -1 & 1 \end{bmatrix} \tag{10.98}$$

Following gives a state variable representation for which the state compiling matrix is diagonal.

$$A' = PAP^{-1} = \begin{bmatrix} 1 & 1 & 0 \\ -\frac{1}{4} & -\frac{1}{2} & 0 \\ \frac{3}{4} & \frac{3}{2} & 1 \end{bmatrix} \begin{bmatrix} -1 & -2 & 0 \\ 1 & 2 & 0 \\ -2 & -1 & -3 \end{bmatrix} \begin{bmatrix} 2 & 4 & 0 \\ -1 & -4 & 0 \\ -1 & -1 & 1 \end{bmatrix} \tag{10.99}$$

$$A' = \begin{bmatrix} 1 & 1 & 0 \\ -\frac{1}{4} & -\frac{1}{2} & 0 \\ \frac{3}{4} & \frac{3}{2} & 1 \end{bmatrix} \begin{bmatrix} 0 & 4 & 0 \\ 0 & -4 & 0 \\ 0 & -1 & -3 \end{bmatrix} = \begin{bmatrix} 0 & 0 & 0 \\ 0 & 1 & 0 \\ 0 & 0 & -3 \end{bmatrix} \tag{10.100}$$

$$B^{\prime} = PB = \begin{bmatrix} 1 & 1 & 0 \\ -\frac{1}{4} & -\frac{1}{2} & 0 \\ \frac{3}{4} & \frac{3}{2} & 1 \end{bmatrix} \begin{bmatrix} 1 \\ 0 \\ 0 \end{bmatrix} = \begin{bmatrix} 1 \\ -\frac{1}{4} \\ \frac{3}{4} \end{bmatrix} \tag{10.101}$$

$$C^{\prime} = CP^{-1} = \begin{bmatrix} 1 & 0 & 1 \\ 1 & -1 & 0 \end{bmatrix} \begin{bmatrix} 2 & 4 & 0 \\ -1 & -4 & 0 \\ 1 & -1 & 1 \end{bmatrix} = \begin{bmatrix} 1 & 3 & 1 \\ 3 & 8 & 0 \end{bmatrix} \tag{10.102}$$

## 10.11 COMPLEX CONJUGATE CHARACTERISTIC ROOTS

The complex conjugate characteristic roots can be explained with the following examples.

**Example 10.15**

$$\frac{Y(s)}{U(s)} = \frac{6s^2+20s+8}{(s+2)(s^2+2s+10)} = \frac{-2}{s+2} + \frac{4+j}{s+1+3j} = \frac{4-j}{s+1-3j} \tag{10.103}$$

$$sX_1(s) = -2x_1(s) + U(s)$$

$$sX_2(s) = (-1+j3)X_2(s) \tag{10.104}$$

$$sX_3(S) = (-1-j3)X_3(s)$$

$$Y(s) = -2x_1(\mathrm{s}) + (4+\mathrm{j})x_2(s) + (4-j)x_3(s)$$

Or

$$\begin{bmatrix} \dot{x}_1 \\ \dot{x}_2 \\ \dot{x}_3 \end{bmatrix} \begin{bmatrix} -2 & 0 & 0 \\ 0 & -1+j3 & 0 \\ 0 & 0 & -1-j3 \end{bmatrix} \begin{bmatrix} X_1 \\ X_2 \\ X_3 \end{bmatrix} + \begin{bmatrix} 1 \\ 1 \\ 1 \end{bmatrix} U \tag{10.105}$$

$$Y = -2 \quad 4+j \quad 4-j \begin{bmatrix} X_1 \\ X_2 \\ X_3 \end{bmatrix} \tag{10.106}$$

Since dealing with complex members is not possible, let's combine the conjugate partial fraction items.

$$\frac{4+j}{s+1+j}+\frac{4-j}{s+1-j3}=\frac{8s+14}{5^2+25+10}$$

This portion of the system may be represented in phase variable form. The state equations for this alternative arrangement are given by

$$sX_1(s)=-2X_1(s)+U(s)$$

$$sX_2(s)=X_3(s)$$

$$sX_3(s)=-10X_2(s)-2X_3(s)+U(s) \tag{10.107}$$

$$Y(s)=-2X_1(\mathrm{s})+14X_2(s)+8X_3(s)$$

$$\begin{bmatrix}\dot{X}_1\\ \dot{X}_2\\ \dot{X}_3\end{bmatrix}=\begin{bmatrix}-2 & 0 & 0\\ 0 & 0 & 1\\ 0 & -10 & -2\end{bmatrix}\begin{bmatrix}X_1\\ X_2\\ X_3\end{bmatrix}+\begin{bmatrix}1\\ 0\\ 1\end{bmatrix}U \tag{10.108}$$

$$\mathrm{Y}=\begin{bmatrix}-2 & 14 & 8\end{bmatrix}\begin{bmatrix}X_1\\ X_2\\ X_3\end{bmatrix} \tag{10.109}$$

## 10.12 COMPUTATION OF $e^{At}$

We can explain the process by considering certain examples given next.

**Example 10.16:** Compute $e^{At}$ for the matrix $A=\begin{bmatrix}0 & 1\\ 0 & -2\end{bmatrix}$.

***Solution***

**Method 1**

$$e^{At}=1+\sum_{k=1}^{\infty}\frac{A^k t^k}{t!}$$

$$=\begin{bmatrix}1 & 0\\ 0 & 1\end{bmatrix}+\begin{bmatrix}0 & 1\\ 0 & -2\end{bmatrix}t+\begin{bmatrix}0 & 1\\ 0 & -2\end{bmatrix}\frac{t^2}{2!}+\ldots \tag{10.110}$$

$$= \begin{bmatrix} 1 & \frac{1}{2} - \frac{t}{2}[1 - 2t + \frac{(2t)^2}{2!} + \frac{(2t)^2}{3!} + \ldots \\ 0 & 1 - 2t + \frac{(2t)^2}{2!} - \frac{(-2t)^3}{3!} + \ldots \end{bmatrix} \quad (10.111)$$

$$= \begin{bmatrix} 1 & \frac{1}{2}\left(1 - e^{-2t}\right) \\ 0 & e^{-2t} \end{bmatrix} \quad (10.112)$$

**Method 2**

$$A = \begin{bmatrix} 0 & 1 \\ 0 & -2 \end{bmatrix}$$

$$[\lambda I—A] = \begin{bmatrix} \lambda & -1 \\ 0 & \lambda + 2 \end{bmatrix}, \lambda(\lambda+2) = 0, \lambda = 0, \lambda = -2$$

$$P = \begin{bmatrix} 1 & 1 \\ 0 & -2 \end{bmatrix}$$

$$e^{At} = Pe^{-\lambda t}P^{-1} \quad (10.113)$$

$$e^{At} = \begin{bmatrix} 1 & 1 \\ 0 & -2 \end{bmatrix} \begin{bmatrix} e^{t} & 0 \\ 0 & e^{-2t} \end{bmatrix} \begin{bmatrix} 1 & \frac{1}{2} \\ 0 & -\frac{1}{2} \end{bmatrix}$$

$$e^{At} = \begin{bmatrix} 1 & \frac{1}{2}\left(1 - e^{-2t}\right) \\ 0 & e^{-2t} \end{bmatrix} \quad (10.114)$$

**Method 3**

$$[\mathrm{sI} - \mathrm{A}] = \begin{bmatrix} s & 0 \\ 0 & s \end{bmatrix} - \begin{bmatrix} 0 & 1 \\ 0 & -2 \end{bmatrix} = \begin{bmatrix} s & -1 \\ 0 & s + 2 \end{bmatrix}$$

$$[\mathrm{sI} - \mathrm{A}]^{-1} = \begin{bmatrix} \frac{1}{\mathrm{s}} & \frac{1}{\mathrm{s}(\mathrm{s}+2)} \\ 0 & \frac{1}{\mathrm{s}+2} \end{bmatrix} \quad (10.115)$$

$$e^{At} = \mathcal{L}^{-1}\left[sI - A\right] = \begin{bmatrix} 1 & \dfrac{1}{2\left(1-e^{-2t}\right)} \\ 0 & e^{-2t} \end{bmatrix} \tag{10.116}$$

## 10.13 CONTROLLABILITY AND OBSERVABILITY

### 10.13.1 Definition

A system is said to be controllable at time $t_0$ if it is possible using an unconstrained control vector to transfer the system from any initial state, $x(t_0)$ to any other state in an infinite interval of time. A system is said to be observable at $t_0$ if it is in state, $x(t_0)$, and it is possible to determine this state by observing the output over a finite time interval.

### 10.13.2 Linear Independence of Vectors

The vectors $X_1, X_2, X_3, \ldots, X_n$ are said to be linearly independent if:

$$c_1X_1 + c_2X_2 + \ldots + c_nX_n = 0 \tag{10.117}$$

where $c_1, c_2, \ldots, c_n$ are constants.

$c_1 = c_2 = \ldots = c_n = 0$. Conversely, the vectors $X_1, X_2, X_3, \ldots, X_n$ are linearly dependent if and only if $X_c$ can be expressed as a linear combination of $X_j\left(i = 1,2,\ldots,n \, and \, j \neq i\right)$

$$X_i = \sum_{\substack{j=1 \\ j\neq i}}^{n} c_j X_j$$

for some set of constants $c_j$. If the n × n matrix is singular, then vectors are dependent.

### 10.13.3 Complete Controllability of Continuous-Time Systems

$$\dot{X} = AX + BU \tag{10.118}$$

If every state is controllable, then the system is said to be completely state controllable without loss of generality. We can assume that the final state is the origin of the state space and that the initial time is zero $\left(t_0 = 0\right)$.

The solution of the equation is:

$$X(t) = e^{At}X(0) + \int_0^t e^{A(t-\tau)}Bu(\tau)\,d\tau \tag{10.119}$$

Applying the definition of complete state controllability, we have

$$X(t_1) = 0 = e^{At_1}X(0) + \int_0^{t_1} e^{A(t_1-\tau)}Bu(\tau)d\tau \tag{10.120}$$

or

$$X(0) = -\int_0^{t} e^{-A\tau}Bu(\tau)d\tau$$

But $e^{-A\tau}$ can be written as

$$e^{-A\tau} = -\sum_{k=0}^{n-1} A^K B \int_0^{t_1} \propto_K(\tau)A^K \tag{10.121}$$

Let us put

$$\int_0^{t_1} \propto_K(\tau)U(\tau)d\tau = B_K$$

so that

$$X(0) = -\sum_{k=0}^{n-1} A^K B = -\left[B : AB! \cdots A^{n-1}B\right]\begin{bmatrix} B_0 \\ B_1 \\ \vdots \\ B_{n-1} \end{bmatrix} \tag{10.122}$$

If this system is completely controllable, then given any initial state $X(0)$, Eq. 10.122 must be satisfied. This requires the rank of the $n \times n$ matrix where $\left[B : AB : A^2B : \cdots \therefore A^{n-1}B\right]$ be of rank n. Thus the system $\dot{X} = AX + BU$ is completely state controllable if and only if the vectors, $B, AB, A^2B, \cdots, A^{n-1}B$ are linearity independent $\left[B : AB : \cdots A^{n-1}B\right]$ of rank n.

The previous result can be extended to the case where the control vector U is $r$ dimensional.

$$\left[B : AB : A^2B : \cdots \therefore A^{n-1}B\right]P \tag{10.123}$$

Eq. 10.123 is known as the controllability matrix.

**Example 10.17**

Consider the system next.

$$\begin{bmatrix} \dot{x}_1 \\ \dot{x}_2 \end{bmatrix} = \begin{bmatrix} 1 & 1 \\ 2 & -1 \end{bmatrix}\begin{bmatrix} x_1 \\ x_2 \end{bmatrix} + \begin{bmatrix} 0 \\ 1 \end{bmatrix}u$$

*Solution*

$$\left[B : AB\right] = \begin{bmatrix} 0 & 1 \\ 1 & -1 \end{bmatrix}$$ is non-singular. Therefore, it is completely controllable.

### 10.13.4 Observability

Consider the unforced system described by the following equation

$$\dot{X}_1 = AX$$

$$Y = CX \tag{10.124}$$

The system is said to be completely observable if every state of $\dot{X}(t_0)$ can be determined from the observation of $\dot{y}(t)$ over a finite time interval $t_0 \leq t \leq t'$.

The system is, therefore, completely observable if every transition of state eventually affects every element of the output vector. The concept of observability is useful in resolving the problem of reconstructing the measurable state of variables from measurable ones in the minimum possible length of time.

Observability is important to estimate the unmeasurable state variables to construct control signals. Such estimation is possible only if the system is observable.

We consider only unforced systems here since u(t) is known, and so are B and C Therefore, in equation here,

$$y(t) = ce^{At}\dot{x}(0) + c\int_0^t e^{A(t-\tau)}Bu(\tau)d\tau + Du$$

the last two terms are known quantities, and therefore, study of unforced system is sufficient.

Now consider,

$$\dot{X} = Ax$$

$$Y = cx$$

$$Y = ce^{At}x(0)$$

Noting that

$$e^{At} = \sum_{k=0}^{n-1} \propto_k (t) A^K$$

$$Y(t) = \sum_{k=0}^{n-1} \alpha_k(t) A^k x(0) \tag{10.125}$$

or

$$y(t) = \propto_0(t)cx(0) + \propto_1(t)cAx(0) + \ldots + \propto_{n-1}(t)cA^{n-1}x(0)$$

If the system is completely observable, the given $y(t)$ is completely observable over $0 \le t < t$, and x(0) is uniquely determined from the previous equation. It can be shown this requires the rank of $nm \times n$ matrix.

$$\begin{bmatrix} c \\ cA \\ \vdots \\ cA^{n-1} \end{bmatrix} \text{to be rank of n}$$

From this analysis, we can state the condition for complete observability as the previous matrix to be nonsingular or $\left[c^* : A^*c^* : \ldots : \left(A^{n-1}\right)c^*\right]$ is of rank n.

The matrix is called the observability matrix.

**Example 10.18**

Consider

$$\dot{X} = AX + Bu$$

$$Y = Cx$$

***Solution***

$$A = \begin{bmatrix} 0 & 1 & 0 \\ 0 & 0 & 1 \\ -6 & -11 & -6 \end{bmatrix}, B = \begin{bmatrix} 0 \\ 0 \\ 1 \end{bmatrix}, c = \begin{bmatrix} 4 & 5 & 1 \end{bmatrix} \quad (10.126)$$

$$\begin{bmatrix} c \\ cA \\ cA^2 \end{bmatrix} = \begin{bmatrix} 4 & 5 & 1 \\ -6 & -7 & -1 \\ 6 & 5 & 1 \end{bmatrix}$$

Det: $4(7+5) - 5(6+6) + 1(-30+42) = 48 - 60 + 12 = 0$

Hence, the rank of a matrix is less than three. Therefore, the system is not observable.

## 10.14 RELATIONSHIP BETWEEN CONTROLLABILITY, OBSERVABILITY, AND TRANSFER FUNCTION

The transfer function has no pole-zero cancellations, and the system is completely state-controlled and completely observable. This means that the cancelled transfer function does not carry along all the information characterizing the dynamic system.

**Example 10.19:** The observability of the system described by following matrices.

$$A = \begin{bmatrix} -1 & 0 \\ 0 & 2 \end{bmatrix}, C = \begin{bmatrix} 1 & 3 \end{bmatrix}$$

***Solution***

$$\begin{bmatrix} C \\ \cdots \\ CA \end{bmatrix} = \begin{bmatrix} 1 & 3 \\ -1 & -6 \end{bmatrix}$$

Determinant is non-zero. The system is observable.

And

$$A = \begin{bmatrix} -1 & 0 \\ 0 & 2 \end{bmatrix}, C = \begin{bmatrix} 0 & 1 \end{bmatrix}$$

$$\begin{bmatrix} C \\ \cdots \\ CA \end{bmatrix} = \begin{bmatrix} 0 & 1 \\ 0 & -2 \end{bmatrix}$$

Determinant is zero. The system is not observable.

## 10.15 OBTAINING STATE SPACE EQUATION IN CANONICAL FORMS

Consider the system defined by

$$y^{(n)} + a_1 y^{(n+1)} + \ldots + a_{n-1} y + a_n y = b_0 u^{(n)} + b_1 u^{n-1} + \ldots + b_{n-1} u + b_n u$$

where, $u$ is the input and y is the output. This can also be written as;

$$\frac{Y(s)}{U(s)} = \frac{b_0 s^n + b_1 s^{n-1} + \ldots + b_{n-1} s + b_n}{s^n + a_1 s^{n-1} + a_2 s^{n-2} + \ldots + a_{n-1} s + a_n} \tag{10.127}$$

There are three methods of state space representation factor:

a) Direct programming method
b) Nested programming method
c) Partial factor expansion programming method for obtaining diagonal or Jordan canonical form.

### 10.15.1 Direct Programming Method

$$\frac{Y(s)}{U(s)} = b_0 + \frac{(b_1 - ab_0)s^{n-1} + \ldots + (b_{n-1}a_{n-1}b_0)s + (b_n - a_n b_0)}{s^n + a_1 s^{n-1} + \ldots + a_{n-1}s + a_n} \quad (10.128)$$

Which can be modified to

$$Y(s) = b_0 U(s) + \hat{Y}(s)$$

where,

$$\hat{Y}(s) = \frac{(b_1 - a_1 b_0)s^{n-1} + \ldots + (b_{n-1} - a_{n-1}b_0)s + (b_n - a_n b_0)}{s^n + a_1 s^{n-1} + \ldots + a_{n-1}s + a_n} U(s) \quad (10.129)$$

Let us write the previous equation in the following form:

$$\frac{\hat{Y}(s)}{(b_1 - a_1 b_0)s^{n-1} + \ldots + (b_{n-1} - a_{n-1}b_0)s + (b_n - a_n b_0)}$$

$$= \frac{U(s)}{s^n + s^{n-1} + \ldots + a_{n-1}s + a_n} = Q(s) \quad (10.130)$$

From Eq. 10.130, the following two equations can be obtained.

$$s^n Q(s) = -a_1 s^{n-1} Q(s) \ldots - a_{n-1} s Q(s) - a_n Q(s) + U(s)$$

$$\hat{Y}(s) = (b_1 - a_1 b_0)s^{n-1} Q(s) + \ldots + (b_{n-1} - a_{n-1}b_0)sQ(s) + (b_n - a_n b_0)Q(s) \quad (10.131)$$

Now define variables as follows:

$$X_1(s) = Q(s)$$

$$X_2(s) = sQ(s)$$

$$\vdots$$

$$X_{n-1}(s) = s^{n-2} Q(s)$$

$$x_n(s) = s^{n-1} Q(s)$$

Then clearly,

$$sX_1(s) = X_2(s)$$
$$sX_2(s) = X_3(s)$$
$$\vdots$$
$$sX_{n-1}(s) = X_n(s)$$

Which may be written as:

$$\dot{x}_1 = x_2$$
$$\dot{x}_2 = x_3$$
$$\vdots$$
$$\dot{x}_{n-1} = x_n$$

Noting that $s^n Q(s) = sX_n(s)$ can be written as

$$sX_n(s) = -a_1 X_n - \ldots - a_{n-1} X_2(s) - a_n X_1(s) + U(s) \tag{10.132}$$
$$\dot{X}_n = -a_n x_1 - a_{n-1} x_2 - \ldots - a_1 x_n + u$$

Then,

$$\begin{aligned} Y(s) &= b_0 U(s) + (b_1 - a_1 b_0) s^{n-1} Q(s) + \ldots + (b_{n-1} - a_{n-1} b_0) sQ(s) \\ &\quad + (b_n - a_n b_0) Q(s) \\ &= b_0 U(s) + (b_1 - a_1 b_0) X_n(s) + \ldots + (b_{n-1} - a_{n-1} b_0) X_2(s) \\ &\quad + (b_n - a_n b_0) X_1(s) \end{aligned} \tag{10.133}$$

The inverse Laplace transform of this output equation becomes

$$y = (b_n - a_n b_0) x_1 + (b_{n-1} - a_{n-1} b_0) x_2 + \ldots + (b_1 - a_1 b_0) x_n + b_0 u$$

Then the state equation and output equation can be given by Eq. 10.134.

$$\begin{bmatrix} \dot{x}_1 \\ \dot{x}_2 \\ \vdots \\ \dot{x}_{n-1} \\ \dot{x}_n \end{bmatrix} = \begin{bmatrix} 0 & 1 & 0 & \cdots & 0 \\ 0 & 0 & 1 & \cdots & 0 \\ \vdots & \vdots & \vdots & & \vdots \\ 0 & 0 & 0 & \cdots & 1 \\ -a_n & -a_{n-1} & -a_{n-2} & \cdots & -a_1 \end{bmatrix} \begin{bmatrix} x_1 \\ x_2 \\ \vdots \\ x_{n-1} \\ x_n \end{bmatrix} + \begin{bmatrix} 0 \\ 0 \\ \vdots \\ \vdots \\ 1 \end{bmatrix} u \tag{10.134}$$

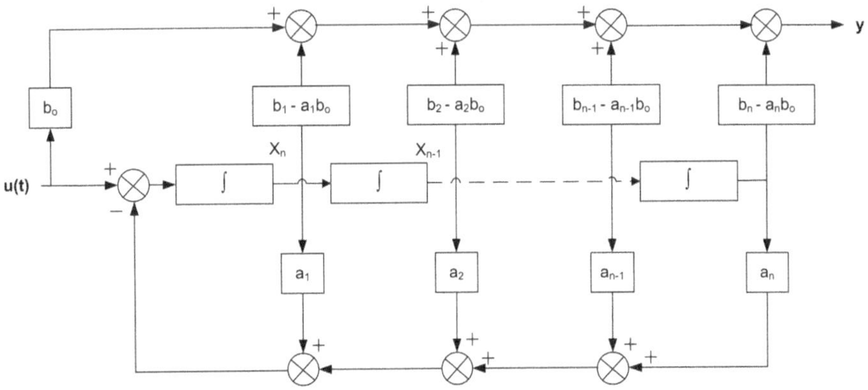


**FIGURE 10.10** Block diagram of controllable canonical form.

$$y = \begin{bmatrix} b_n - a_n b_0 & b_{n-1} - a_n - b_0 & b_1 - a_1 b_0 \end{bmatrix} \begin{bmatrix} x_1 \\ x_2 \\ \vdots \\ x_n \end{bmatrix} + b_0 u$$

Figure 10.10 shows a block diagram of controllable canonical form.

### 10.15.2 Nested Programming Method

Consider

$$s^n\left[Y(s) - b_0U(s)\right] + s^{n-1}\left[a_1Y(s) - b_1U(s)\right] + \cdots$$

$$\cdots + s\left[a_{n-1}Y(s) - b_{n-1}U(s)\right] + \left[a_nY(s) - b_nU(s)\right] = 0$$

$$Y(s) = b_0U(s) + \frac{1}{s}\left[b_1U(s) - a_1Y(s)\right] + \cdots$$

$$\cdots + \frac{1}{s^{n-1}}\left[b_{n-1}U(s) - a_{n-1}Y(s)\right] + \frac{1}{s^n}\left[b_nU(s) - a_nY(s)\right] \tag{10.135}$$

Now define the state variables as follows.

$$X_n(s) = \frac{1}{s}\left[b_1U(s) - a_1Y(s)a + X_{n-1}(s)\right]$$

$$X_{n-1}(s) = \frac{1}{s}\left[b_2U(s) - a_2Y(s) + X_{n-2}(s)\right] \tag{10.136}$$

$$\vdots$$

$$x_2(s) = \frac{1}{s}\left[b_{n-1}U(s) - a_{n-1}Y(s) + x_1(s)\right]$$

$$x_1(s) = \frac{1}{s}\left[b_n U(s) - a_n Y(s)\right]$$

Eq. 10.135 can be written as

$$Y(s) - b_n U(s) + X_n(s) \tag{10.137}$$

By substituting Eq. 10.137 into Eq. 10.136 and multiplying both sides of the equation by $s$, we get:

$$\begin{aligned}
s(s) &= X_{n-1}(s) - a_1 X_n(s) + (b_1 - a_1 b_0)U(s) \\
sX_{n-1}(s) &= X_{n-2}(s) - a_2 X_n(s) + (b_2 - a_2 b_0)U(s) \\
&\vdots \\
sX_2(s) &= X_1(s) - a_{n-1} X_n(s) + (b_{n-1} - a_{n-1} b_0)U(s) \\
sX_1(s) &= -a_n X_n(s) + (b_n - a_n b_0)U(s)
\end{aligned} \tag{10.138}$$

Taking inverse Laplace transform

$$\begin{aligned}
\dot{x}_1 &= -a_n x_n + (b_n - a_n b_0)u \\
\dot{x}_2 &= x_1 - a_{n-1} x_n + (b_{n-1} - a_{n-1} b_0)u \\
&\vdots \\
\dot{x}_{n-1} &= x_{n-2} - a_2 x_n + (b_2 - a_2 b_0)u \\
\dot{x}_n &= x_{n-1} - a_1 x_n + (b_1 - a_1 b_0)u
\end{aligned} \tag{10.139}$$

And $y = x_n + b_0 u$ from Eq. 10.137.

In the matrix form,

$$\begin{bmatrix} \dot{x}_1 \\ \dot{x}_2 \\ \vdots \\ \dot{x}_{n-1} \\ \dot{x}_n \end{bmatrix} = \begin{bmatrix} 0 & 0 & \cdots & 0 & -a_n \\ 1 & 0 & \cdots & 0 & -a_{n-1} \\ \cdot & \cdot & \cdots & \cdots & \cdots \\ \cdot & \cdot & \cdots & \cdots & \cdots \\ 0 & 0 & \cdots & 1 & -a_1 \end{bmatrix} \begin{bmatrix} x_1 \\ x_2 \\ \cdot \\ \cdot \\ x_n \end{bmatrix} + \begin{bmatrix} b_n - a_n b_0 \\ b_{n-1} - a_{n-1} b_0 \\ \cdot \\ \cdot \\ b_1 a_1 b_0 \end{bmatrix} u$$

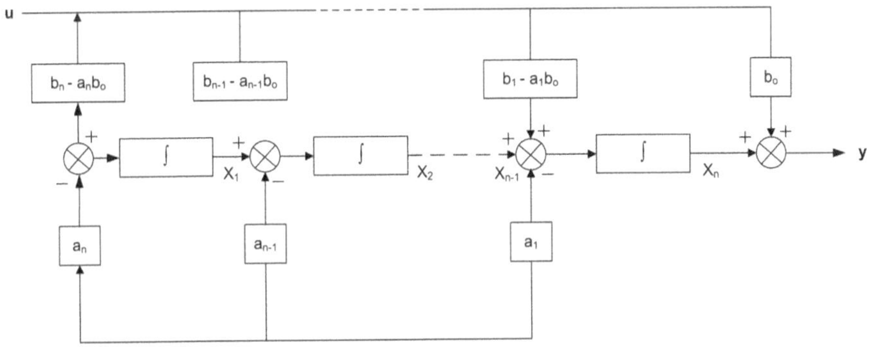


**FIGURE 10.11** Block diagram of an observable canonical form.

$$Y = \begin{bmatrix} 0 & 0 & \cdots & 0 & 1 \end{bmatrix} \begin{bmatrix} x_1 \\ x_2 \\ \cdot \\ \cdot \\ x_n \end{bmatrix} + b_0 u \tag{10.140}$$

Figure 10.11 shows a block diagram representation of an observable canonical form.

### 10.15.3 Partial Fraction Expansion Programming Method

Consider the transfer function

$$\frac{Y(S)}{U(S)} = \frac{b_0 s^n + b_1 s^{n-1} + \cdots + b_{n-1} s + b_n}{(s - p_1)(s - p_2)\cdots(s - p_n)} = b_0 + \frac{c_1}{s - p_1} + \frac{c_2}{s - p_2} + \cdots + \frac{c_n}{s - p_n}$$

$$Y(s) = b_0 U(s) + \frac{c}{s - p_1} U(s) + \frac{c_2}{s - p_2} U(s) + \cdots + \frac{c_n}{s - p_n} U$$

$$\begin{aligned} x_1(s) &= \frac{1}{s - p_1} U(s) \\ x_2(s) &= \frac{1}{s - p_2} U(s) \\ &\vdots \\ x_n(s) &= \frac{1}{s - p_n} U(s) \end{aligned} \tag{10.141}$$

Which may be written as

$$\begin{aligned} sX_1(s) &= p_1X_1(s)+U(s) \\ sX_2(s) &= p_2X_2(s)+U(s) \\ &\vdots \\ sX_n(s) &= p_nX_n(s)+U(s) \end{aligned} \tag{10.142}$$

In time domain,

$$\begin{aligned} \dot{x}_1 &= p_1x_1+u \\ \dot{x}_2 &= p_2x_2+u \\ &\vdots \\ \dot{x}_n &= p_nx_n+u \end{aligned} \tag{10.143}$$

In terms of state variable

$X_1(s), X_2(s), \cdots, X_n(s)$ can be written as:

$$Y(s) = b_0U(s)+C_1X_1(s)+C_2X_2(s)+\cdots++C_nX_n(s)$$

or

$$y = c_1x_1+c_2x_2+\cdots+c_nx_n+b_ou$$

which is the output equation.

$$\begin{bmatrix} \dot{x}_1 \\ \dot{x}_2 \\ \vdots \\ \cdot \\ \dot{x}_n \end{bmatrix} = \begin{bmatrix} P_1 & 0 & 0 & 0 \\ 0 & P_2 & 0 & 0 \\ \cdot & \cdot & \ddots & 0 \\ 0 & 0 & 0 & P_n \end{bmatrix} \begin{bmatrix} x_1 \\ x_2 \\ \cdot \\ \cdot \\ x_n \end{bmatrix} + \begin{bmatrix} 1 \\ 1 \\ \cdot \\ \cdot \\ 1 \end{bmatrix} u$$

$$Y = \begin{bmatrix} C_1 & C_2 & \cdots & C_n \end{bmatrix} \begin{bmatrix} x_1 \\ x_2 \\ \cdot \\ \cdot \\ x_n \end{bmatrix} + b_0u \tag{10.144}$$

The state space representation given by the previous equations is said to be in the diagonal canonical form. Figure 10.12 shows the block diagram of partial fraction method.

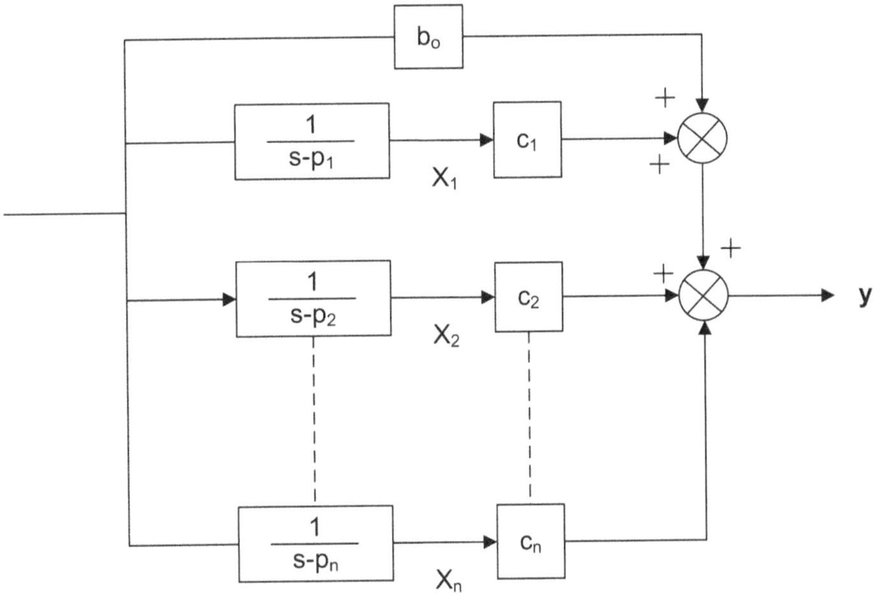


**FIGURE 10.12** Block diagram of partial fraction method.

## 10.16 DESIGN OF CONTROL SYSTEM BY POLE PLACEMENT

$$\dot{x}(t) = Ax(t) + Bu(t) \tag{10.145}$$

$$y(t) = Cx(t)$$

Consider only a single input, single output system. $u(t)$ and $y(t)$ are scalars.
Also consider a control system input, $r(t) = 0$.

$$u(t) = f[x(t)]$$

This equation is called the control law. In pole placement design, control law is specified as

$$u(t) = Kx(t)$$

where $K = 1 \times n$ vector of control gains. This control law allows all poles of the closed loop system to be placed in any desirable locations. The rule can be expressed as

$$u(t) = -k_1x_1(t) - k_2x_2(t) \cdots k_nx_n(t) \tag{10.146}$$

It can be seen in Figure 10.13 that the system feedback to the plant input is a weighted sum of all the system states. Since the close loop system is equal to zero, we consider

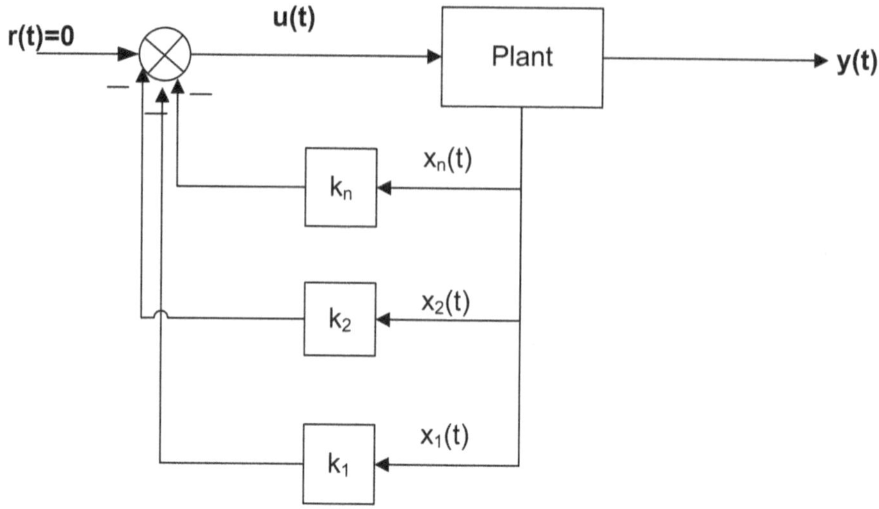


**FIGURE 10.13** Block diagram of pole placement method.

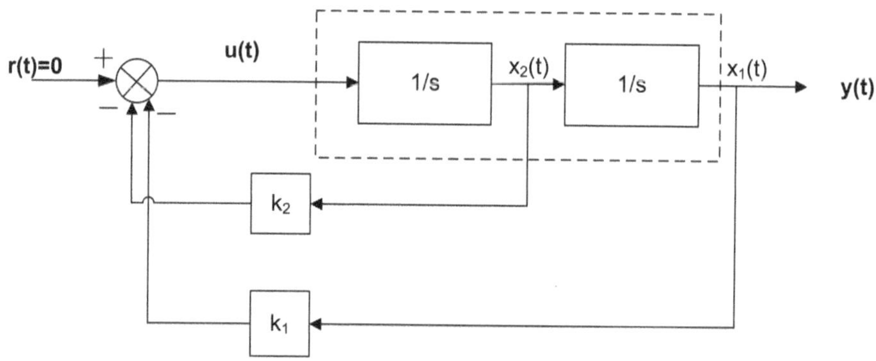


**FIGURE 10.14** Block diagram of rigid satellite system.

the proposed system to maintain system output at zero. However, disturbances will tend to protect the system, and feedback will tend to bring it to zero. Such a system is called a Regulator control system.

Consider a rigid satellite system as shown in Figure 10.14.

The state equation is,

$$\dot{x}(t) = \begin{bmatrix} 0 & 1 \\ 0 & 0 \end{bmatrix} x(t) + \begin{bmatrix} 0 \\ 1 \end{bmatrix} u(t) \tag{10.147}$$

$$u(t) = -k_1 x_1(t) - k_2 x_2(t) = [-k_1 \quad -k_2] X(t) = -kX(t) \tag{10.148}$$

The state equation for a closed loop system becomes,

$$\dot{x}(t) = AX(t) - BKX(t)$$
$$= (A - BK)X(t) = A_f X(t)$$

$A_f = (A - BK)$ is the system matrix for closed loop system. The Cahn-Hilliard (Ch.equation) for closed loop system is then

$$|sI - A_f| = |sI - A + BK| = 0$$

Suppose that design specifications require that zeros of the Ch-equation be $\lambda - \lambda_1, \lambda_2, \cdots \lambda_n$. Then the desired ch.equation $\propto_c (s)$ is:

$$\propto_c (s) = s^n + \propto_{n-1} s^{n-1} + \cdots + \propto_1 s + \propto_0$$
$$= (s + \lambda_1)(s + \lambda_2) \cdots (s + \lambda_n) = 0$$

The pole placement design procedure results in a gain factor k such that,

$$|sI - A + BK| = \propto_c (s) = s^n + \propto_{n-1} s^{n-1} + \cdots + \propto_1 s + \propto_0 \tag{10.149}$$

In this equation, there are n unknown $(k_1, k_2, \cdots, k_n)$. Equation coefficients results in equation with n unknowns.

The solution can be obtained by a digital computer.

$$\begin{bmatrix} 0 & 1 \\ 0 & 0 \end{bmatrix} x(t) - \begin{bmatrix} 0 \\ 1 \end{bmatrix} [k_1 \quad k_2] x(t)$$

$$\begin{bmatrix} 0 & 1 \\ 0 & 0 \end{bmatrix} x(t) - \begin{bmatrix} 0 & 0 \\ k_1 & k_2 \end{bmatrix} x(t) = \begin{bmatrix} 0 & 0 \\ -k_1 & -k_2 \end{bmatrix} x(t) \tag{10.150}$$

Hence,

$$\dot{x}(t) = \begin{bmatrix} 0 & 0 \\ k_1 & k_2 \end{bmatrix} \mathrm{x}(t) = \lambda_f x(t)$$

$$[SI - \lambda_f] = \begin{bmatrix} s & -1 \\ k & s + k_2 \end{bmatrix} = s^2 + k_2 s + k_1 = 0$$

Let the desired Ch.Equation be with roots $x_1$ and $x_2$.

$$\propto_c (s) = (s + \lambda_1)(s + \lambda_2) = s^2 + (\lambda_1 + \lambda_2)s + \lambda_1\lambda_2 = 0 \tag{10.151}$$

$$k_1 = \lambda_1\lambda_2$$
$$k_2 = \lambda_1 + \lambda_2$$

For this example, we see that the proper choice of feedback gains allows the root of the Ch.Equation for the system model to be placed anywhere in the s-plane.

In general,

$$\dot{X} = AX + Bu(t)$$

$$k = [k_1, k_2, \ldots, k_n]$$

$$u(t) = -KX(t) \tag{10.152}$$

### 10.16.1 Design Steps for Pole Placement

Consider

$$\dot{X} = AX + Bu$$

$$u = -kx \text{ (control signal)} \tag{10.153}$$

The following steps can determine the feedback gain matrix k that forces the eigenvalues of $A - BK$ to be $\lambda_1, \lambda_2, \ldots, \lambda_n$.

If $\lambda_i$ is a complex eigenvalue, then its conjugate must also be eigenvalues of $A - BK$.

Step 1: Check the controlling conditions for the system before it is time to proceed to step 2.

Step 2: From the characteristic polynomial for matrix A,

$$|SI - A| - s^N + a_1 s^{n-1} + \ldots + a_{n-1} s + a_2 \tag{10.154}$$

Determine the value $a_1, a_2, \ldots, a_n$.

Step 3: Determine the transformation matrix T that transfers the system state equation into the controllable canonical form.

The transformation matrix T is given by:

$$T = MW$$

where,

$$\mu = \begin{bmatrix} B & AB & A^{n-1}B \end{bmatrix}$$

and

$$w = \begin{bmatrix} a_{n-1} & a_{n-2} & \cdots & a_1 & 1 \\ a_{n-2} & a_{n-3} & \cdots & 1 & 0 \\ \vdots & \vdots & \cdots & \vdots & \vdots \\ a_1 & 1 & \cdots & 0 & 0 \\ 1 & 0 & \cdots & 0 & 0 \end{bmatrix} \tag{10.155}$$

Step 4: Using the desired eigenvalues and closed-loop poles, write the desired characteristic polynomial

$$(s-\lambda_1)(s-\lambda_2)\dots(s-\lambda_n)=s^n+\propto_1 s^{n-1}+\dots+\propto_{n-1} s+\propto_n$$

and determine the values of $\propto_1,\propto_2,\dots$ (10.156)

Step 5: The required state feedback gain matrix k can be determined from the following:

$$k=kT^{-1}=\left[\propto_n-a_n \quad \propto_{n-1}-a_{n-1} \quad \dots \quad \propto_2-a_2 \quad \propto_1-a_1\right]T^{-1} \tag{10.157}$$

If the system is of lower order $(n \le 3)$, it can be solved analytically. For example, if $n=3$

$$k=\begin{bmatrix}k_1 & k_2 & k_3\end{bmatrix}$$

and

$$\left[SI-A+BK\right]=(s-\lambda_1)(s-\lambda_2)(s-\lambda_3) \tag{10.158}$$

Since both sides of this Ch.equation are polynomial in s, by equating coefficients of like powers of s on both sides, it is possible to determine the values of $k_1, k_2$, and $k_3$.

**Example 10.20:** Consider the system defined by

$$\dot{X}=AX+BU,\; A=\begin{bmatrix}0 & 1\\ 20.6 & 0\end{bmatrix},\; B=\begin{bmatrix}0\\ 1\end{bmatrix}$$

$$|sI-A|=\begin{vmatrix}s & -1\\ -20.6 & s\end{vmatrix}=s^2-20.6=0$$

$$s=\pm 4.539$$

The system is unstable. Using feedback control $U=-Kx$, it is desired to have the closed loop poles at $-1.8\pm j2.4$. Determine the state feedback matrix k.

***Solution***

Figure 10.15 shows the block diagram for Example 10.20.
Check controllability.

$$\mu=\begin{bmatrix}B & AB\end{bmatrix}\begin{bmatrix}0 & 1\\ 1 & 0\end{bmatrix}$$

The rank is 2. So pole placement is possible.

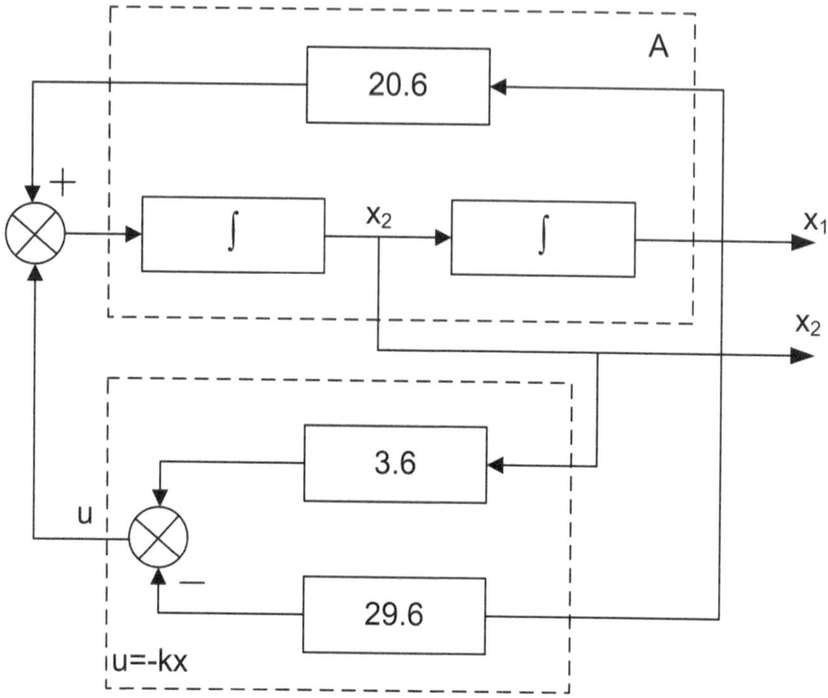


**FIGURE 10.15** Block diagram for Example 10.20.

**Method 1**: Since the state equation is in controllable canonical form, T = I. From the characteristic equations $a_1 = 0, a_2 = -20.6$.
The desired Ch.equation is:

$$(s-\lambda_1)(s-\lambda_2) = (s+1.8-j2.4)(s+1.8+j2.4)$$

$$s^2 + 3.6s + 9 = s^2 + \propto_1 s + \propto_2$$

Hence, $\propto_1 = 3.6, \propto_2 = 9$.
Using equations

$$|\propto_n - a_n \quad \propto_{n-1} - a_{n-1} \quad \dots \quad \propto_2 - a_2 \quad \propto_1 - a_1| T^{-1}$$

As T = I

$$[9+20.6 \quad 3.6-0]\mathrm{I} = |29.6 \quad 3.6|\mathrm{I} = |29.6 \quad 3.6|$$

$$K_1 = 29.6,\ K_2 = 3.6$$

**Method 2:** We use direct substitutions $[k = k_1, k_2]$ into the desired Ch.polynomial. The Ch.polynomial for the desired system is:

$$[sI - A + Bk] = \begin{bmatrix} s & 0 \\ 0 & s \end{bmatrix} - \begin{bmatrix} 0 & 1 \\ 20.6 & 0 \end{bmatrix} + \begin{bmatrix} 0 \\ 1 \end{bmatrix} \begin{bmatrix} k_1 & k_2 \end{bmatrix}$$

$$= \begin{bmatrix} s & -1 \\ -20.6 + k_1 & s + k_2 \end{bmatrix} = s^2 + k_2 s - 20.6 + k_1$$

The characteristic polynomial must be equal to:

$$(s + \lambda_1)(s + \lambda_2) = (s + 1.8 + j2.4)(s + 1.8 - j2.4)$$

$$= s^2 + 3.6s + 9$$

$$k_2 = 3.6 - 20.6 + k_1 = 9$$

$$k_2 = 29.6$$

$$K = [29.6 \quad 3.6]$$

**Example 10.21:** Design of servo system.

Figure 10.16 shows the block diagram of a servo system.

***Solution***

The equations for a servo system when the plant has an integrator is:

$$\dot{X} = AX + Bu \tag{10.159}$$

$$Y = Cx$$

Where, u and Y are scalars.

Properly choosing state variables allows one to choose the output to be equal to one of state variables. Figure 10.17 shows a simplified block diagram of Example 10.21.

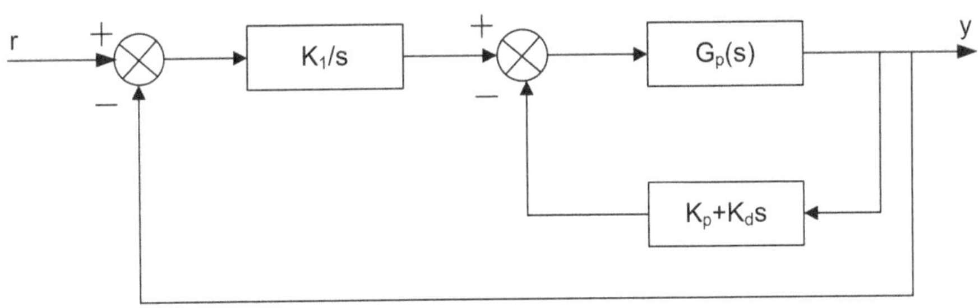


**FIGURE 10.16** Block diagram of Example 10.21 with PID control of the plant as $K = k_p + k_d s$.

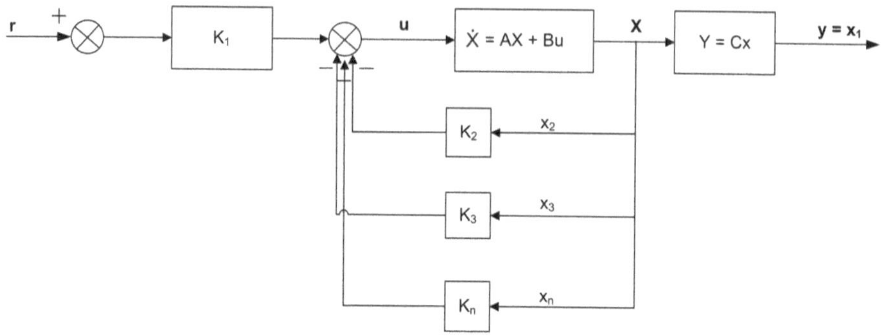


**FIGURE 10.17** Simplified block diagram of Example 10.21.

Consider the type 1 servo system, where $y = x_1, r$ is step input. Following state feedback scheme is used:

$$u = -\left[0, k_2, k_3 \ldots k_n\right]\begin{bmatrix} x_1 \\ x_2 \\ \vdots \\ x_n \end{bmatrix} + k_1\left(r_1 - x_1\right) \tag{10.160}$$

$$= -kX + k_1 r, k = \begin{bmatrix} k_1 & k_2 & \cdots & k_n \end{bmatrix}$$

Assume that r is applied at $t = 0$, then for $t > 0$, the system dynamics can be described as:

$$\dot{X} = AX + Bu = \left(A - Bk\right)x + Bk_1 r$$

The designed system will be an asymptomatically state system, that is, $y(\infty)$ will reach a constant value r, and $u(\infty)$ will approach zero.

Notice at steady state.

$$\dot{X}(\infty) = (A - Bk)X(\infty) + BK_1 r(\infty), \qquad r(\infty) = r(step) \tag{10.161}$$

By subtracting the steady state value from the dynamic equations,

$$\dot{X}(t) - \dot{X}(\infty) = (A - Bk)\left[x(t) - x(\infty)\right]$$

Now, define $X(t) - X(\infty) = e(t)$.

$$\dot{e} = (A - Bk)e \text{ (error dynamics)}$$

At steady state,

$$\dot{X}(\infty) = 0 = (A - Bk)x(\infty) + Bk_1 r \tag{10.162}$$

Since the desired eigenvalues of A-Bk are in LHP, the inverse of a-Bk exists and $x(\infty)$ can be determined as,

$$x(\infty) = -(A - Bk)^{-1} Bk_1 r \tag{10.163}$$

Also, $U(\infty) = -kx(\infty) + k_1 r = 0.$

**Example 10.22**

Consider

$$\frac{Y(S)}{U(S)} = \frac{1}{s(s+1)(s+2)}$$

It is derived to design a type 1 servo with closed-loop poles of $-2 \pm j3.464$ and $-10$. The reference input r is a step function.

***Solution***

Define state variables $x_1, x_2, x_3$.

$$x_1 = y$$

$$x_2 = \dot{x}_1$$

$$x_3 = \dot{x}_2$$

$$\dot{X} = AX + Bu \tag{10.164}$$

$$y = cx$$

$$A = \begin{bmatrix} 0 & 1 & 0 \\ 0 & 0 & 1 \\ 0 & -2 & -3 \end{bmatrix} \quad B = \begin{bmatrix} 0 \\ 0 \\ 1 \end{bmatrix} \quad C = \begin{bmatrix} 1 & 0 & 0 \end{bmatrix}$$

The control signal is given by

$$u = -(k_2 x_2 + k_3 x_3) + k_1 (r - x_1) = kx + k_1 r$$

where, $k = \begin{bmatrix} k_1 & k_2 & k_3 \end{bmatrix}$. (10.165)

Now, determine the state feedback gain matrix 'k' using the pole placement approach. Let's examine the controllability matrix.

$$\mu = \begin{bmatrix} B & AB & A^2B \end{bmatrix} = \begin{bmatrix} 0 & 0 & 1 \\ 1 & 1 & -3 \\ 1 & -3 & 7 \end{bmatrix}$$

It does not vanish as the rank = 3. Hence, the plant is completely controllable. The characteristic equation for the system is:

$$|SI - A| = \begin{bmatrix} s & -1 & 0 \\ s & s & -1 \\ 0 & 2 & s+3 \end{bmatrix} = s^3 + 3s^2 + 2s \tag{10.166}$$

$$= s^3 + a_1 s^2 + a_2 s + a_3$$

$$a_1 = 3, a_2 = 2, a_3 = 0$$

For the determination of matrix k by the pole placement method,

$$k = [\propto_3 - a_3 \quad \propto_2 - a_2 \quad \propto_1 - a_1]T^{-1}$$

Since the state equation is already in the controllable canonical form $T = I$,

$$\dot{X} = AX + Bu, u = -kx + k_1 r$$

$$\dot{X} = AX + B(-kx + k_1 r) \tag{10.167}$$

$$\dot{X} = (A - Bk)x + Bk_1 r$$

Thus, as $t \to \infty, x(t) \to x(\infty)$

At state space,

$$\dot{X}(\infty) = (A - Bk)X(\infty) + Bk_1 r$$

$$\dot{X}(t) - \dot{X}(\infty) = (A - Bk)(x(t) - x(\infty)) \tag{10.168}$$

$$x(t) - x(\infty) = e(t)$$

$$\dot{e}(t) = (A - Bk)e(t)$$

Since the desired eigenvalues of $(A - Bk)$ are

$$\lambda_1 = -2 + j3.464, \lambda_2 = -2 - j3.464, \lambda_3 = -10$$

The characteristic equation is:

$$(s - \lambda_1)(s - \lambda_2)(s - \lambda_3) = s^3 + 14s^2 + 56s + 160$$

$$= s^3 + \propto_1 s^2 + \propto_2 s + \propto_3$$

$$\propto_1 = 14, \propto_2 = 56, \propto_3 = 160$$

Then,

$$k = [\propto_3 - a_3 \quad \propto_2 - a_2 \quad \propto_1 - a_1]T^{-1}$$

$$= \begin{bmatrix} 160-0 & 56-2 & 14-3 \end{bmatrix} I \quad (10.169)$$

$$= \begin{bmatrix} 160 & 54 & 11 \end{bmatrix}$$

$$[A-Bk] = \begin{bmatrix} 0 & 1 & 0 \\ 0 & 0 & 1 \\ 0 & -2 & -3 \end{bmatrix} - \begin{bmatrix} 0 \\ 0 \\ 1 \end{bmatrix} \begin{bmatrix} 160 & 54 & 11 \end{bmatrix}$$

$$= \begin{bmatrix} 0 & 1 & 0 \\ 0 & 0 & 1 \\ -160 & -56 & -14 \end{bmatrix}$$

The state equation for the designed system is:

$$\begin{bmatrix} \dot{x}_1 \\ \dot{x}_2 \\ \dot{x}_3 \end{bmatrix} = \begin{bmatrix} 0 & 1 & 0 \\ 0 & 0 & 1 \\ -160 & -54 & -14 \end{bmatrix} \begin{bmatrix} x_1 \\ x_2 \\ x_3 \end{bmatrix} + \begin{bmatrix} 0 \\ 0 \\ 160 \end{bmatrix} \quad (10.170)$$

And output $y$ is:

$$y = \begin{bmatrix} 1 & 0 & 0 \end{bmatrix} \begin{bmatrix} x_1 \\ x_2 \\ x_3 \end{bmatrix} \quad (10.171)$$

Notice that $\dot{X}(\infty) = 0$. Hence, $(A-Bk)x(\infty) = -Bk_1 r$.

$$(A-Bk)^{-1} = \begin{bmatrix} 0 & 1 & 0 \\ 0 & 0 & 1 \\ -160 & -54 & -14 \end{bmatrix}^{-1} = \begin{bmatrix} -\frac{7}{20} & -\frac{7}{80} & -\frac{1}{160} \\ 1 & 0 & 0 \\ 0 & 1 & 0 \end{bmatrix}$$

$$x(\infty) = -(A-Bk)^{-1} Bk_1 r = \begin{bmatrix} -\frac{7}{20} & -\frac{7}{80} & -\frac{1}{160} \\ 1 & 0 & 0 \\ 0 & 1 & 0 \end{bmatrix} \begin{bmatrix} 0 \\ 0 \\ 1 \end{bmatrix} (160) r$$

$$\begin{bmatrix} \frac{1}{160} \\ 0 \\ 0 \end{bmatrix} (160) r = \begin{bmatrix} 1 \\ 0 \\ 0 \end{bmatrix} r = \begin{bmatrix} r \\ 0 \\ 0 \end{bmatrix}$$

$x_1(\infty) = y(\infty) = r$. The step response has no steady state error.

$$u(\infty) = -kx(\infty) + k_1 r(\infty) = -kx(\infty) + k_1 r$$

We have

$$u(\infty) = -\begin{bmatrix}160 & 54 & 11\end{bmatrix}\begin{bmatrix}x_1(\infty) \\ x_2(\infty) \\ x_3(\infty)\end{bmatrix} + 160r$$

$$= -\begin{bmatrix}160 & 54 & 11\end{bmatrix}\begin{bmatrix}r \\ 0 \\ 0\end{bmatrix} + 160r \tag{10.172}$$

So $u$ is zero at steady state.

**Example 10.23:** Consider a transfer function as under:

$$\frac{Y(S)}{U(S)} = \frac{1}{s(s+1)}$$

***Solution***

$$\dot{X}_1 = X_2$$

$$\dot{X}_2 = -X_2$$

$$A = \begin{bmatrix}0 & 1\end{bmatrix}, B = \begin{bmatrix}0 \\ 1\end{bmatrix}, c = \begin{bmatrix}1 & 0\end{bmatrix}$$

$$\begin{bmatrix}B & AB\end{bmatrix} = \begin{bmatrix}0 & 1 \\ 1 & -1\end{bmatrix} > 0, \text{ controllable}$$

$$\begin{bmatrix}SI - A\end{bmatrix} = \begin{bmatrix}s & -1 \\ 0 & s+1\end{bmatrix} = s(s+1) = s^2 + s$$

$$a_1 = 1, a_2 = 0$$

It is desired that close loop poles be at $\lambda_1 = -1, \lambda_2 = -2$.

Hence, characteristic equations are:

$$(s-\lambda_1)(s-\lambda_2) = (s+1)(s+2) = s^2 + 3s + 2$$

$$\propto_1 = 3, \propto_2 = 2$$

$$\begin{bmatrix}\propto_2 - a_2 & \propto_1 - a_1\end{bmatrix} T^{-1}, T = I$$

$$[2-0 \quad 3-1]I = [2 \quad 2]$$

$$k_1 = 2, k_2 = 2$$

$$[A-Bk] = \begin{bmatrix} 0 & 1 \\ 0 & -1 \end{bmatrix} - \begin{bmatrix} 0 \\ 1 \end{bmatrix} [2 \quad 2]$$

$$= \begin{bmatrix} 0 & 1 \\ -2 & -3 \end{bmatrix}$$

$$\begin{bmatrix} \dot{x}_1 \\ \dot{x}_2 \end{bmatrix} \begin{bmatrix} 0 & 1 \\ -2 & -3 \end{bmatrix} \begin{bmatrix} x_1 \\ x_2 \end{bmatrix} + \begin{bmatrix} 0 \\ 2 \end{bmatrix} r$$

$$y = [1 \quad 0] \begin{bmatrix} x_1 \\ x_2 \end{bmatrix}$$

$$[A-Bk] = \begin{bmatrix} 0 & 1 \\ -2 & -3 \end{bmatrix}$$

$$[A-Bk]^{-1} = \begin{bmatrix} -\frac{3}{2} & -\frac{1}{2} \\ 1 & 0 \end{bmatrix}$$

$$(A-Bk)^{-1} Bk_1 r$$

$$\begin{bmatrix} x_1(\infty) \\ x_2(\infty) \end{bmatrix} = -\begin{bmatrix} -\frac{3}{2} & -\frac{1}{2} \\ 1 & 0 \end{bmatrix} \begin{bmatrix} 0 \\ 1 \end{bmatrix} 2r$$

$$= -\left[ -\frac{1}{2} \times 2r \right] = r$$

$$u(\infty) = -kx(\infty) + k_1 r(\infty)$$

$$= -kx(\infty) + k_1 r$$

$$u(\infty) = -[2 \quad 2] \begin{bmatrix} r \\ 0 \end{bmatrix} + 2r$$

$$= -2r + 2r = 0 \quad (10.173)$$

**Problems**

P10.1. Find the transfer function when

$$A = \begin{bmatrix} -1 & 1 \\ 0 & -2 \end{bmatrix}, B = \begin{bmatrix} 0 \\ 1 \end{bmatrix}, C = [1 \quad 1]$$

P10.2. A system is described by the following transfer function, $G(s)=\frac{3}{s^4+2s^3+3s+2}$. Find the state and output equation of the system.

P10.3. Check the controllability of the following state space model

$$\begin{bmatrix}\dot{x}_1\\ \dot{x}_2\end{bmatrix}=\begin{bmatrix}1 & 1\\ 2 & -1\end{bmatrix}\begin{bmatrix}x_1\\ x_2\end{bmatrix}+\begin{bmatrix}0\\ 1\end{bmatrix}u$$

P10.4. Check the observability of the following system

$$\dot{X}=AX+Bu$$
$$Y=Cx$$

where

$$A=\begin{bmatrix}0 & 1 & 0\\ 0 & 0 & 1\\ -6 & -11 & -6\end{bmatrix}, B=\begin{bmatrix}0\\ 0\\ 1\end{bmatrix}, c=\begin{bmatrix}4 & 5 & 1\end{bmatrix}$$

P10.5. Find the transition matrix for $A=\begin{bmatrix}0 & 1\\ -2 & -3\end{bmatrix}$ using Cayley Hamilton theorem.

P10.6. Find $f(A)=A^{10}$ for $A=\begin{bmatrix}0 & 1\\ -3 & -4\end{bmatrix}$ using Cayley Hamilton theorem.

P10.7. Find the eigenvalues of $A=\begin{bmatrix}0 & 1 & 0\\ 0 & 0 & 1\\ -6 & 11 & -6\end{bmatrix}$.

P10.8. Compute $e^{At}$ for the matrix $A=\begin{bmatrix}0 & 1\\ 0 & -2\end{bmatrix}$.

P10.9. Consider the system defined by

$$\dot{X}=AX+BU \quad A=\begin{bmatrix}0 & 1\\ 20.6 & 0\end{bmatrix} \quad B=\begin{bmatrix}0\\ 1\end{bmatrix}$$

$$|sI-A|=\begin{vmatrix}s & -1\\ -20.6 & s\end{vmatrix}=s^2-20.6=0, \quad s=\pm 4.539$$

The system is unstable. By using feedback control $U=-Kx$ it is desired to have the closed loop poles at $-1.8\pm j2.4$. Determine state feedback matrix k.

# 11 Modern Methods for Analysis of Control Systems

## 11.1 LYAPUNOV STABILITY ANALYSIS

Lyapunov stability analysis can be done using two methods.

**Explicit solution method:** This traditional approach involves explicitly solving the system's differential equations to analyze its stability.
**Direct method:** This method bypasses the need for explicit solutions, allowing stability to be assessed directly through system properties.

The following terms are important for understanding the concept of Lyapunov stabilization.

**System:**

$\bar{X} = f(\bar{x},t)$, where $f(\bar{x},t)$ is an n-vector whose elements are functions of $x_1, x_2, \ldots, x_n$ and $t$.

**Solution:**

$$\phi(t;\bar{x}_0,t_0) \text{ and } \phi(t_0;\bar{x}_0,t_0) = \bar{x}_0$$

**Equilibrium state:**

In the system, state $\bar{X} = f(X_c,t) = 0$ for all $t$ is called the equilibrium state of $f(x,t) = A$

For linear systems, only one equilibrium state exists if A is nonsingular, and many states exist if A is singular. For nonlinear systems, there may be many equilibrium states.

**Stability in the sense of Lyapunov:**

Let $\varphi_i(t)$ be a solution to the set of equations

$$X_i = f_i(x_i;t)$$

Define a new set of variables. $q_i = x_i - \varphi_i$. If they are subtracted in $(A)$, a new set of equations results, which has an equilibrium position at $q$. The equilibrium position

DOI: 10.1201/9781003484202-11

is said to be stable in the sense of Lyapunov if for every $\in > 0$, there exists $\delta > 0$, such that $q(t) < \in$ whenever $q(t) < \delta$ for all $t > 0$. If $q$ approaches zero at $t \to \infty$, the equilibrium is called asymptotically stable in the sense of Lyapunov.

The diagram representing stable and unstable equilibrium is shown in Figure 11.1.

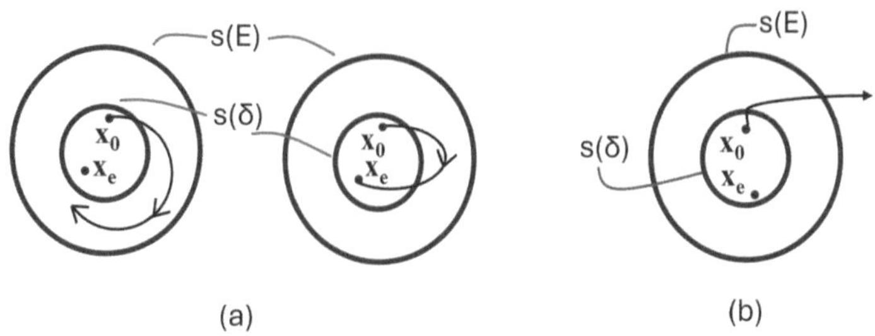


**FIGURE 11.1** (a) Stable equilibrium, (b) unstable equilibrium.

**Euclidean norm**

$$\| X - X_e \| \leq k,$$

Let $s(\delta)$ consider all points such that

$$\| X - X_e \| \leq \delta$$

Let $s(\in)$ consist of all such points such that $\| \phi(t, x_0, t_0) - X_c \| \leq \in$ for all $t \leq t_0$

**Positive definiteness of scalar functions**

A scalar function $V(x)$ is said to be positive definite in the region $\Omega$ (which includes the origin of the state space) if $V(x) > 0$ for all non-zero states x in the region $\Omega$ and $V(0) = 0$

A time-varying function $V(x,t)$ is said to be positive definite in a region $\Omega$ (which includes the origin) if it is bound from below by a time-invariant positive definite function $V(x)$ such that

$$V(x,t) > V(x) \qquad \text{for all } t \geq t_0$$
$$V(0,t) = 0 \qquad \text{for all } t \geq t_0$$

**Negative definiteness of scalar functions**

A scalar function $V(x)$ is said to be negative definite if $-V(x)$ is positive definite.

**Positive semi-definiteness of scalar function**

A scalar function $V(x)$ is said to be positive semidefinite if it is positive in all states in the region $\Omega$ except at the origin and certain other states where it is zero. Negative semidefiniteness of scalar function: a scalar function $V(x)$ is said to be negative semidefinite if $-V(x)$ is positive definite.

**Indefiniteness of scalar function**

A scalar function $V(x)$ is said to be indefinite if in the region $\Omega$ it assures both positive and negative values, no matter how small the region $\Omega$ is.

Example

1. $V(x) = x_1^2 + x_2^2$ Positive definite
2. $V(x) = (x_1 + x_2)^2$ Positive semidefinite
3. $V(x) = -x_1^2 - (3x_{11} + 2x_2)^2$ Negative definite
4. $V(x) = x_1 x_2 + x_2^2$ Indefinite
5. $V(x) = x_1^2 + \dfrac{2x_2^2}{1 + x_2^2}$ Positive definite

**Quadratic form:**

$$V(x) = X^T PX = [x_1, x_2, \ldots, x_n] \begin{bmatrix} p_{11} & p_{12} & \cdots & p_{1n} \\ p_{21} & p_{22} & \cdots & p_{2n} \\ \vdots & \vdots & \ddots & \vdots \\ p_{n1} & p_{n2} & \cdots & p_{nn} \end{bmatrix} \begin{bmatrix} x_1 \\ x_2 \\ \vdots \\ x_n \end{bmatrix}$$

Show the following quadratic form in positive definite

$$V(x) = X^T PX = \begin{bmatrix} x_1 & x_2 & x_3 \end{bmatrix} \begin{bmatrix} 10 & 1 & -2 \\ 1 & 4 & -1 \\ -2 & -1 & 1 \end{bmatrix} \begin{bmatrix} x_1 \\ x_2 \\ x_3 \end{bmatrix}$$

Applying Sylvester's condition

$$10 > 0, \begin{vmatrix} 10 & 1 \\ 1 & 4 \end{vmatrix} > 0 \begin{vmatrix} 10 & 1 & -2 \\ 1 & 4 & -1 \\ -2 & -1 & 1 \end{vmatrix} > 0$$

Example

$$\dot{x}_1 = x_2 - x_1\left(x_1^2 + x_2^2\right)$$

$$\dot{x}_2 = -x_1 - x_2\left(x_1^2 + x_2^2\right)$$

Let $V(x) = 2x\dot{x}_1 + 2x_2\dot{x}_2$

$$= 2x_1(x_2 - x_1(x_1^2 + x_2^2) + 2x_2(-x_1 - x_2(x_1^2 + x_2^2))$$

$$= 2x_1x_2 - 2x_1^2(x_1^2 + x_2^2) - 2x_1x_2(x_1^2 + x_2^2)$$

$V(X) = -2(x_1^2 + x_2^2)^2$ is negative definite

Theorem: Suppose a system is defined as

$$\dot{X} = f(x,t)$$

Where $f(0,t) = 0$ for all $t \geq t_0$

If there exists a scalar function $V(x,t)$ having continuous first partial derivatives and satisfying the following conditions:

1. $V(x,t)$ is positive definite.
2. $V(x,t)$ is negative definite.
3. $V(\phi(t,x_0,t_0),t)$ does not vanish identically in $t \geq t_0$ for any $t_0$ and any $x_0 \neq 0$, where $\phi(t,x_0,t_0)$ denotes the trajectory or solution starting from $x_0$ as $t_0$, then the equilibrium state at the origin of the system is uniformly asymptomatically stable in the large.

### 11.1.1 Definition of Lyapunov Function

Let $V(x)$ be a function with the following properties:

1. $V(x)$ is continuous together with its first derivatives in a certain open region $\Omega$ about the origin.
2. $V(0) = 0$
3. Outside the origin (and always in $\Omega$) $V(x)$ is positive.
4. $\dot{V}(x) < 0$ in $\Omega$ $\dot{V}(x)$ is total time derivative of $V(x)$

Then $V(x)$ is called the Lyapunov function

Theorem: Suppose a system is described by

$$\dot{X} = f(x,t) \text{ were } f(0,t) = 0 \text{ for all } t \geq t_0$$

If there exists a scalar function $W(\overline{x},t)$ having continuous first partial derivatives and satisfying the following condition

1. $W(\overline{x},t)$ is a positive definite in some regions above the origin.
2. $\dot{W}(\overline{x},t)$ is a positive definite in the same origin.

Then the equilibrium state at the origin is unstable.

### 11.1.2 Lyapunov Stability Analysis of the Linear Time-Invariant System

Let $\dot{X} = AX$

The condition for the asymptotic stability of the origin of the system is that all eigenvalues of the matrix A have negative real parts, or equivalently, the roots of the characteristic equation lie in the left half of the complex plane.

The characteristic equation can be written as

$$|SI - A| = s^n + a_1 s^{n-1} + \ldots + a_{n-1} s + a_n \tag{11.1}$$

having negative parts.

Here assumption is that $A$ is nonsingular and the origin $x = 0$ is the only equilibrium state. Let us choose a possible Lyapunov function as

$$V(x) = \dot{X}^* PX + X^* P\dot{X}$$

Where P is the positive definite Hermitian matrix

$$\dot{V}(X) = \dot{X}Px + X^* P\dot{X}$$

$$= \left(AX^*\right)PX + X^* PAX$$

$$= X^* A^* PX + X^* PAX$$

$$= X^*\left(A^* P + PA\right)X$$

Since $V(X)$ was chosen to be positive definite, we require, for asymptotic stability that $\dot{V}(X)$ be negative definite. Therefore, we require

$$\dot{V}(X) = -X^* QX \tag{11.2}$$

Where $Q = -\left(A^* P + PA\right)$, which is positive definite.

Therefore, it is sufficient for Q to be positive definite, which can be verified using Sylvester's criterion. According to this criterion, a matrix is positive definite if and only if the determinants of all its leading principal minors are positive.

$$A^* P + PA = -Q \tag{11.3}$$

Theorem:

Consider the system

$$\dot{X} = AX$$

A necessary and sufficient condition for the asymptotic stability of the equilibrium state X=0 in the large is that, for any positive definite Hermitian (or real) matrix Q, there exists a positive definite Hermitian (or real symmetric) matrix P such that

$$A^{*}P + PA = -Q \tag{11.4}$$

The scalar function $X^{*}PX$ is a Lyapunov function for this system.

Comments

1. If the system invalidates the real state vector, then $A^{T}P + PA = -Q$.
2. If $\dot{V}(X) = -X^{*}QX$ does not vanish identically along any trajectory, the $Q$ may be chosen to be positive semidefinite.

**Example 11.1.** Consider

$$\dot{x}_1 = x_2$$

$$\dot{x}_2 = -x_1 - x_2$$

$$A = \begin{bmatrix} 0 & 1 \\ -1 & -1 \end{bmatrix}$$

Let us assume $LF \quad V(X) = X^{T}PX$

$$A^{T}P + PA = -1$$

$$= \begin{bmatrix} 0 & -1 \\ 1 & -1 \end{bmatrix}\begin{bmatrix} p_{11} & p_{12} \\ p_{21} & p_{22} \end{bmatrix} + \begin{bmatrix} p_{11} & p_{12} \\ p_{21} & p_{22} \end{bmatrix}\begin{bmatrix} 0 & 1 \\ -1 & -1 \end{bmatrix} = \begin{bmatrix} -1 & 0 \\ 0 & -1 \end{bmatrix}$$

$$= \begin{bmatrix} -p_{21} & -p_{22} \\ -p_{11} - p_{21} & -p_{12} - p_{22} \end{bmatrix} + \begin{bmatrix} -p_{12} & -p_{11} - p_{12} \\ -p_{22} & -p_{21} - p_{22} \end{bmatrix} = \begin{bmatrix} -1 & 0 \\ 0 & -1 \end{bmatrix}$$

$$= \begin{bmatrix} p_{21} - p_{22} & -p_{22} + p_{11} - p_{12} \\ -p_{11} - p_{21} - p_{22} & -p_{12} - p_{22} + p_{21} - p_{22} \end{bmatrix} = \begin{bmatrix} -1 & 0 \\ 0 & -1 \end{bmatrix}$$

$$-p_{21} - p_{12} = -1, \qquad 2p_{12} = 1$$

$$-p_{12} - p_{22} = 0, \qquad p_{12} = p_{21} = \frac{1}{2}$$

$$2p_{12} - 2p_{22} = 1, \qquad p_{22} = 1$$

$$p_{11} = \frac{3}{2}$$

$$\begin{bmatrix} p_{11} & p_{12} \\ p_{21} & p_{22} \end{bmatrix} = \begin{bmatrix} \frac{3}{2} & \frac{1}{2} \\ \frac{1}{2} & 1 \end{bmatrix} \qquad \frac{3}{2} > 0$$

$$\begin{vmatrix} \frac{3}{2} & \frac{1}{2} \\ \frac{1}{2} & 1 \end{vmatrix} > 0$$

$p$ is positive definite

$$V(X) = X^T PX = \frac{1}{2}\left(3x_1^2 + 2x_1x_2 + 2x_2^2\right)$$

$$\dot{V}(X) = -\left(x_1^2 + x_2^2\right)$$

**Example 11.2**

$$\dot{x}_1 = x_2 - ax_1\left(x_1^2 + x_2^2\right)$$

$$\dot{x}_2 = -x_1 - ax_2\left(x_1^2 + x_2^2\right)$$

Let the Lyapunov function be $V(X) = C_1X_1^2 + C_2X_2^2$

$$\begin{aligned}\dot{V}(X) &= 2c_1x_1\dot{x}_1 + 2c_2x_2\dot{x}_2 \\ &= 2c_1x_1\left(x_2 - ax_1\left(x_1^2 + x_2^2\right)\right) + 2c_2x_2\left(-x_1 - ax_2\left(x_1^2 + x_2^2\right)\right) \\ &= 2c_1x_1x_2 - 2c_1ax_1^2\left(x_1^2 + x_2^2\right) - 2c_1x_2x_1 - 2c_2ax_2^2\left(x_1^2 + x_2^2\right)\end{aligned}$$

When $c_1 = c_2$

$$-2ac\left(x_1^2 + x_2^2\right)\left(x_1^2 + x_2^2\right)$$

$$-2ac\left(x_1^2 + x_2^2\right)^2$$

**Example 11.3:** Consider a system represented by A, find the coefficients of P.

$$A = \begin{bmatrix} -3 & 2 \\ -1 & -1 \end{bmatrix} \qquad A^T = \begin{bmatrix} -3 & -1 \\ 2 & -1 \end{bmatrix}$$

$$A^T P + PA = \begin{bmatrix} -3 & -1 \\ 2 & -1 \end{bmatrix}\begin{bmatrix} p_{11} & p_{12} \\ p_{21} & p_{22} \end{bmatrix} + \begin{bmatrix} p_{11} & p_{12} \\ p_{21} & p_{22} \end{bmatrix}\begin{bmatrix} -3 & 2 \\ -1 & -1 \end{bmatrix} = \begin{bmatrix} -1 & 0 \\ 0 & -1 \end{bmatrix}$$

$$= \begin{bmatrix} -3p_{11}-p_{21} & -3p_{12}-p_{22} \\ 2p_{11}-p_{21} & 2p_{12}-p_{22} \end{bmatrix} + \begin{bmatrix} -3p_{11}-p_{12} & 2p_{11}-p_{12} \\ -3p_{21}-p_{22} & 2p_{21}-p_{22} \end{bmatrix} = \begin{bmatrix} -1 & 0 \\ 0 & -1 \end{bmatrix}$$

$$-3p_{11}-p_{21}-3p_{11}-p_{12}=-1 \qquad p_{12}=p_{21}$$

$$-3p_{12}-p_{22}+2p_{11}-p_{12}=0 \qquad -6p_{11}-2p_{12}=-1$$

$$2p_{11}-p_{21}-3p_{21}-p_{22}=0 \qquad -p_{22}+2p_{11}-4p_{12}=0$$

$$2p_{12}-p_{22}+2p_{21}-p_{22}=-1 \qquad 4p_{12}-2p_{22}=-1$$

$$P = \begin{bmatrix} \frac{7}{40} & -\frac{1}{40} \\ -\frac{1}{40} & \frac{18}{40} \end{bmatrix} \qquad p_{11}=\frac{7}{40}, p_{12}=p_{21}=-\frac{1}{40}, p_{22}=\frac{18}{40}$$

**Example 11.4:** Consider the system

$$\dot{X} = AX + Bu$$

$$y = cx \qquad \text{where } A = \begin{bmatrix} 1 & 2 \\ -4 & -3 \end{bmatrix}, B = \begin{bmatrix} 1 \\ 2 \end{bmatrix}, C = \begin{bmatrix} 1 & 1 \end{bmatrix}$$

Transform the system into a controllable canonical form.

$$\frac{Y(S)}{U(S)} = G(S) = C(SI-A)^{-1}B = \begin{bmatrix} 1 & 1 \end{bmatrix} \begin{bmatrix} S-1 & -2 \\ 4 & S+3 \end{bmatrix}^{-1} \begin{bmatrix} 1 \\ 2 \end{bmatrix}$$

$$= \frac{1}{S^2+2S+S} \begin{bmatrix} 1 & 1 \\ 0 & 0 \end{bmatrix} \begin{bmatrix} S+3 & 2 \\ -4 & S-1 \end{bmatrix} \begin{bmatrix} 1 \\ 2 \end{bmatrix}$$

$$= \frac{3S+1}{S^2+2S+S} = \frac{b_0 s^2 + b_1 s + b_2}{s^2 + a_1 s + a_2}$$

$$b_0 = 0 \qquad a_1 = 2$$

$$b_1 = 3 \qquad a_2 = 5$$

$$b_2 = 1$$

The controllable canonical form for state equation is

$$\begin{bmatrix} \dot{x}_1 \\ \dot{x}_2 \\ \vdots \\ \dot{x}_n \end{bmatrix} = \begin{bmatrix} 0 & 1 & 0 & \dots & 0 \\ 0 & 0 & 1 & \dots & 0 \\ \vdots & \vdots & \vdots & \ddots & \vdots \\ -a_n & -a_{n-1} & \dots & \dots & -a_1 \end{bmatrix} \begin{bmatrix} x_1 \\ x_2 \\ \vdots \\ x_n \end{bmatrix} + \begin{bmatrix} 0 \\ 0 \\ \vdots \\ 1 \end{bmatrix} u$$

And $Y = \begin{bmatrix} b_n - a_n b_0 & b_{n-1} - a_{n-1} b_0 & \cdots & b_1 - a_1 b_0 \end{bmatrix}$

$$y = \begin{bmatrix} 1 - 5x_0 & 3 - 2x_0 \end{bmatrix},\ Y = \begin{bmatrix} 1 & 3 \end{bmatrix} \begin{bmatrix} x_1 \\ x_2 \end{bmatrix}$$

### 11.1.3 Quadratic Optimal Control Systems

Consider

$$\dot{X} = AX + BU$$

$$X - n \text{ vector}$$

$$U - r \text{ vector}$$

$$A - n \times n$$

$$B - n \times r$$

It is desired to choose the control vector $u(t)$ such that a given performance index (PI) is minimized. It can be proved that $J = \int_0^\infty L(x,u)\,dt$, where $L(x,u)$ is a quadratic function or Hermitian function of $x$ and $u$ will load to linear control law $u(t) = -kx(t)$

Where k is $r \times n$ matrix or

$$\begin{bmatrix} u_1 \\ u_2 \\ \vdots \\ u_r \end{bmatrix} = -\begin{bmatrix} k_{11} & k_{12} & \cdots & k_{1n} \\ k_{21} & k_{22} & \cdots & k_{2n} \\ \vdots & \vdots & \ddots & \vdots \\ k_{r1} & k_{r2} & \cdots & k_{rn} \end{bmatrix} \begin{bmatrix} x_1 \\ x_2 \\ \vdots \\ x_n \end{bmatrix} \tag{11.5}$$

Therefore, the design of the optimal regulator control system is the determination of elements of k. In the following, we consider the problem of determining the vector $u(t)$ for the system with the following performance index.

$$J = \int_0^\infty \left( X^* QX + u^* RU \right) dt \tag{11.6}$$

Where $Q$ is positive definite or (positive semidefinite) Hermitian or real matrix and $U$ is unconstrained. The approach to designing such a system is based on the method of Lyapunov. We shall use complex (Hermitian) performance indices and real will become the special case. In other words,

$$\int_0^\infty \left( X^* QX + u^* Ru \right) dt \quad \text{is complex.}$$

$$\int_0^{\infty}\left(X^T QX + u^T Ru\right)dt \quad \text{is real.} \tag{11.7}$$

In many systems, there is a direct relationship between Lyapunov functions and quadratic performance indices, which are commonly used in the design of optimal control systems.

### 11.1.4 Parameter Optimization

We can solve such a problem using the second method of Lyapunov, for example, consider

$$\dot{X} = AX$$

If all eigenvalues of A have negative real parts, the origin x = 0 is asymptotically stable. We assume that A depends on an adjustable parameter $(s)$. It is desired to minimize the following

$$J = \int_0^{\infty} X^* QXdt$$

The problem thus becomes of determining values of adjustable parameters to minimize the performance index.

Let us assume.

$$X^* QX = \frac{d}{dt}\left(X^* PX\right)$$

$p$ is a positive definite Hermitian or real matrix.

$$\begin{aligned} X^* OX &= -\dot{X}PX - X^* P\dot{X} \\ &= -X^* A^* PX - X^* PAX \\ &= -X^* \left(A^* P + PA\right) X \end{aligned}$$

By the second method of Lyapunov, we know that for a given $Q$ there exists a $P$ such that,

$$A^* P + PA = -Q$$

Hence, we can determine the elements of $P$ from this equation.

Thus $PI(J)$ can be evaluated as,

$$J = \int_0^\infty X^* QX dt = -X^* PX |_0^\infty \tag{11.8}$$

$$= -X^*(\infty) PX(\infty) - X^*(0) PX(0)$$

Since all eigenvalues of $A$ have negative real parts, we have $x(\infty) = 0$
Therefore,

$$J = X^*(0) PX(0)$$

Thus $J$ can be obtained in terms of initial conditions $x(0)$ and $P$ which is related to $P$ and $Q$ by

$$A^* P + PA = -Q$$

$J$ can be minimized by minimizing $X^*(0) PX(0)$ with respect to parameters in question. Since $x(0)$ is given and $Q$ is also given, $P$ is a function of elements of $A$. Hence the minimization process will result in the optimal value of the adjustable parameters.

**Example 11.5:** Determine the value of damping ratio $\xi > 0$ so that when the system shown in Figure 11.2 is subjected to $u(t)$, the following $PI$ is minimized.

$$PI = J = \int_0^\infty \left(e^2 + u\dot{e}^2\right) dt \qquad (u > 0)$$

$$\ddot{c} + 2\xi\dot{c} = r$$

$$\ddot{e} + 2\xi\dot{e} + e = \ddot{r} + 2\xi\dot{r}$$

$$\ddot{r}(0) = 0$$

$$\dot{r}(0) = 0$$

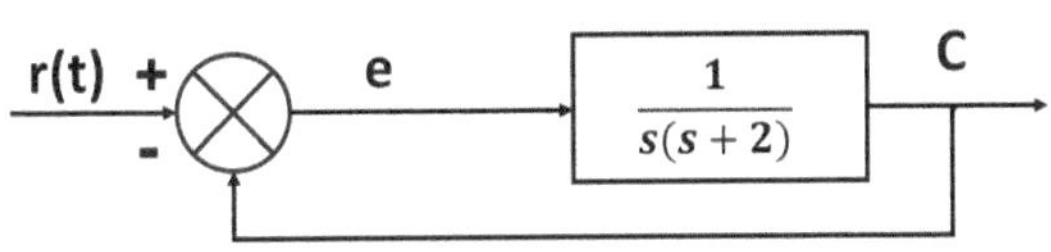


**FIGURE 11.2** Block diagram of Example 11.5.

Define:

$$X_1 = e$$

$$X_2 = \dot{e} = \dot{X}_1$$

$$\dot{X}_2 = -2\xi X_2 - X_1$$

$$\dot{X} = AX \qquad A = \begin{bmatrix} 0 & 1 \\ -1 & -2\xi \end{bmatrix}$$

Then

$$J = \int_0^\infty \left(e^2 + u\dot{e}^2\right) dt$$

$$= \int_0^\infty \left(X_1^2 + uX_2^2\right) dt$$

$$= \int_0^\infty [X_1 \quad X_2] \begin{bmatrix} 1 & 0 \\ 0 & u \end{bmatrix} \begin{bmatrix} x_1 \\ x_2 \end{bmatrix} dt$$

$$= \int_0^\infty X^T QX dt$$

Since $A$ is a stable matrix

$$J = X^T(0+) PX(0+)$$

Where $P$ is determined from,

$$ATP + PA = -Q$$

$$\begin{bmatrix} 0 & -1 \\ 1 & -2\xi \end{bmatrix} \begin{bmatrix} p_{11} & p_{12} \\ p_{21} & p_{22} \end{bmatrix} + \begin{bmatrix} p_{11} & p_{12} \\ p_{21} & p_{22} \end{bmatrix} \begin{bmatrix} 0 & 1 \\ -1 & -2\xi \end{bmatrix} = \begin{bmatrix} -1 & 0 \\ 0 & -u \end{bmatrix}$$

So that

$$-2p_{12} = -1$$

$$p_{11} - 2\xi p_{12} - p_{22} = 0$$

$$2p_{12} - 4\xi p_{22} = -u$$

$$P = \begin{bmatrix} p_{11} & p_{12} \\ p_{21} & p_{22} \end{bmatrix} = \begin{bmatrix} \xi + \dfrac{1+u}{4\xi} & \dfrac{1}{2} \\ \dfrac{1}{2} & \dfrac{1+u}{4\xi} \end{bmatrix}$$

Then $J = X^T(0+)PX(0+)$

$$= \left(\xi + \frac{1+u}{4\xi}\right)x_1^2(0) + x_1(0)x_2(0) + \frac{1+u}{4\xi}x_2^2(0)$$

$$x_1(0) = 1, \qquad x_2(0) = 0$$

$$J = \xi + \frac{1+u}{4\xi^2} = 0 \quad 4\xi^2 = 1+u \quad \xi = \sqrt{\frac{1+u}{2}}$$

### 11.1.5 Quadratic Optimal Control Problem

Consider the system, $\dot{X} = AX + Bu$. To find the matrix corresponding to the optimal control vector, $u(t) = -kX(t)$, the following procedure is adopted.

To minimize the performance index

$$J = \int_0^\infty \left(X^* QX + u^* Ru\right)dt$$

Where $u^* Ru$ is the expenditure of energy of the control signal.

Q—Positive definite (positive semidefinite) Hermitian or real symmetric matrix.

R—Positive definite, Hermitian, or real symmetric matrix.

$u(t)$ is assumed to be uncontrolled. If unknown elements of k are determined, then $u(t) = -kx(t)$ is optimal for any initial state $x(0)$.

$$A^* P + PA - K^* B^* P - PBK + K^* T^* TK + Q = 0$$

The minimization of $J$ for $K$ requires the minimization of

$$X^* \left[TK - \left(T^*\right)^{-1} B^* P^*\right]\left[TK - \left(T^*\right)^{-1} B^* P\right]X$$

Since this equation is nonnegative, the minimum occurs where it is zero or when,

$$TK = \left(T^*\right)^{-1} B^* P$$

$$K = T^{-1}\left(T^*\right)^{-1} B^* P = R^{-1} B^* P \tag{11.9}$$

Thus $u(t) = -kx(t) = -R^{-1}B^* Px(t)$

The matrix $P$ in Eq (11.9) must satisfy the following reduced equation.

$$A^* P + PA - PBR^{-1}B^* P + Q = 0 \tag{11.10}$$

This equation is called the reduced matrix Riccati equation. The design steps are,

1. Solve Eq (11.10) for matrix $P$.
2. Substitute this matrix $P$ in Eq (11.9). The resulting matrix K is the optimal one.

**Example 11.6:** Consider the example shown in Figure 11.3.

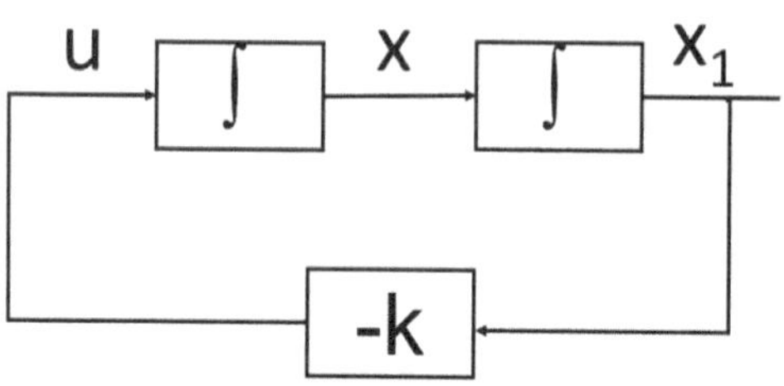


**FIGURE 11.3** Block diagram of example 11.6.

$$u(t) = -kx(t)$$

And let

$$J = \int_0^\infty \left(X^T QX + u^2\right) dt \qquad \text{where } Q = \begin{bmatrix} 1 & 0 \\ 0 & \mu \end{bmatrix}, \mu \geq 0$$

Another depiction of Example 11.6 is shown in Figure 11.4.

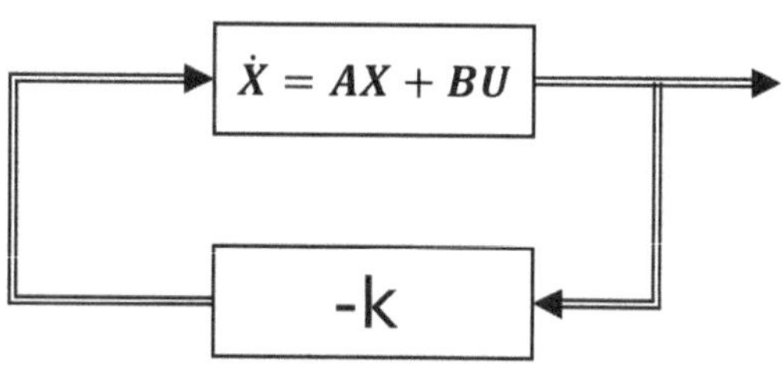


**FIGURE 11.4** Block diagram of Example 11.6, another depiction.

$$\dot{X} = AX - BKX = (A - BK)X$$

$$J = \int_0^\infty \left(X^* QX + X^* KRKX\right) dt$$

$$= \int_0^\infty X^* \left(Q + K^* RK\right) X dt$$

Set $X^*\left(Q+K^*RK\right)=-\frac{d}{dt}\left(X^*PX\right)$

Then, we obtain

$$X^*\left(Q+K^*RK\right)X=-\dot{X}^*PX-X^*P\dot{X}$$

$$=-X^*\left[\left(A-BK\right)^*P+P\left(A-BK\right)\right]X$$

$$\left(A-BK\right)^*P+P\left(A-BK\right)=-\left(Q+K^*RK\right)$$

By the second method of Lyapunov, if $A-BK$ is a stable matrix, there exists a positive definite matrix $P$ that satisfies

$$\left(A-BK\right)^*P+P\left(A-BK\right)=-\left(Q+K^*RK\right) \tag{11.11}$$

Since $X\left(\infty\right)\rightarrow 0$, then performance index $J$ can be written as

$$J=X^*\left(0\right)PX\left(0\right)$$

To obtain the solution to the quadratic optimal control problem, we proceed as follows,

$$R=T^*T$$

Where $T$ is a nonsingular matrix, then Eq. 11.11 can be written as

$$\dot{X}=AX+Bu$$

$$A=\begin{bmatrix}0 & 1\\ 0 & 0\end{bmatrix} \qquad B=\begin{bmatrix}0\\ 1\end{bmatrix}$$

Noting that $\quad Q=\begin{bmatrix}1 & 0\\ 0 & \sqrt{\mu}\end{bmatrix}\begin{bmatrix}1 & 0\\ 0 & \sqrt{\mu}\end{bmatrix}$

The rank is

$$\begin{bmatrix}Q^{\frac{1}{2}}\\ Q^{\frac{1}{2}}A\end{bmatrix}=\begin{bmatrix}1 & 0\\ 0 & \sqrt{\mu}\\ 0 & 1\\ 0 & 0\end{bmatrix} \quad \text{to be 2.}$$

Thus $A-BK$ is a stable matrix

The reduced Riccati equation

$$A^*P+PA-PBR^{-1}B^*P+Q=0$$

Since $A$ and $Q$ are symmetric, $P$ is real symmetric. Hence the last equation can be written as,

$$\begin{bmatrix} 0 & 0 \\ 1 & 0 \end{bmatrix}\begin{bmatrix} p_{11} & p_{12} \\ p_{21} & p_{22} \end{bmatrix}+\begin{bmatrix} p_{11} & p_{12} \\ p_{21} & p_{22} \end{bmatrix}\begin{bmatrix} 0 & 1 \\ 0 & 0 \end{bmatrix}-\begin{bmatrix} p_{11} & p_{12} \\ p_{21} & p_{22} \end{bmatrix}\begin{bmatrix} 0 \\ 1 \end{bmatrix}[1][0 \quad 1]\begin{bmatrix} p_{11} & p_{12} \\ p_{21} & p_{22} \end{bmatrix}$$
$$+\begin{bmatrix} 1 & 0 \\ 0 & \mu \end{bmatrix}=\begin{bmatrix} 0 & 0 \\ 0 & 0 \end{bmatrix}$$

$$\begin{bmatrix} 0 & 0 \\ p_{11} & p_{12} \end{bmatrix}+\begin{bmatrix} 0 & p_{11} \\ 0 & p_{12} \end{bmatrix}-\begin{bmatrix} p_{11}^2 & p_{12}p_{22} \\ p_{21}p_{22} & p_{22}^2 \end{bmatrix}+\begin{bmatrix} 1 & 0 \\ 0 & \mu \end{bmatrix}=\begin{bmatrix} 0 & 0 \\ 0 & 0 \end{bmatrix}$$

For which, we obtain

$$1-p_{12}^2=0$$

$$p_{11}-p_{12}p_{22}=0$$

$$\mu+2p_{12}-p_{22}^2=0$$

$$P=\begin{bmatrix} p_{11} & p_{12} \\ p_{21} & p_{22} \end{bmatrix}=\begin{bmatrix} \sqrt{\mu+2} & 1 \\ 1 & \sqrt{\mu+2} \end{bmatrix}$$

The optimal feedback gain matrix k is

$$K=R^{-1}B^{*}P$$

$$=[1][0 \quad 1]\begin{bmatrix} p_{11} & p_{12} \\ p_{21} & p_{22} \end{bmatrix}=[p_{12} \quad p_{22}]=\begin{bmatrix} 1 & \sqrt{\mu+2} \end{bmatrix}$$

Thus, the optimal control signal

$$u=-kx=-x_1-\sqrt{\mu+2}\,x_2$$

The block diagram of optimal control system for Example 11.6 is shown in Figure 11.5.

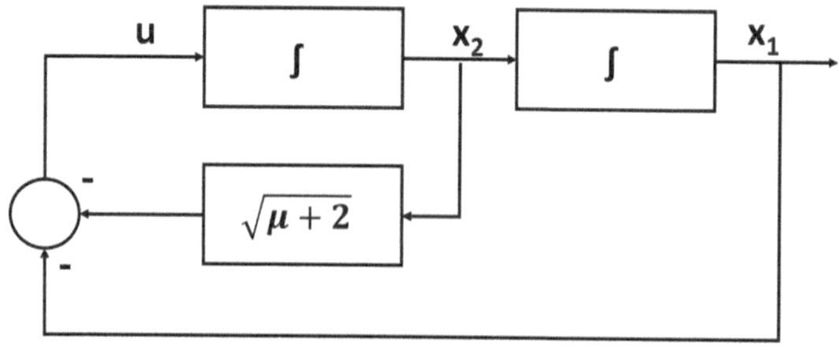


**FIGURE 11.5** Block diagram of optimal control system of Example 11.6.

### 11.1.6 Alternate Solutions of State Space Equation

Consider

$$x(t)=\varphi(t-t_0)x(t_0)+\int_{t_0}^{t}\phi(t-\tau)Br(\tau)d\tau \tag{11.12}$$

Laplace transform

$$X(s)=[SI-A]^{-1}X(t_{0.})+[AI-A]^{-1}BR(s) \tag{11.13}$$

Once Eq. 11.13 is determined for a specific system, the inverse Laplace transform is found to obtain Eq. 11.12. Mason's formula can be used to find the Eq. 11.12.

$$\frac{x_1(s)}{x_2(s)}=\mu(s)=\sum_k\frac{\mu k(s)\Delta k(s)}{\Delta(s)}$$

Where $\mu, \Delta k$, and $\Delta$ have the usual meaning

**Example 11.7:** Consider

$$\begin{bmatrix}\dot{X}_1\\ \dot{X}_2\end{bmatrix}=\begin{bmatrix}0 & 1\\ -\frac{k}{\mu} & -\frac{f}{\mu}\end{bmatrix}x+\begin{bmatrix}0\\ \frac{1}{\mu}\end{bmatrix}r(t)\quad \frac{k}{\mu}=2$$

$$\begin{bmatrix}\dot{X}_1\\ \dot{X}_2\end{bmatrix}=\begin{bmatrix}0 & 1\\ -2 & -3\end{bmatrix}\begin{bmatrix}x_1\\ x_2\end{bmatrix}+\begin{bmatrix}0\\ 1\end{bmatrix}r(t)\quad \frac{f}{\mu}=3,\frac{1}{\mu}=1$$

The signal flow graph of Example 11.7 is shown in Figure 11.6.

Using Mason's rule

$$\Delta(s)=1-\left(-\frac{3}{s}-\frac{2}{s^2}\right)=1+\left(\frac{3}{s}+\frac{2}{s^2}\right)$$

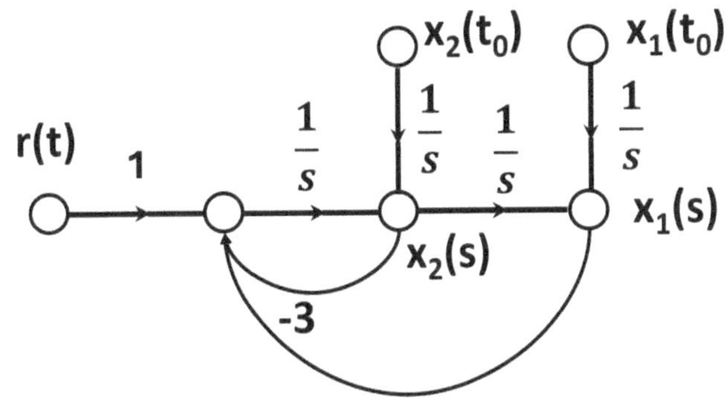


FIGURE 11.6 Signal flow graph of Example 11.7.

$$=\frac{s^2+3s+2}{s^2}=\frac{(s+1)(s+2)}{s^2}$$

$$X_1(s)=\frac{1*\left(1+\frac{3}{s}\right)s^2}{(s+1)(s+2)}*\frac{x_1(t_0)}{s}+\frac{\frac{1}{s}*\frac{x_2(t_0)}{s}}{(s+1)(s+2)}+\frac{1}{s^2}*\frac{R(s)}{(s+1)(s+2)}*s^2$$

$$X(s)=\frac{s+3}{(s+1)(s+2)}x_1(t_0)+\frac{x_2(t_0)}{(s+1)(s+2)}s^2+\frac{R(s)}{(s+1)(s+2)}$$

$$X_2(S)=1*-\frac{2}{s^2}*\frac{s^2}{(s+1)(s+2)}x_1(t_0)+1*\frac{\frac{\frac{x_2(t_0)}{s}}{(s+1)(s+2)}}{s^2}+\frac{R(s)}{s}*\frac{s^2}{(s+1)(s+2)}$$

$$=-\frac{2x_1(t_0)}{(s+1)(s+2)}+\frac{x_2(t_0)s}{(s+1)(s+2)}+\frac{sR(s)}{(s+1)(s+2)}$$

The inverse Laplace transform of the previous equation is

$$x_1(t)=\left(2e^{-t}-e^{\ 2t}\right)x_1(t_0)+\left(e^{-t}-e^{-2t}\right)x_2(t_0)+\mathcal{L}^{-1}\left[\frac{R(s)}{(s+1)(s+2)}\right]$$

$$x_2(t)=\left(-2e^{-t}+2e^{-2t}\right)x_1(t_0)+\left(-e^{-t}+2e^{-2t}\right)x_2(t_0)+\mathcal{L}^{-1}\left[\frac{R(s)}{(s+1)(s+2)}\right]$$

When the initial conditions are zero and $R(S)=\frac{1}{s}$

$$x_1(t)=\mathcal{L}^{-1}\left[\frac{\frac{1}{s}e^{-st_0}}{(s+1)(s+2)}\right]=\frac{1}{2}e^{-(t-t_0)}+\frac{1}{2}e^{-2(t-t_0)}$$

$$x_2(t)=\mathcal{L}^{-1}\left[\frac{s*\frac{1}{s}e^{-st_0}}{(s+1)(s+2)}\right]=e^{-(t-t_0)}-e^{-2(t-t_0)}$$

**Example 11.8:** $\frac{d^3c}{dt^3}+\frac{3d^2c}{dt^2}+\frac{2dc}{dt}=r(t)$

The signal flow graph of Example 11.8 is shown in Figure 11.7.

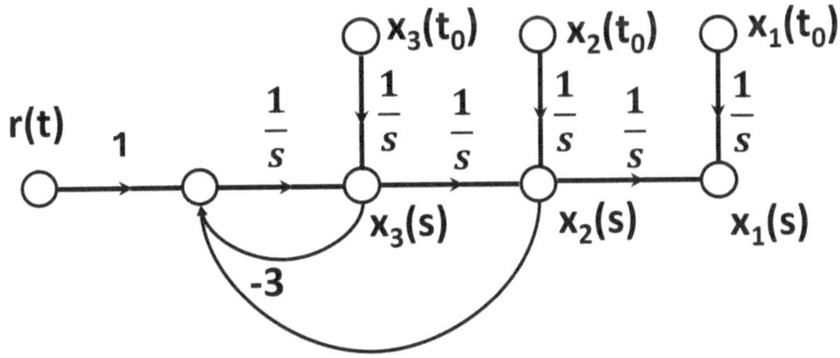


**FIGURE 11.7** Signal flow graph of Example 11.8.

The state variable $X^T = [x_1, x_2, x_3] = [c, \dot{c}, \ddot{c}]^T$

$$\frac{dx_1}{dt} = x_2, \frac{dx_2}{dt} = x_3, \frac{dx_3}{dt} = -2x_2 - 3x_3 + r$$

Rewriting in matrix form $\dot{X} = \begin{bmatrix} 0 & 1 & 0 \\ 0 & 0 & 1 \\ 0 & -2 & -3 \end{bmatrix} X + \begin{bmatrix} 0 \\ 0 \\ 1 \end{bmatrix} r(t)$

$$\Delta(S) = \frac{\frac{(s+1)(s+2)}{s^2} * \frac{x_1(t_0)}{s}}{(s+1)(s+2)} + \frac{s^{-2}\left(1+\frac{3}{s}\right)}{\frac{(s+1)(s+2)}{s^2}} x_2(t_0) + \frac{s^{-2} * \frac{x_3(t_0)}{s}}{(s+1)(s+2)} s^{-2}$$

$$+ \frac{s^{-3}R(S)}{\frac{(s+1)(s+2)}{s^2}}$$

$$= x_1(t_0) + \frac{(x+3)x_2(t_0)}{s(s+1)(s+2)} + \frac{x_3(t_0)}{s(s+1)(s+2)} + \frac{R(s)}{s(s+1)(s+2)}$$

$$X_2(S) = 0 + \frac{x_2(t_0)s^{-1}}{(s+1)(s+2)s^{-2}} + \frac{x_3(t_0)s^{-1} * s^{-1}}{(s+1)(s+2)s^{-2}} + \frac{R(S)s^{-2}}{(s+1)(s+2)s^{-2}}$$

$$= 0 + \frac{s+3}{(s+1)(s+2)} x_2(t_0) + \frac{x_3(t_0)}{(s+1)(s+2)} + \frac{R(S)}{(s+1)(s+2)}$$

$$x_3(S) = 0 + \frac{-2s^{-1}x_2(t_0)s^{-1}}{(s+1)(s+2)s^{-2}} + \frac{1 * x_3(t_0)s^{-1}}{(s+1)(s+2)s^{-2}} + \frac{R(S)s^{-1}}{(s+1)(s+2)s^{-2}}$$

$$= 0 + \frac{-2x_2(t_0)}{(s+1)(s+2)} + \frac{sx_3(t_0)}{(s+1)(s+2)} + \frac{sR(S)}{(s+1)(s+2)}$$

When the loop is closed.

$$\dot{X} = \begin{bmatrix} 0 & 1 & 0 \\ 0 & 0 & 1 \\ -1 & -2 & -3 \end{bmatrix} + \begin{bmatrix} 0 \\ 0 \\ 1 \end{bmatrix} r$$

And the equation can be formulated accordingly.

**Example 11.9:** This example relates to a state variable feedback system.

We have a system described by

$$\begin{bmatrix} \dot{x}_1 \\ x_2 \end{bmatrix} = \begin{bmatrix} 0 & 1 \\ 0 & 0 \end{bmatrix} \begin{bmatrix} x_1 \\ x_2 \end{bmatrix} + \begin{bmatrix} 0 \\ 1 \end{bmatrix} u(t)$$

$$A = \begin{bmatrix} 0 & 1 \\ 0 & 0 \end{bmatrix} \quad \text{and} \quad B = \begin{bmatrix} 0 \\ 1 \end{bmatrix}$$

We choose a feedback control system so that

$$u(t) = -k_1 X_1 - k_2 X_2$$

Therefore, the control signal is a linear function of two state variables. The sign of feedback is negative to provide negative feedback.

$$\dot{x}_1 = x_2$$

$$\dot{x}_2 = -k_1 X_1 - k_2 X_2$$

Or $\quad \dot{X} = HX = \begin{bmatrix} 0 & 1 \\ -k_1 & -k_2 \end{bmatrix} X$

Note that $x_1$ represents position and transfer function $G(S) = \frac{1}{\mu s^2}$ where $\mu = 1$, Let $k_1 = 1$ and determine a suitable value of $k_2$ so that the performance index is minimized then

$$H^T P + PH = -I$$

$$\begin{bmatrix} 0 & -1 \\ 1 & -k_2 \end{bmatrix} \begin{bmatrix} p_{11} & p_{12} \\ p_{21} & p_{22} \end{bmatrix} + \begin{bmatrix} p_{11} & p_{12} \\ p_{21} & p_{22} \end{bmatrix} \begin{bmatrix} 0 & 1 \\ -1 & -k_2 \end{bmatrix} = \begin{bmatrix} -1 & 0 \\ 0 & -1 \end{bmatrix}$$

$$-p_{12} - p_{12} = -1 \qquad p_{12} = \frac{1}{2}$$

$$p_{11} - k_2 p_{12} - p_{22} = 0 \qquad p_{22} = \frac{1}{k_2}$$

$$p_{12} - k_2 p_{22} + p_{12} - k_2 p_{22} = -1 \qquad p_{11} = \frac{k_2^2 + 2}{2k_2}$$

The integral performance index is then

$$J = X^T(0)PX(0)$$

And we will consider the case where each state is initially displaced one unit from equilibrium so that $x(0) = [1 \quad 1]$. Therefore,

$$J = [1 \quad 1]\begin{bmatrix} p_{11} & p_{12} \\ p_{21} & p_{22} \end{bmatrix}\begin{bmatrix} 1 \\ 1 \end{bmatrix}$$

$$= [1 \quad 1]\begin{bmatrix} p_{11} + p_{12} \\ p_{21} + p_{22} \end{bmatrix} = p_{11} + p_{12} + p_{21} + p_{22}$$

Substituting the values of the elements of P,

We have,

$$J = \frac{k_2^2 + 2}{2k_2} + 1 + \frac{1}{k_2} = \frac{k_2^2 + 2k_2 + 4}{2k_2}$$

To minimize as a function of $k_2$,

$$\frac{\partial J}{\partial k_2} = \frac{2k_2(2k_2 + 2) - 2(k_2^2 + 2k_2 + 4)}{(2k_2)^2}$$

$$k_2^2 = 4, \qquad k_2 = 2$$

where $J$ is minimized.

The minimized value of $J$ is obtained by substituting $K_2 = 2$

$$J = \frac{4 + 4 + 4}{4} = 3$$

The system matrix H obtained for the compensated system is then

$$H = \begin{bmatrix} 0 & 1 \\ -1 & -2 \end{bmatrix}$$

The signal flow graph of the compensated system is shown in Figure 11.8.

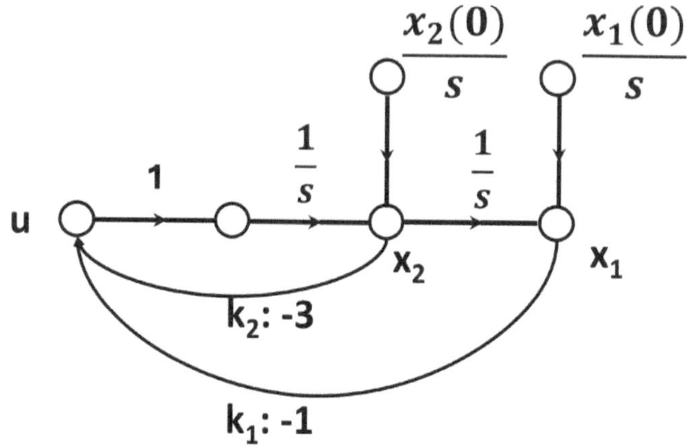


**FIGURE 11.8** Signal flow graph of compensated system.

The characteristic equation of the compensated system is

$$\det[\lambda I - H] = \det\begin{bmatrix} \lambda & -1 \\ 1 & \lambda+2 \end{bmatrix} = \lambda^2 + 2\lambda + 1$$

Because this is a second-order system, we note that the characteristic equation is of the form $\left(s^2 + 2\xi\omega_n s + \omega_n^2\right)$, and therefore, the damping ratio of the compensated system is $\xi$=1. This compensated system is considered an optimal system that results in a minimum value for $PI$.

### 11.1.7 Determination of an Optimal System

In the previous system, both the feedback gains $k_1$ and $k_2$ are unknown for simplification, and we can fix them as $k_1 = k_2 = k$

$$\dot{X} = HX = \begin{bmatrix} 0 & 1 \\ -k & -k \end{bmatrix} X$$

To determine $P$ matrix

$$H^T P + PH = -I$$

Once we solve these simultaneous equations, we obtain

$$p_{12} = 2k, \qquad p_{22} = \frac{k+1}{2k^2}, \qquad p_{11} = \frac{1+2k}{2k}$$

We assume a scenario where the system is displaced one unit from equilibrium initially so that $X^T(0) = [1 \quad 0]$ then the *PI*

$$J = \int_0^\infty X^T X dt = X^T(0) P X(0) = p_{11}$$

Thus the *PI* to be minimized.

$$J = p_{11} = \frac{1+2k}{2k} = 1 + \frac{1}{2k}$$

The minimum value of *J* is obtained where *K* approaches infinity, the result is $J_{min} = 1$

$$u(t) = -k[x_1(t) + x_2(t)]$$

To be very large. Therefore, we must introduce a constraint on $u(t)$ so that the gain $k$ is not made too large. Thus, if we establish $|u(t)| \le 50$, we require that the maximum acceptable value of $k$

$$k_{max} = \frac{|u|_{max}}{x_1(0)} = 50$$

The min value of $J = 1 + \frac{1}{2k_{max}} = 1.01$

**Example 11.10:** This example relates to the third-order system design

$$\frac{d^3y}{dt^3} + \frac{5d^2y}{dt^2} + \frac{3dy}{dt} + 2y = u$$

$$Y[1 \quad 0 \quad 0]X$$

where matrix $k$ is given as

$$k = [k_1 \quad k_2 \quad k_3]$$

And $u = -kX$

Then $\dot{X} = AX - BKX = (A - BK)X$

The state feedback matrix is

$$[A-BK]=\begin{bmatrix} 0 & 1 & 0 \\ 0 & 0 & 1 \\ -2-k & -2-k_2 & -5-k_3 \end{bmatrix}$$

And the characteristic equation is

$$det[A-BK]=s^3+(5+k_3)s^2+(3+k_2)s+2+k_1$$

To obtain a rapid response with a low overshoot, the desired characteristic equation can be chosen as

$$(s^2+2\xi\omega_n s+\omega_n^2)(s+\xi\omega_n)$$

We choose $\xi=0.8$ for minimal overshoot and $\omega_n$ it meets the setting time requirement. If we want a settling time (2% criterion) equal to 1 sec, then $T_s=\dfrac{4}{\xi\omega_n}=\dfrac{4}{0.8\omega_n}=1$

If we choose $\omega_n=6$, the desired characteristic equation is $(s^2+9.6s+36)(s+4.8)=s^3+14.4s^2+82.1s+172.8$. Therefore, we require $k_3=9.4,\quad k_2=79.1,$ $k_1=170.1$

### 11.1.8 Ackermann's Formula

Ackermann's formula is effective in finding the state variable feedback matrix K for a SISO (single input, single output) system.

$$K=[k_1,k_2,\ldots,k_n]$$

Where $u=-KX$

Given the desired characteristic equation

$$q(s)=s^n+\propto_1 s^{n-1}+\ldots+\propto_n$$

The state feedback gain matrix is given by

$$k=[0\ \ 0\ \ \ldots\ \ 1]Pe^{-1}q(A)$$

Where

$$q(A)=A^n+\propto_1 A^{n-1}+\ldots+\propto_{n-1} A+\propto_n I$$

And $P_c$ is the controllability matrix

**Example 11.11:** Consider $G(S)=\frac{1}{s^2}$. To determine the feedback gain to place the closed-loop poles at $-1\pm j$, we require:

$$q(s)=s^2+2s+2 \qquad \propto_1=2, \qquad \propto_2=2$$

The matrix equation for the system is

$$\dot{X}=\begin{bmatrix}0 & 1\\0 & 0\end{bmatrix}X+\begin{bmatrix}0\\1\end{bmatrix}u$$

The controllability matrix is

$$P_c=\begin{bmatrix}B & AB\end{bmatrix}=\begin{bmatrix}0 & 1\\1 & 0\end{bmatrix}$$

We obtain $k=\begin{bmatrix}0 & 1\end{bmatrix}\begin{bmatrix}0 & 1\\1 & 0\end{bmatrix}q(A)$

$$p_c^{-1}=\frac{1}{-1}\begin{bmatrix}0 & -1\\-1 & 0\end{bmatrix}=\begin{bmatrix}0 & 1\\1 & 0\end{bmatrix}$$

And $q(A)=\begin{bmatrix}0 & 1\\0 & 0\end{bmatrix}^2+2\begin{bmatrix}0 & 1\\0 & 0\end{bmatrix}+2\begin{bmatrix}1 & 0\\0 & 1\end{bmatrix}=\begin{bmatrix}2 & 2\\0 & 2\end{bmatrix}$

Thus,

$$K=\begin{bmatrix}0 & 1\end{bmatrix}\begin{bmatrix}0 & 1\\1 & 0\end{bmatrix}\begin{bmatrix}2 & 2\\0 & 2\end{bmatrix}$$

$$=\begin{bmatrix}0 & 1\end{bmatrix}\begin{bmatrix}0 & 2\\2 & 2\end{bmatrix}=\begin{bmatrix}2 & 2\end{bmatrix}$$

## 11.2 ROBUST CONTROL SYSTEM

Recent control design methodologies emphasize robustness in stability, performance, and resilience to uncertainty. These approaches recognize that physical systems and the environments in which they operate cannot be modeled with complete accuracy, are subject to unpredictable changes, and may experience significant disturbances. While classical control techniques can also yield robust control systems, they often rely on models that are inherently imperfect representations of the actual physical system. Such models are prone to inaccuracies due to factors including:

1 Parameter variations
2 Unmodeled dynamics
3 Unmodeled time delays
4 Shifts in equilibrium or operating points
5 Sensor noise
6 Unanticipated disturbances.

The goal of robust system design is to ensure consistent system performance despite model inaccuracies and changes. A system is considered robust when its performance remains acceptable even in the presence of model variations or uncertainties.

The block diagram of a control system is shown in Figure 11.9.

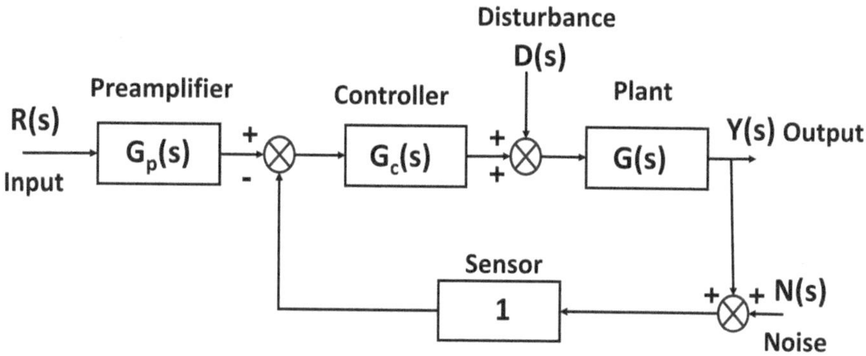


**FIGURE 11.9** Block diagram of a control system.

A control system is considered robust when:

1. It exhibits low sensitivity to disturbances and uncertainties.
2. It remains stable across a range of parameter variations.
3. Its performance continues to meet the specified requirements despite changes in system parameters.

For small perturbations in parameters, the differential sensitivities can be used as a measure of robustness. System sensitivity is defined as:

$$S{\propto}^{T}=\frac{\partial\left(\frac{T}{T}\right)}{\partial\left(\frac{\propto}{\propto}\right)}$$, where $\propto$ is a parameter and $T$ is the system transfer function.

Root sensitivity: $s_{\propto}^{ri}=\dfrac{\partial ri}{\partial\left(\frac{\propto}{\propto}\right)}$

When the zero of $T(s)$ is independent of the parameter $\propto$,

$$s_{\propto}^{T} = -\sum_{i=1}^{n} s_{\propto}^{ri} \frac{1}{s+ri}$$

For example, $T(s) = \frac{1}{s+(\propto+1)}$

$$S_{\propto}^{T} = \frac{-\propto}{s+\propto+1}$$

Furthermore, root is $r = (\propto+1)$ and

$$-s_{\propto}^{ri} = -\propto$$

Therefore, $s_{\propto}^{T} = -s_{r}^{r} * \frac{1}{s+\propto+1}$

### 11.2.1 Sensitivity of a Controlled System

Consider $G(S) = \frac{1}{s^2}$ and a PD controller $G(S) = b_1 + b_2 s$. Then the sensitivity w.r.t changes in $G(S)$ is

$$S_{G}^{T} = \frac{1}{1+GG(S)} = \frac{s^2}{s^2+b_2 s+b_1}$$

And $T(S) = \frac{b_2 s+b_1}{s^2+b_2 s+b_1}$

### 11.2.2 Analysis of Robustness

Consider the block diagram of the feedback control system as shown in Figure 11.10.

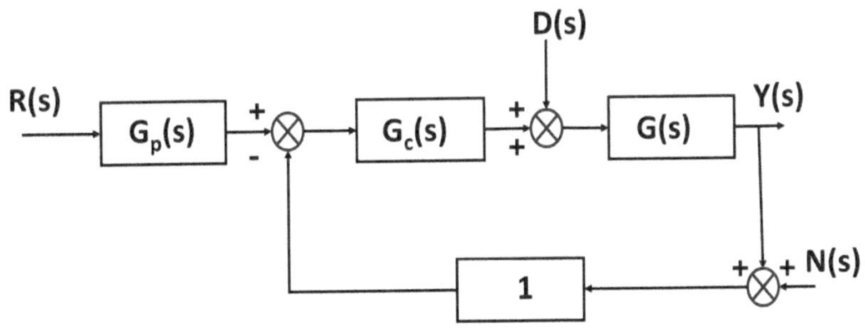


**FIGURE 11.10** Block diagram of feedback control system.

System goals include maintaining small tracking error $\left[e(t)=r(t)-y(t)\right]$ for an input $r(t)$ and keeping the output $y(t)$ small for a disturbance $d(t)$. The sensor noise $n(t)$ must be small for $r(t)$ so that $|r| \gg |n|$. The sensitivity function is:

$$S_s = \frac{1}{1+G_c(s)G(s)}$$

The closed-loop transfer function is:

$$T(S) = \frac{G_c(s)G(s)}{1+G_c(s)G(s)}$$

When $G_p(s)=1,\ S(s)+T(s)=1$

Thus, it is important to make $S(s)$ small. For a physically realizable system, the loop gain $L(s)=G_c(s)G(s)$ must be small for high frequencies. This means that $S(j\omega)$ approaches 1 at high frequencies.

An additive perturbation characterizes the set of possible plants as follows (assuming $G_c(s)=1$)

$G_a(s)=G(s)+A(s)$ where $G(s)$ is the nominal plant and $A(s)$ is the perturbation that is bounded in magnitude. It is assumed that $G_a(s)$ and $G(s)$ have the same number of poles in the RHP if any. Then the system stability will not change if:

$$|A(j\omega)| < |1+G(j\omega)|, \text{ for all } \omega$$

This assumes stability but not dynamic performance.

A multiplicative perturbation gives the plant:

$$G_m(s) = G(s)\left[1+\mu(s)\right]$$

The perturbation is bounded in magnitude and again it is assumed that $G_m(s)$ and $G(s)$ have the same number of poles in RHP. Then the system stability will not change if

$$|\mu(j\omega)| < \left|1+\frac{1}{G(j\omega)}\right|, \text{ for all } \omega$$

The previous equation is referred to as the robust stability condition. It serves as a test for robustness under multiplicative perturbations. This type of perturbation is commonly used because it satisfies the following key properties:

1 It is small at low frequencies, where the nominal plant model is typically well-identified.
2 It is large at high frequencies, where the nominal model is generally less accurate or uncertain.

**Example 11.12:** A system with multiplicative perturbation considers the previous block diagram of Figure 11.10 with $G_p(S) = 1, G_c = k$ and $G(s) = \frac{170,000(s+0.7)}{s(s+3)(s^2+10s+10,000)}$.

The system is unstable for $k = 1$ but is stable for $k = 0.5$. Now consider the effect of an unmodeled pole at $50\,rad/sec$. In this case, the multiplicative perturbation is determined from.

$$[1+\mu(s)] = \frac{50}{s+50}$$

Or $\mu(s) = \frac{-s}{s+50}$

Bound is then

$$|\mu(j\omega)| = \left|\frac{-j\omega}{j\omega+50}\right|$$

The criterion that $|\mu(j\omega)| < \left|1+\frac{1}{G(j\omega)}\right|$ is not satisfied. Thus the system may not be stable. If we use a lag compensator $G_c(s)$, where

$$G_c(s) = \frac{0.15(s+25)}{s+2.5}$$

The loop transfer function is $1+G_c(s)G(s)$ and thus we reshape the function $G_cG(j\omega)$ in the frequency range $2 < \omega < 25$ (use bode plot to observe). Then we have the altered magnitude.

$$\left|1+\frac{1}{G_cG(j\omega)}\right|$$

Here the robustness inequality is satisfied, and the system remains stable.

### 11.2.3 System with Uncertain Parameters

Many systems have constant parameters that are uncertain within a range. For example, consider a system with characteristics equation,

$$s^2 + a_{n-1}s^{n-1} + a_{n-2}s^{n-2} + \ldots + a_0 = 0$$

With known coefficients within bounds

$$\propto_i \le a_i \le \beta_i \qquad i = 0,1,\ldots,n$$

Where $a_n = 1$

To determine the stability of the system, one might need to examine all possible combinations of parameters. Fortunately, it is often sufficient to analyze a limited set of worst-case polynomials. In fact, only four key polynomials need to be considered, as they are both coefficient-based and easily defined.

For a third-order system with characteristics equation

$$s^3 + a_2 s^2 + a_1 s + a_0 = 0$$

Then the four polynomials are

$$q_1(S) = s^3 + \propto_2 s^2 + \beta_1 s + \beta_0$$

$$q_2(S) = s^3 + \beta_2 s^2 + \propto_1 s + \propto_0$$

$$q_3(S) = s^3 + \beta_2 s^2 + \beta_1 s + \propto_0$$

$$q_4(S) = s^3 + \propto_2 s^2 + \propto_1 s + \beta_0$$

One of the four polynomials represents the worst case and may indicate unstable performance or at least the worst performance for the system.

**Example 11.13:** This example relates to a third-order system with uncertain coefficients and explains the previous sub-section.

Let us consider a system as under

$$8 \le a_0 \le 60 \Rightarrow \propto_0 = 8, \beta_0 = 60$$

$$12 \le a_1 \le 100 \Rightarrow \propto_1 = 12, \beta_1 = 100$$

$$7 \le a_2 \le 25 \Rightarrow \propto_2 = 7, \beta_2 = 25$$

The four polynomials are

1. $q_1(s) = s^3 + 7s^2 + 100s + 60$
2. $q_2(s) = s^3 + 25s^2 + 12s + 8$
3. $q_3(s) = s^3 + 25s^2 + 100s + 8$
4. $q_4(s) = s^3 + 7s^2 + 12s + 60$

After using the Routh-Hurwitz Criterion, we find

1. $q_1(s)$ is stable
2. Stable

3. Stable
4. Stable

**Example 11.14:** Given $G(S)=\dfrac{4.5}{s(s+1)(s+2)}$.

The polynomial characteristics equation is then

$$q(s)=s^3+3s^2+2s+4.5=0$$

The system is nominally stable according to the Routh-Hurwitz criterion. However, if the system has uncertain coefficients

$$4 \le a_0 \le 5 \qquad \propto_0 = 4, \beta_0 = 5$$

$$1 \le a_1 \le 3 \qquad \propto_1 = 1, \beta_1 = 3$$

$$2 \le a_2 \le 4 \qquad \propto_2 = 2, \beta_2 = 4$$

Then the resulting polynomials are

1. $q_1(s)=s^3+2s^2+3s+5=0$

$$\begin{bmatrix} s^3 & 1 & 3 \\ s^2 & 2 & 5 \\ s^1 & 1 & 0 \\ s^0 & 5 & \end{bmatrix} \quad \text{stable.}$$

2. $q_2(s)=s^3+4s^2+1s+4=0$

$$\begin{bmatrix} s^3 & 1 & 1 \\ s^2 & 4 & 4 \\ s^1 & \propto\in & 0 \\ s^0 & 4 & \end{bmatrix} \quad \text{marginally stable.}$$

3. $q_3(s)=s^3+4s^2+3s+4=0$

$$\begin{bmatrix} s^3 & 1 & 3 \\ s^2 & 4 & 4 \\ s^1 & 2 & \\ s^0 & 4 & \end{bmatrix} \quad \text{stable.}$$

4 $q_4(s) = s^3 + 2s^2 + 1s + 5 = 0$

$$\begin{bmatrix} s^3 & 1 & 1 \\ s^2 & 2 & 5 \\ s^1 & -\frac{3}{2} & \\ s^0 & 5 & \end{bmatrix} \quad \text{unstable.}$$

Therefore $q_4(s)$ is unstable for the worst case, where $\propto_2$ is minimized and $p_0$ is maximum. This occurs when the plant has changed to

$$G(s) = \frac{5}{s(s+1)(s+1)}$$

### 11.2.4 Design of Robust Control System

The design process involves two main tasks: (i) determining the structure of the controller, and (ii) tuning its parameters to achieve optimal performance. This process assumes complete knowledge of the plant, which is typically modeled as a linear time-invariant (LTI) continuous system. The controller's structure is chosen to ensure that the system's response meets specified performance criteria.

One common objective in control system design is to ensure that the output of the controlled system exactly and instantaneously follows the input. In other words, the system's transfer function should ideally be unity

$$T(s) = \frac{C(s)}{R(s)} = 1$$

In other words, the system should ideally be represented on a Bode plot with a 0 dB gain over an infinite bandwidth and zero phase shift. In practice, this is not achievable, as all systems contain inductive and capacitive components that store energy in some form, resulting in a finite bandwidth. One key design objective is to keep the magnitude response as flat and as close to unity as possible over a wide bandwidth for a given plant and controller combination. Another important goal in control system design is to minimize the effect of disturbances on the system's output. Thus we wish to minimize $\frac{Y(s)}{D(s)}$ over a range of frequency. Consider the control system shown in Figure 11.11.

$$\frac{T(s)}{R(s)} = \frac{G_c G_1 G_2(s)}{1 + G_c G_1 G_2(s)}$$

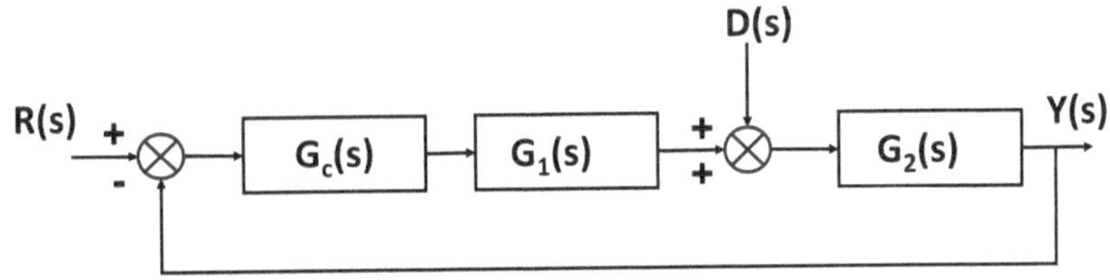


**FIGURE 11.11** A typical feedback control system.

And $\frac{Y(s)}{D(s)} = \frac{G_2(s)}{1 + G_c G_1 G_2(s)}$

The characteristics equation for both is

$$1 + G_c(s)G_1(s)G_2(s) = 1 + L(s) = 0$$

Furthermore, we recall that the sensitivity of $T(s)$ with respect to $G(s)$ is $G(s) = G_1 G_2(s)$

$$S_G^T = \frac{1}{1 + G_c G_1 G_2(s)}$$

The characteristic equation governs the sensitivity factor of the system. The previous equation illustrates that for low sensitivity s, a high value of the loop gain L(jω) is required. However, it is well-known that excessively high gain can lead to instability or degraded responsiveness of the transfer function T(s). Therefore, the designer aims to achieve the following:

1. A transfer function T(s) with a wide bandwidth and faithful reproduction of the reference input R(s).
2. A sufficiently large loop gain L(s) to minimize sensitivity s.
3. Large loop gain $L(s)$ attained primarily by $G_c(s)G_1(s)$ since $\frac{Y(s)}{D(s)} \cong \frac{1}{G_c G_1(s)}$

In the context of robust system design in the frequency domain, the goal is to find an appropriate compensator Gc(s) such that the closed-loop sensitivity is within a specified tolerance. Sensitivity minimization involves determining a compensator that ensures the closed-loop sensitivity is as close as possible to the minimal achievable sensitivity. Similarly, the gain margin problem focuses on finding a compensator that achieves a prescribed gain margin, while gain margin maximization aims to identify the compensator that provides the maximal attainable gain margin. For the frequency domain specifications, the following conditions are required for the Bode plot (Figure 11.12) of the system $G_c G(j\omega)$.

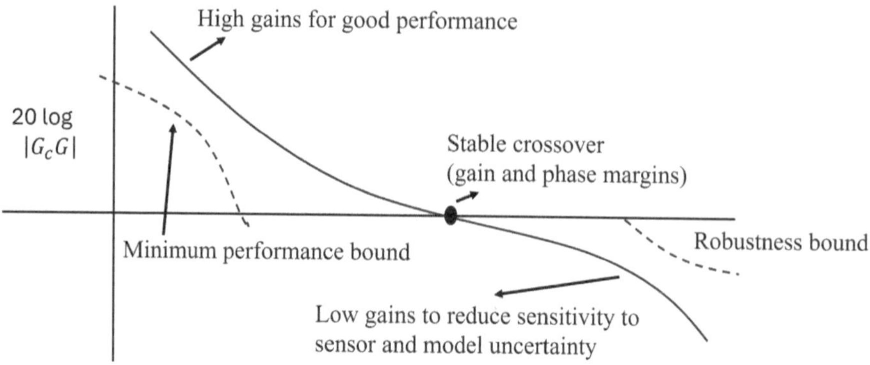


**FIGURE 11.12** A typical Bode plot.

1. Relative stability $G_cG(j\omega)$ must have, for an adequate range of $\omega$, not more than $-20\,db\,/\,decade$ slope at or near the cross over frequency.
2. Steady-state accuracy is achieved by the low-frequency gain.
3. Accuracy over a bandwidth $\omega_B$ by not allowing $|G_cG(\omega)|$ to fall below a prescribed level.
4. Disturbance rejection by high gain for $G_c(j\omega)$ over the system bandwidth.

Using the root sensitivity concept, we can state that we require that $S_{\infty}^{T}$ be minimized while attaining $T(s)$ with dominant roots that will provide the appropriate response and minimize the effect of $D(s)$. Again, we see that the goal is to have the gain of the loop primarily attained by $G_c(s)$. As an example, let $G_c(s) = K,\ G_1(s) = 1,\ G_2(s) = \dfrac{1}{s(s+1)}$.

Thus system has two roots, and we select a gain K so that $\dfrac{Y(s)}{D(s)}$ is minimized, $S_K^T$ minimized, and $T(s)$ has desirable dominant roots.

The sensitivity is

$$S_K^T = \frac{dn}{dk} * \frac{k}{r} = \left.\frac{ds}{dk}\right|_{s=r} * \frac{k}{r}$$

The characteristic equation is

$$s(s+1) + k = 0$$

Therefore, $\dfrac{dk}{ds} = -(2s+1)$ since $k = -s(s+1)$.

We then obtain

$$S_K^T = \frac{-1}{(2s+1)} * \frac{-s(s+1)}{8}\bigg|_{s=r} = \frac{-0.5 + j\omega + 1}{-1 + 2j\omega + 1} = \frac{0.5 + jk}{2j\omega}$$

When $\xi < 1$, the roots are complex, and $r = 0.5 + j\omega'$. Then

$$\left|S_K^T\right| = \left(\frac{0.25 + \omega^2}{4\omega^2}\right)^{1/2}$$

The magnitude of the sensitivity is plotted as shown in Figure 11.13.

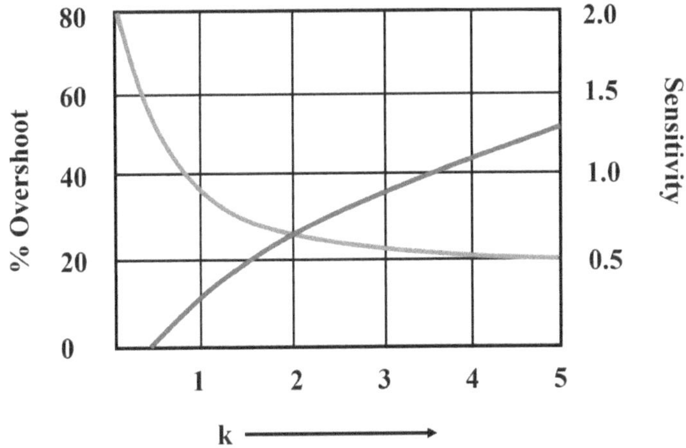


**FIGURE 11.13** A typical Bode plot of the magnitude of sensitivity.

The percent overshoot to a step input is shown in the diagram. It is desirable to minimize sensitivity, but it is recommended to limit the sensitivity factor k to 1.5 or less. The design procedure is as follows:

1. **Sketch the root locus** of the compensated system with Gc(s) chosen to position the dominant root at the desired location.
2. **Maximize the gain** of Gc(s) to minimize the effect of disturbances.
3. **Determine** $S_{\infty}^{r}$ and achieve the minimum sensitivity consistent with the required transient response.

**Example 11.15** (Sensitivity and compensation)

Let $G(s) = \frac{1}{s^2}, H(s) = 1$, and $G_c(s)$ are to be selected by frequency response methods. Therefore, the compensator is to be selected to achieve an appropriate gain and phase margin while minimizing sensitivity and the effect of the disturbance. Thus, we choose

$$G_c(s) = \frac{k\left(\frac{s}{z} + 1\right)}{\frac{s}{p} + 1}$$

Choose $k = 10$ to reduce the effect of the disturbance. To attain a phase margin of $45^0$, we select $z = 2.0$ and $p = 12$.

The compensated diagram is shown in Figure 11.14. The closed-loop bandwidth $\omega_b = 1.6\omega_c$. Thus we will increase the bandwidth by using the compensator and improve the fidelity of reproduction of the input signals.

$$\left|S_G^T\right| = \left|\frac{1}{1+G_cG(j\omega)}\right|_{\omega c}$$

To estimate $\left|S_G^T\right|$, Nichol's chart is used. Which gives

$$\left|T(\omega)\right| = \left|\frac{G_cG(\omega)}{1+G_cG(\omega)}\right|$$

### 11.2.5 Sensitivity with Lead Compensator

Let us consider $G(s) = \frac{1}{s^2}, H(s) = 1$, and $G_c(s) = \frac{8.1(s+1)}{s+3.6}$ for the system. The block diagram of a control system with lead compensator is shown in Figure 11.15.

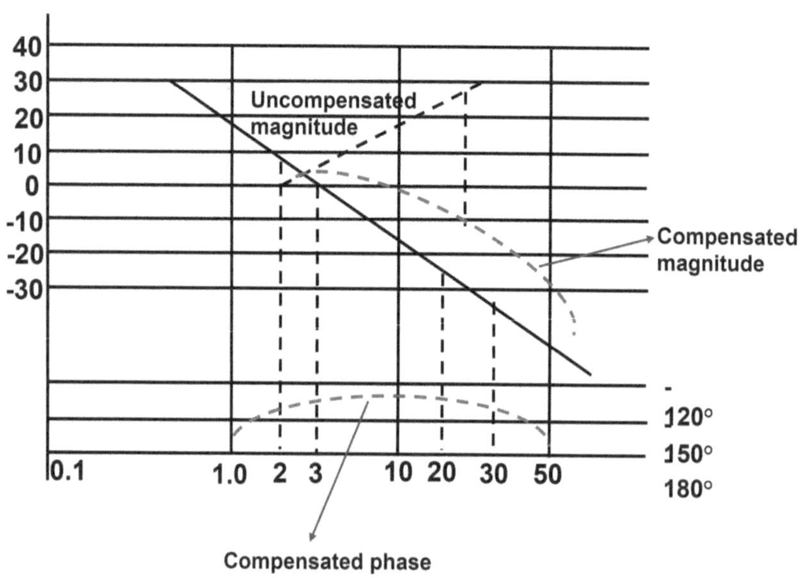


**FIGURE 11.14** Bode diagram of Example 11.15.

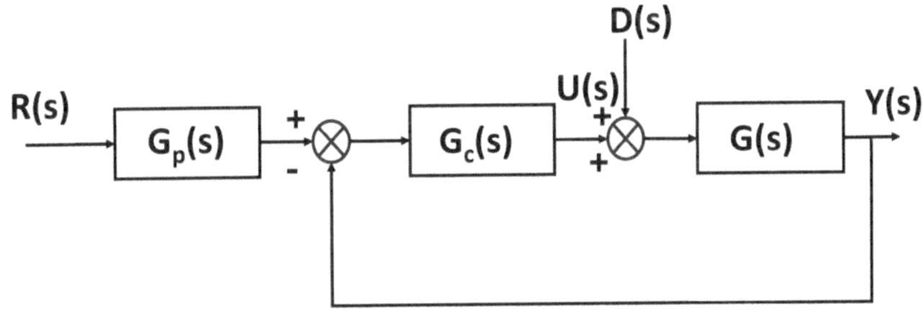


**FIGURE 11.15** Block diagram for a control system with lead compensator.

By root locus, the dominant roots are $s = -1 \pm j2$. Since the gain is 8.1, the effect of disturbance is reduced, and the response time meets the specifications. The sensitivity at the root $r$ by assuming that the system with dominant roots may be approximated by second order system

$$T(s) = \frac{k}{s^2 + 2\xi\omega_n s + k} = \frac{k}{s^2 + 2s + k}$$

Since $\xi\omega_n = 1$. The characteristic equation is thus

$$s^2 + 2s + k = 0$$

Then $\frac{dk}{ds} = -2(s+1)$, since $k = -(s^2 + 2s)$

Therefore, $S_k^T = \frac{-1}{2(s+1)} * \frac{s^2+2s}{s}\bigg|_{s=r} = \frac{s+2}{2s+2}\bigg|_{s=r}$

Where $r = -1 + j2$, thus

$$\frac{s+2}{2s+2}\bigg|_{-1+j2} = \frac{-1+j2+2}{-2+j4+2} = \frac{1+j2}{j4} = \frac{\sqrt{5}}{4} = 0.56$$

If we raise the gain to 10, $r \cong -1.1 \pm j2.2$

$$\left|S_k^T\right| = 0.53$$

Thus as $k$ increases, the sensitivity decreases but the transient performance deteriorates.

### 11.2.6 PID Controller

$$G_c(s) = k_1 + \frac{k_2}{s} + k_3 s$$

The popularity of PID controllers can be attributed to their robust performance across a wide range of operating conditions, as well as their functional simplicity, which allows engineers to use them in a straightforward and intuitive manner. To implement a PID controller, the appropriate parameters must be determined for the given process.

Consider the PID controller,

$$G_c(s) = k_1 + \frac{k_2}{s} + k_3 s = \frac{k_3 s^2 + k_1 s + k_2}{s}$$

$$= \frac{k_3\left(s^2 + as + b\right)}{s}$$

$$= \frac{k_3\left(s+z_1\right)\left(s+z_2\right)}{s}, \qquad a = \frac{k_1}{k_3}, \qquad b = \frac{k_2}{k_3}$$

Therefore, the PID controller introduces one pole at the origin and two zeros, which can be located anywhere in LHP.

If we have a system

$$G(s) = \frac{1}{(s+2)(s+5)}$$

For a PID controller with complex zeros, the root locus can be plotted as shown in Figure 11.16.

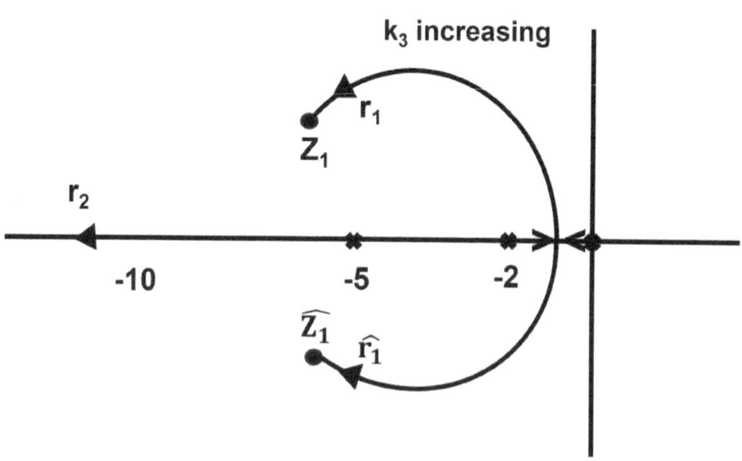


FIGURE 11.16 Root locus using PID controller.

The closed-loop transfer function is

$$T(s) = \frac{G(s)G_c(s)G_p(s)}{1+G_c(s)G(s)}$$

$$\frac{k_3(s+z_1)(s+\hat{z}_1)}{(s+r_2)(s+r_1)(S+\hat{r}_1)}G_p(s)$$

$$\cong \frac{k_3G_p(s)}{(s+r_2)}$$

Because the complex roots and zeroes are almost equal.

Setting $G_p(s)=1$

$$T(s) = \frac{k_3}{s+r_2} = \frac{k_3}{s+k_3}$$

when $k_3 \gg 1$. If $k_3$ is 100, the system has a fast response and zero steady-state error.

### 11.2.7 The Design of Robust PID Controlled System

The design procedure consists of three steps

1. Select the $\omega_n$ of the closed-loop system by specifying the settling time.
2. Determine the three coefficients using the appropriate optimum equation (table) and $\omega_n$ of step 1 to obtain $G_c(s)$.

The optimum coefficients of $T(s)$ are based on the ITAE criterion for a step input.

$$ITAE = \int_0^T t\left|e(t)\right| dt$$

Table 11.1 shows the optimum coefficients of T(s).

**TABLE 11.1**
**Table Showing the Optimum Coefficients of T(s)**

| Degree of Polynomial | Equation |
|---|---|
| 1 | $s+\omega_n$ |
| 2 | $s^2+1.4\omega_n s+\omega_n^2$ |
| 3 | $s^3+1.7\omega_n s^2+2.15\omega_n^2 s+\omega_n^3$ |
| 4 | $s^4+2.1\omega_n s^3+3.4\omega_n^2 s^2+2.7\omega_n^3+\omega_n^4$ |

3. Determine a filter $G_p(s)$, 100 that the closed-loop system transfer function, $T(s)$, does not have any zeros, as required by the following equation:

$$T(s)=\frac{Y(s)}{R(s)}=\frac{b_0}{s^n+b_{n-1}s^{n-1}+\ldots+b_1s+b_0}$$

**Example 11.16:** A control system for temperature settings is given in Figure 11.17.

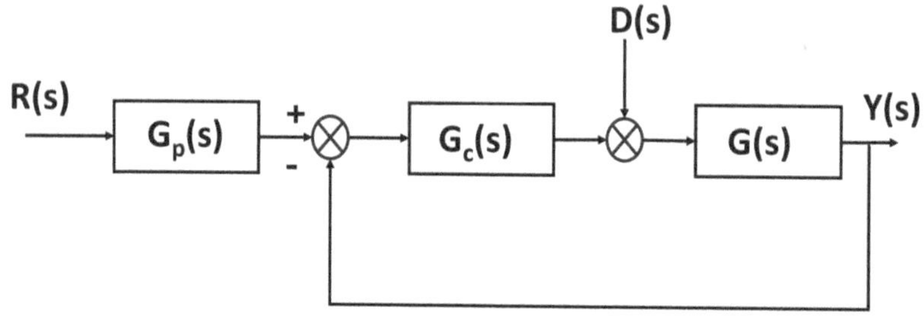


**FIGURE 11.17** Block diagram for Example 11.16.

$$G(s)=\frac{1}{(s+1)^2}$$

If $G_c(s)=1$, the steady-state error is found to be 50% and the settling time with a 2% criterion is 3.2 sec for a step input. We desire to obtain an optimum ITAE performance for a step input and a settling time of less than 0.5 sec using the PID controller

$$G_c(s)=\frac{k_3s^2+k_1s+k_2}{s}$$

The closed-loop transfer function without prefiltering $G_p=1$ is

$$T_1(s)=\frac{Y(s)}{R(s)}=\frac{G_cG(s)}{1+G_cG(s)}=\frac{k_3s^2+k_1s+k_2}{s^3+(2+k_3)s^2+(1+k_1)s+k_2} \qquad \text{A}$$

The optimum coefficient from the table

$$\left(s^3+1.7\omega_ns^2+2.15\omega_n^2s+\omega_n^3\right) \qquad \text{B}$$

We need to select $\omega_n$ om order to meet the settling time requirement. Since $T_s=\frac{4}{\xi\omega_n}$ and $\xi$ is unknown but near $\xi=0.8$, we set $\omega_n=10$. Then to equate the denominator of Equation (A) to Equation (B), we obtain

$$k_1 = 214, \qquad k_2 = 1000, \qquad k_3 = 15.5$$

$$T_1(s) = \frac{15.5s^2 + 214s + 1000}{s^3 + 17.5s^2 + 215s + 1000}$$

$$= \frac{15.5(s + 6.9 + j4.1)(s + 6.9 - j4.1)}{s^3 + 17.5s^2 + 215s + 1000}$$

The response to step input gives a 32% overshoot. We select a prefilter $G_p(s)$ so that we achieve an ITAE response with

$$T(s) = \frac{G_c G G_p(s)}{1 + G_c G(s)} = \frac{1000}{s^3 + 17.5s^2 + 215s + 1000}$$

Therefore, we require

$$G_p(s) = \frac{64.5}{(s^2 + 12.8s + 64.5)}$$

In order to eliminate the zeros and have the overall numbers to 1,000. The resulting system has a small overshoot, a settling time of less than ½ sec, and zero steady-state error. Table 11.2 shows the system characteristics for the system given in Example 11.16.
Becomes

$$(A - BK)^T P + P(A - BK) = -(I - K^T K)$$

And equation

$$K = T^{-1}(T^*) B^* P = R^{-1} B^* P$$

**TABLE 11.2**
**System Characteristics for the System Given in Example 11.16**

| Controller | $G(s) = 1$ | PID and $G_p(s) = 1$ | PID with $G_p(s)$ Prefilter |
|---|---|---|---|
| % overshoot | 0 | 31.7 | 1.9 |
| Settling time | 3.2 | 0.2 | 0.45 |
| ss error | 50.1% | 0.0% | 0.00% |

Becomes

$$K = R^{-1}B^T P = B^T P$$

Where $P$ is determined from the reduced matrix Riccati equation

$$A^T P + PA - PBB^T P + I = 0$$

Solving for $P$ requiring that it is positive

$$P = \begin{bmatrix} 2 & 1 \\ 1 & 1 \end{bmatrix}$$

The optimal feedback gain matrix K becomes

$$K = B^T P = \begin{bmatrix} 0 & 1 \end{bmatrix} \begin{bmatrix} 2 & 1 \\ 1 & 1 \end{bmatrix} \begin{bmatrix} 1 & 1 \end{bmatrix}$$

Thus optimal control signal $u$ is given by

$$u = -KX = -x_1 - x_2$$

## 11.3 SLIDING MODE CONTROLLER

Sliding mode control (SMC) is a robust control strategy widely used for nonlinear systems and systems with uncertainties. It belongs to the family of variable structure control methods, characterized by their ability to adapt the structure of the control law dynamically based on the system's state. SMC is particularly valued for its robustness to modeling inaccuracies and external disturbances, making it a preferred approach in applications ranging from robotics and aerospace systems to power electronics and industrial automation.

### 11.3.1 Principle of Sliding Mode Control

The central idea of SMC is to design a control law that forces the system's state to converge to and remain on a predefined manifold, known as the sliding surface, in the state space. Once the system reaches this surface, it "slides" along it toward the desired equilibrium point, maintaining a specified behavior. The sliding surface is typically defined to satisfy specific performance criteria, such as stability, tracking accuracy, or disturbance rejection.

#### 11.3.1.1 Fundamental Principles behind Sliding Mode Control

**a) System dynamics and control objective**

Consider a dynamic system with the general form of equations:

$$\dot{x} = f(x) + g(x)u + d(t)$$

x is the state vector
u is the control input

f(x) and g(x) represent the system's natural dynamics and how control input influences the system, respectively

d(t) is an unknown disturbance that may affect the system

The primary goal of sliding mode control is to make the system robust to disturbances d(t), uncertainties in the system model, and any unmodeled dynamics.

**b) The sliding surface**

The core of SMC is the sliding surface, which defines the desired behavior of the system. One common choice for the sliding surface is a linear function of the state error, e= $x$-$x_d$ where $x_d$ is the desired trajectory. For a simple system, the sliding surface can be:

$$S(x) = e = x - x_d$$

In this case, the surface S(x) = 0 corresponds to the state being equal to the desired state $x_d$, and the objective is to drive the system towards this surface. For higher-order systems, the sliding surface is typically defined using error and its derivatives, which allows for better control over the system's dynamics.

$$S(x) = \frac{d^{n-1}e}{dt^{n-1}} + \lambda_1 \frac{d^{n-1}e}{dt^{n-1}} + \ldots + \lambda_{n-1}e$$

Where $\lambda_i$ are tuning parameters that ensure desired performance.

**c) Sliding mode dynamics**

Once the system reaches the sliding surface (S(x) = 0), the system dynamics slide along the surface. The idea is that the behavior of the system, once on the surface, becomes insensitive to disturbances and uncertainties. In other words, when the system is on the sliding surface, the dynamics can be controlled in a way that is independent of external disturbances.

For example, in the sliding mode, the state dynamics can be written as:

$$\dot{x} = f(x) + g(x)u$$

This means that, once the system reaches the sliding surface, the effect of disturbances d(t) is eliminated or at least greatly reduced.

**d) The sliding mode control law**

The control law is designed to achieve two objectives:

1. Reaching the sliding surface: The system should be driven towards the sliding surface. This is often achieved by using a control law that forces the system state to converge to the surface in finite time.
2. Maintaining the sliding mode: Once the system reaches the sliding surface, the control law ensures that the system stays on the surface.

The control law typically consists of two components:

- Equivalent control: This component ensures that once the system is on the sliding surface, it stays there. It is determined by solving for the control input that results in zero dynamics along the sliding surface

$$u_{eq} = -g(x)^{-1}\left[f(x) + \frac{\partial S}{\partial x} f(x)\right]$$

- Switching control: This component drives the system toward the sliding surface in the first place. The switching control forces the state to approach the sliding surface in finite time. A common form of switching control is:

$$u_{sw} = -k.sgn\left(S(x)\right)$$

Where k > 0 is a constant and sgn(S(x)) is the signum function, which produces a control force that switches the direction of the system state depending on whether the system is above or below the sliding surface.

The total control law is then the sum of the equivalent control and the switching control:

$$u = u_{eq} + u_{sw}$$

**e) Reaching and Sliding phases**

SMC can be described as having two phases:

1. Reaching phase: During this phase, the system state is driven toward the sliding surface $S(x)=0$. The switching control, $u_{sw}$, is responsible for ensuring that the state converges to the surface in finite time.

2. Sliding phase: Once the system reaches the sliding surface, the equivalent control, $u_{eq}$, ensures that the system remains on the surface. In this phase, the dynamics are governed by the sliding surface equation, making the system robust to disturbances and uncertainties.

**Example 11.17:** Consider a first-order system described by:

$$\dot{x} = -x + u + d(t)$$

Where, x is the state, u is the control input, and d(t) is an unknown disturbance bounded as $|d(t)| \leq 0.5$. Design a sliding mode control law to stabilize the system such that x→0.

**Solution:**

The sliding surface is chosen as:

$$S(x) = x$$

When S(x) = 0, the state reaches the desired equilibrium x = 0.

On the sliding surface, we require:

$$S(\dot{x}) = \dot{x} = 0$$

Substituting the value of $\dot{x}$, we get

$$0 = -x + u + d(t)$$

Ignoring the disturbance term, d(t), the equivalent control is:

$$u_{eq} = x$$

To ensure the system reaches the sliding surface, we use the reaching condition:

$$S(\dot{x}) = -\eta.sgn(S(x))$$

Where η > 0 is a positive constant.

This leads to

$$-x + u + d(t) = -\eta.sgn(S(x))$$

Solving for u, we get:

$$u = x - d(t) - \eta.sgn(S(x))$$

Since d(t) is unknown but bounded,

$$u = x - \eta.sgn(x)$$

Where η > 0 0.5 to account for the disturbance.

The final sliding mode control law is then represented as:

$$u = x - (\eta + 0.5).sgn(x)$$

**Example 11.18:** For a second-order system:

$$\ddot{x} + 2\dot{x} + +3x = u + d(t)$$

Design a sliding mode controller to stabilize the system at x = 0, given that the disturbance $|d(t)| \leq 1$.

**Solution:**

Let $x_1 = x$, $x_2 = \dot{x}$

Then, $\dot{x}_1 = x_2$, $\dot{x}_2 = -3x_1 - 2x_2 + u + d(t)$

The system becomes:

$$\dot{x} = \begin{bmatrix} \dot{x}_1 \\ \dot{x}_2 \end{bmatrix} = \begin{bmatrix} x_2 \\ -3x_1 - 2x_2 + u + d(t) \end{bmatrix}$$

Defining the sliding surface as:

$$S(x) = c_1x_1 + x_2$$

Where $c_1 > 0$ is a design parameter.

On the sliding surface we have:

$$\dot{S}(x) = \frac{d}{dt}(c_1x_1 + x_2) = c_1\dot{x}_1 + \dot{x}_2$$

Substituting the necessary values:

$$\dot{S}(x) = c_1x_2 - 3x_1 - 2x_2 + u + d(t)$$

Setting the value of $\dot{S}(x) = 0$, we get the equivalent control as:

$$u_{eq} = -c_1 x_2 + 3x_1 + 2x_2$$

To ensure the system reaches the sliding surface,

$$S(\dot{x}) = -\eta.sgn(S(x))$$

Substituting for $S(\dot{x})$,

$$c_1 x_2 - 3x_1 - 2x_2 + u + d(t) = -\eta.sgn(S(x))$$

Solving for u, we get:

$$u = -\eta.sgn(S(x)) - c_1 x_2 + 3x_1 + 2x_2 - d(t)$$

Since $|d(t)| \leq 1$, the control law becomes:

$$u = -\eta.sgn(S(x)) + 3x_1 + (2 - c_1)x_2$$

With $\eta > 1$ for disturbance rejection.

### 11.3.2 Key Properties of Sliding Mode Control

- Robustness:
- SMC is inherently robust against matched uncertainties (those affecting the same channels as the control input). Once the system enters the sliding phase, the dynamics are decoupled from certain types of disturbances and model inaccuracies.
- Finite-Time Convergence:
- The control law guarantees finite-time convergence to the sliding surface, ensuring rapid response.
- Reduced-Order Dynamics:
- On the sliding surface, the system dynamics are of reduced order, simplifying the analysis and design of the control system.
- Global Stability:
- Proper design of the sliding surface ensures global asymptotic stability of the closed-loop system.

### 11.3.3 Challenges and Practical Considerations

Despite its advantages, SMC faces some challenges in practical implementation:

- Chattering Phenomenon:
- The discontinuous nature of the control law can lead to high-frequency oscillations, known as chattering. Chattering can cause wear and tear in

mechanical systems or excessive power consumption in electronic systems. Mitigating chattering often requires modifications to the control law, such as replacing the discontinuous term with a smooth approximation (e.g., using a sigmoid or saturation function).

- Unmatched Uncertainties:
- While SMC is robust to matched uncertainties, handling unmatched uncertainties (those that do not affect the same channels as the control input) requires additional design considerations.
- Implementation Complexity:
- Accurate knowledge of system dynamics and the design of the sliding surface are critical for effective implementation. Inaccuracies in these can degrade performance.

### 11.3.4 Applications of Sliding Mode Control

Sliding mode control has been successfully applied in various domains, including:

- Robotics: SMC is used for trajectory tracking, force control, and adaptive control of robotic manipulators.
- Aerospace Systems: It is employed for attitude control, altitude stabilization, and fault-tolerant flight control.
- Power Electronics: SMC ensures robust performance in DC-DC converters, inverters, and motor drives.
- Automotive Systems: Applications include anti-lock braking systems (ABS), traction control, and vehicle stability control.

## 11.4 DATA-DRIVEN CONTROLLERS: A BRIEF INTRODUCTION

The advent of data-driven approaches has revolutionized modern control systems, shifting the paradigm from traditional model-based designs to dynamic, adaptive solutions that leverage data. Unlike conventional controllers, which rely on predetermined mathematical models, data-driven controllers utilize the vast amounts of information generated by sensors, IoT devices, and operational systems. These controllers dynamically adjust their behavior based on real-time feedback and learned patterns, making them exceptionally suited for complex and rapidly changing environments. Industries such as robotics, autonomous vehicles, and industrial automation are benefiting immensely from this transformation, as data-driven controllers enhance precision, adaptability, and robustness in handling intricate system dynamics.

Machine learning (ML)-based approaches form the foundation of data-driven controllers by enabling systems to learn and optimize performance using data. Unlike static models, ML-based controllers are adaptive and can predict system behavior under uncertainty, making them indispensable for dynamic environments. Techniques such as reinforcement learning empower these controllers to interact with their environment, iteratively improving performance through trial and error. For instance, ML controllers in robotic systems can learn optimal motion trajectories over time, enhancing efficiency in pick-and-place tasks or navigation. Moreover, supervised learning aids in system modeling and state prediction, while unsupervised learning clusters operational states to identify faults or optimize performance. However, ML-based controllers require extensive datasets for training and continuous updates to maintain accuracy and relevance in changing conditions.

Deep learning (DL)-based approaches build on ML by employing multi-layered neural networks that excel in modeling complex and high-dimensional relationships within data. These controllers are particularly effective in applications involving unstructured data, such as images, videos, and time-series information. For instance, convolutional neural networks (CNNs) enable image-guided control tasks, such as visual servoing in robotics or obstacle avoidance in drones, by processing and interpreting visual inputs. Similarly, recurrent neural networks (RNNs) and their variants, such as long short-term memory (LSTM) networks, are used for predictive control in time-dependent processes, such as power grid stability or adaptive traffic management. Deep learning enables data-driven controllers to uncover intricate patterns that traditional methods cannot, making them invaluable for tasks demanding high precision and adaptability. However, the computational demands of DL models can be a limitation, requiring advanced hardware for training and deployment.

At the apex of data-driven methodologies are artificial intelligence (AI)-based approaches, which integrate ML, DL, and cognitive technologies to create controllers capable of reasoning, planning, and autonomous decision-making. These controllers are not limited to pattern recognition; instead, they incorporate higher-order functionalities such as logic-based reasoning and contextual awareness. AI-based controllers are instrumental in autonomous systems, where they enable end-to-end automation of tasks. For example, in industrial robotics, AI-powered systems can adapt to varying product specifications without human intervention, optimizing assembly lines for efficiency and flexibility. Furthermore, AI controllers utilize cognitive computing to interpret and act on unstructured inputs, such as voice commands in collaborative robots, facilitating seamless human-machine interaction. These systems excel in scenarios requiring real-time decision-making under uncertainty, such as autonomous navigation through hazardous environments or adaptive mission planning in aerospace.

## Problems

P11.1 The transfer function of a system is given by:

$$G(S) = \frac{1}{(s+1)(s+2)} = \frac{1}{s^2+3s+2}$$

a) Derive the state-space representation of the system.
b) Determine if the system is stable by checking the eigenvalues of matrix A.
c) Propose a candidate Lyapunov function for the system and analyze its stability using the Lyapunov method.

P11.2. Consider the system defined by

$$\begin{bmatrix} \dot{x}_1 \\ \dot{x}_2 \end{bmatrix} = \begin{bmatrix} -1 & 1 \\ 0 & 2 \end{bmatrix} \begin{bmatrix} x_1 \\ x_2 \end{bmatrix} + \begin{bmatrix} 1 \\ 0 \end{bmatrix} u$$

Show that the system cannot be stabilized by the state feedback control scheme $u = -k$ whatever the value of k.

P11.3. Consider the system shown in Figure P11.3.

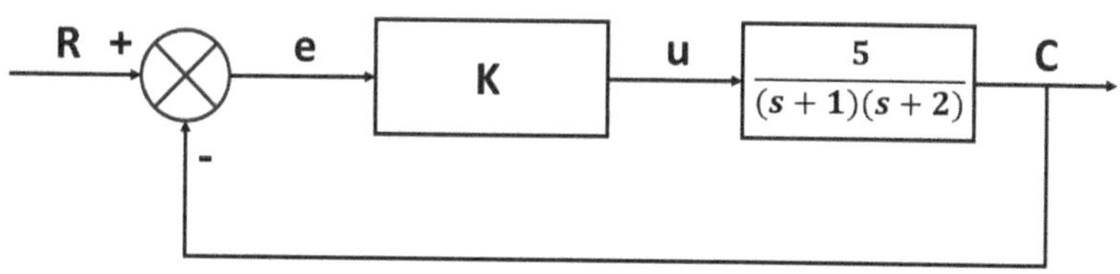


Determine the value of gain $k$, so that the damping ratio $\xi$ of the closed-loop system is equal to 0.5, then determine $\omega_n$ of the closed-loop system. Assuming that $e(0) = 1, \dot{e}(0) = 0$

Evaluate $\int_0^{\infty} e^2(t)\,dt$.

P11.4. Determine the transition matrixes of the following system matrixes.

$$A = \begin{bmatrix} 1 & 2 \\ 3 & 2 \end{bmatrix}$$

# Index

## A



## B



## C



## D



## E



## F



## G

## H



## I



## L



## M



## N



## O



## P



## R



## S

## T

FSC www.fsc.org
MIX
Papier | Fördert gute Waldnutzung
FSC® C083411

Kolibri 360 GmbH
Ferdinand-Jühlke-Straße 7
99095 Erfurt, Deutschland
produktsicherheit@kolibri360.de